ECONOMICS
OF
AGRICULTURAL
DEVELOPMENT

Economics of Agricultural Development examines the causes, severity, and effects of persistent poverty, rapid population growth, and malnutrition in developing countries. It discusses potential solutions to these problems, and considers the implications of globalization for agriculture, poverty, and the environment.

The authors provide a broad view of development and world food issues, on the means for utilizing agricultural surpluses to further overall economic development, and on issues related to trade and capital flows. The roles of governments and of international institutions in promoting broad-based development are also explored.

Economics of Agricultural Development covers topics related to the sustainability of the natural resource environment, gender roles in relation to agriculture and resource use, and the importance of macroeconomic policies as related to development and trade. The successes and failures of such policies, and the implications for what might be done in the future to encourage more rapid agricultural and economic development, are also considered.

George W. Norton is Professor of Agricultural and Applied Economics at Virginia Polytechnic Institute and State University, Blacksburg, USA. **Jeffrey Alwang** is Professor of Agricultural and Applied Economics at Virginia Polytechnic Institute and State University, Blacksburg, USA. **William A. Masters** is Professor of Agricultural Economics at Purdue University, West Lafayette, USA.

ECONOMICS OF

AGRICULTURAL

DEVELOPMENT

WORLD FOOD SYSTEMS
AND RESOURCE USE

GEORGE W. NORTON
JEFFREY ALWANG
Virginia Polytechnic Institute and State University

WILLIAM A. MASTERS
Purdue University

Routledge
Taylor & Francis Group

NEW YORK AND LONDON

First published 2006
by Routledge
270 Madison Ave., New York, NY 10016

Simultaneously published in the USA and Canada
by Routledge
2 Park Square, Milton Park, Abingdon, Oxon OX14 4RN

Routledge is an imprint of the Taylor & Francis Group, an informa business

© 2006 George W. Norton, Jeffrey Alwang, and William A. Masters
Typeset in the United States by Pocahontas Press, Inc.
Printed and bound in Great Britain by
TJ International Ltd, Padstow, Cornwall

British Library Cataloguing in Publication Data
A catalogue record for this book is available from the British Library

Library of Congress Cataloging in Publication Data
A catalog record for this book has been requested.

ISBN 10: 0–415–77045–9 (hbk)
ISBN 10: 0–415–77046–7 (pbk)

ISBN 13: 978–415–77045–3 (hbk)
ISBN 13: 978–0–415–77046–0 (pbk)

Contents

Preface

Persistent poverty, rapid population growth, and malnutrition in developing countries are among the most serious issues facing the world today. *Economics of Agricultural Development* examines the causes, severity, and effects of these problems. It identifies potential solutions, and considers the implications of globalization for agriculture, poverty, and the environment. It identifies linkages in the world food system, and stresses how agricultural and economic situations in poor countries affect industrialized nations and vice versa. It focuses on the role that agriculture can play in improving economic and nutritional well-being and how that role might it be enhanced.

Much has been learned about the roles of technology, education, international trade and capital flows, agricultural and macroeconomic policies, and rural infrastructure in stimulating agricultural and economic development. In some cases, the same factors can contribute to economic growth and lead to price and income instability or environmental risk. These lessons and other issues are examined in the book using the basic tools of economic analysis. The need is stressed for improved information flows to help guide institutional change in light of social, cultural, and political disruptions that occur in the development process.

The challenge in studying the economics of agricultural development is to build a broad view of the problem, and to bring economic theory to bear on specific challenges faced by the rural sector and on means for utilizing agricultural surpluses to further overall economic development. The goal of this book is to help students and other interested practitioners gain an understanding of the agricultural development problem, including the environmental and human consequences of different development paths, and the influence of international trade and capital flows. It is designed to help students develop skills that will enhance their capability to analyze world food and development problems.

This book interprets for students the economics of development and trade, including the importance of extending economic theory to account for institutions, imperfect information, and the willingness of people to exploit others and to act collectively. This extension provides important insights for development policy and helps explain why some countries develop while others are left behind. The role of the government in promoting broad-based development is explored. The book also covers topics related to sustainability of the environment, gender roles in relation to agriculture and resource use, and the importance of macroeconomic policies as related to development and trade.

Intended Audience

Economics of Agricultural Development is designed as a comprehensive text for the first course on the economics of world food issues and agricultural development. The book is aimed primarily at undergraduate students, with the only prerequisite a course in introductory economics. Students in undergraduate world food and agricultural development courses frequently represent a wide variety of majors. Economic jargon is kept to a minimum and explained where necessary, and the book sequentially builds a base of economic concepts that are used in later chapters to analyze specific development problems. A second audience for the book is those who work for public and private international development organizations.

Organization of the Book

Agricultural development is important for rural welfare and for overall economic development. Part One of the book considers the many dimensions of the world food-income-population problem in both a human and an economic context. After the severity and dimensions of the problem have been established, Part Two examines the economic transformation experienced by countries as they develop, sources of economic growth, and theories of economic development, including the role of agriculture in those theories. Part Three provides students with an overview of traditional agriculture, agricultural systems, and their determinants in developing countries, with particular attention to issues such as environmental sustainability and gender roles. Part Four then identifies agricultural development theories and the technical and institutional elements required for improving the agricultural sector. It stresses the need to build on and modify current agricultural development theories. Finally, Part Five considers the importance of the international environment, including trade and trade policies, macroeconomic policies, capital flows, and foreign assistance. The concluding chapter integrates various development components addressed in the book and discusses future prospects for agricultural development.

Acknowledgments

This textbook has benefited from the contributions of numerous individuals. It has benefited from the feedback of students in our classes at Virginia Tech and Purdue, and from comments from students of Laura McCann at the University of Missouri. We thank Laura and Laurian Unnevehr for reviewing an earlier draft. The encouragement and assistance of our colleagues at Virginia Tech and Purdue are gratefully acknowledged. We especially thank Brad Mills, David Orden, Dan Taylor, S.K. DeDatta, Anya McGuirk, Herb Stoevener, Jerry Shively, Sally Thompson, and Wally Tyner. The book has benefited greatly from discussions and interactions on development issues over many years with Phil Pardey, Stan Wood, Paul B. Siegel, Terry Roe, Bill Easter, Dan Sisler, Brady Deaton, Mesfin Bezuneh, and numerous graduate students.

We also thank Robert Langham and Taiba Batool of Routledge Press for their editorial assistance, and we especially want to thank Mary Holliman of Pocahontas Press for her invaluable editorial and production assistance. We thank Daren McGarry, Steve Aultman, Jessica Bayer, and Jacob Ricker-Gilbert for their assistance on figures and illustrations.

An earlier book by Norton and Alwang, published by McGraw Hill in 1992, proved popular in undergraduate classes in the United States and abroad, including a Spanish edition sold in Spain and Latin America. We thank Laura, Laurian, Nicole, Ben, Tim, Sophie, Rafael, Rob, Mary, Jerry, and others who stuck with us after it went out of print and used draft chapters in your classes while waiting for us to publish this new version. We hope you find it was worth the wait.

George W. Norton
Jeffrey Alwang
William A. Masters

PART 1

Dimensions of World Food and Development Problems

Rural family in Colombia.

Introduction

Most hunger is caused by a failure to gain access to the locally available food or to the means to produce food directly.
— C. Peter Timmer, Walter P. Falcon, and Scott R. Pearson[1]

This Chapter
1. Examines the basic dimensions of the world food situation
2. Discusses the meaning of economic development
3. Considers changes that occur during agricultural and economic development.

OVERVIEW OF THE WORLD FOOD PROBLEM

One of the most urgent needs in the world today is to reduce the pervasive problems of hunger and poverty in developing countries. Despite many efforts and some successes, millions of people remain ill-fed, poorly housed, underemployed, and afflicted by a variety of illnesses. These people regularly suffer the pain of watching loved ones die prematurely, often from preventable causes. In many countries, the natural resource base is also being degraded, with potentially serious implications for the livelihoods of future generations.

Why do these problems persist, how severe are they, and what are their causes? What does the globalization of goods, services, and capital mean for agriculture, poverty, and environment around the world? How does the situation in poor countries feed back on industrialized nations, and vice versa? An understanding of the fundamental causes of the many problems in poorer countries is essential if solutions are to be recognized and implemented. What role does agriculture play and how might it be enhanced? What can rich countries do to help? How do the policies in developed countries affect developing countries? These are some of the questions addressed in this book. Globalization will

---

[1] C. Peter Timmer, Walter P. Falcon, and Scott R. Pearson, *Food Policy Analysis* (Baltimore, Md.: Johns Hopkins University Press, 1983), p. 7.

continue, and a key issue is how to manage it to the betterment of developing and developed countries alike.

Much has been learned over the past several years about the roles of technology, education, international trade and capital flows, agricultural and macroeconomic policies, and rural infrastructure in stimulating agricultural and economic development. In some cases, these same factors can be two-edged: they contribute to economic growth on the one hand, but lead to price and income instability or environmental risk on the other. These lessons and other potential solutions to development problems are examined herein from an economic perspective. The need is stressed for improved information flows to help guide institutional change in light of the social, cultural, and political disruptions that occur in the development process.

World Food and Income Situation

Are people hungry because the world does not produce enough food? No. In the aggregate, the world produces a surplus of food. If the world's food supply were evenly divided among the world's population, each person would receive substantially more than the minimum amount of nutrients required for survival. The world is not on the brink of starvation. Population has roughly doubled over the past 40 years, and food production has grown even faster.

If total food supplies are plentiful, why do people die every day from hunger-related causes? At its most basic level, hunger is a poverty problem. Only the poor go hungry. They go hungry because they cannot afford food or cannot produce enough of it themselves. The very poorest groups tend to include: families of the unemployed or of underemployed landless laborers; the elderly, handicapped, and orphans; and persons experiencing temporary misfortune due to weather, agricultural pests, or political upheaval. Thus, hunger is for some people a chronic problem and for others a periodic or temporary problem. Many of the poorest live in rural areas.

Hunger is an individual problem related to the distribution of food and income within countries, and a national and international problem related to the geographic distribution of food, income, and population. Roughly one-fifth of the world's population (about one billion people) lives on less than $1 per day; about one-half lives on less than $2 per day. These people are found primarily in Asia and Africa. The largest number of the poor and hungry live in Asia, although severe hunger and poverty are found in Sub-Saharan Africa and in parts of Latin America. Great strides have been made in reducing global poverty; between 1981 and 2001, the proportion of the world's population living

Many farm workers earn about two dollars per workday.

on less than $1 per day fell from around 40 percent to near 20 percent. However, more remains to be done to alleviate poverty-related problems.[2]

While hunger and poverty are found in every region of the world, Sub-Saharan Africa is the only major region where per capita food production has failed to at least trend upward for the past 30 years. As Figure 1-1 shows, per capita food production in Africa has stagnated since 1980 and had experienced a downward trend for several years before that time. Latin America, and particularly Asia, have experienced relatively steady increases. The result has been significant progress in reducing hunger and poverty in the latter two regions, while per capita calorie availability remains below minimum nutritional standards in many Sub-Saharan countries. Low agricultural productivity (farm output divided by farm inputs), wide variations in yields due to natural, economic, and political causes, and rapid population growth have combined to create a precarious food situation in these countries.

Annual variation in food production is a serious problem, particularly in Sub-Saharan Africa (Fig. 1-1). This variation has caused periodic famines in individual countries, particularly when production

---

[2] In 2001, the proportion of the population living on less than $1 per day was still almost 50 percent in Sub-Saharan Africa and 30 percent in South Asia, according to World Bank statistics.

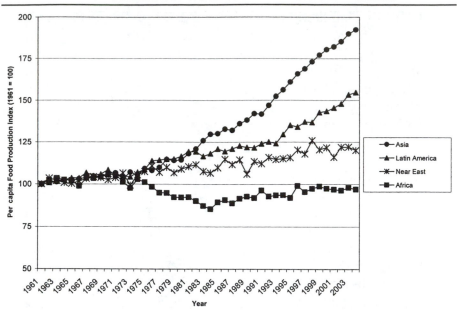

Figure 1-1. Index of per capita food production (Source: FAOSTAT data, 2005).

problems have been compounded by political upheaval or wars that have hindered international relief efforts. Production variability causes wide price swings that reduce food security for millions who are on the margin of being able to purchase food. If the world is to eliminate hunger, it must distinguish among solutions needed for short-term famine relief, those needed to reduce commodity price instability (or its effects), and those needed to reduce long-term or chronic poverty problems.

Malnutrition

Hunger is most visible to people in developed countries when a drought or other disaster results in images in the news of children with bloated bellies and bony limbs enduring the pain of extreme hunger. Disturbing as such images are, in a sense they mislead. The less conspicuous but more pernicious problem, in terms of people suffering and dying, is chronic malnutrition. While accurate figures of the number of malnourished in the world are not available, and even good estimates depend on the definition used, a conservative estimate is that roughly 800 million people suffer from chronic or severe malnutrition associated with food deprivation. More than 10 million people, many of them young children, die each year from causes related to inadequate food consumption. Increasing per capita food production has allowed more of the world's population to eat better. But for those in the lower income groups, the situation remains difficult.

Health

People born in developing countries live, on average, 14 years less (in Sub-Saharan Africa, 27 years less) than those born in developed countries. Health problems, often associated with poverty, are responsible for most of the differences in life expectancies. Mortality rates for children under age five are particularly high, often 10 to 20 times higher than in developed countries (Fig. 1-2). Though countries with high rates of infant mortality are found in all regions, Sub-Saharan African countries are particularly afflicted. The band of high infant mortality stretching from the Atlantic coast across Africa to Somalia on the Indian Ocean covers some of the poorest and most undernourished populations in the world.

Poverty affects health by limiting people's ability to purchase food, housing, medical services, and even soap and water. Inadequate public sanitation and high prevalence of communicable diseases are also closely linked with poverty. A major health problem, particularly among children, is diarrhea, usually caused by poor water quality. According to the World Bank, 5 to 10 million children die each year from causes related to diarrhea. Respiratory diseases account for an additional 4 to 5 million deaths, and malaria another million. Many people have never been vaccinated against such common — but preventable — diseases as rubella and measles, although vaccination rates for children under one year old have improved over the past 20 years. Basic health services are almost totally lacking in many areas; on average, ten times as many people per doctor and per nurse are found in low-income countries as in developed countries.

A major health problem that continues to grow rapidly in the developing world is acquired auto-immune deficiency syndrome (AIDS). The disease is particularly difficult to contain in many African countries because of the ease of its heterosexual spread, lack of education about the disease, limited use of protective birth-control devices, and in some cases absence of government commitment to address the problem. Estimates are that a quarter to one third of the adult populations in certain countries, such as Zimbabwe and Botswana, are HIV positive. Effects are felt in lost productivity and increased poverty, in addition to its effects on direct human suffering. As serious as the problem currently is in Africa, the region likely to be devastated most by AIDS in the future is Asia. Worldwide, an estimated 40 million people were living with HIV/AIDS in 2003.

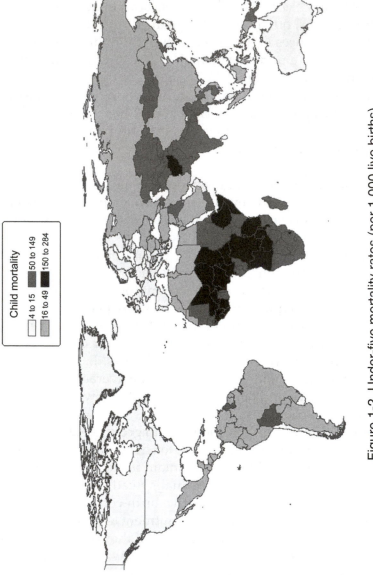

Figure 1-2. Under-five mortality rates (per 1,000 live births)
(Source: World Bank, *World Development Indicators*, 2005).

Children in Honduras.

Population Growth

How important is population growth to the food-poverty-population problem? It is very important, and will continue to be so at least for the next 40 to 50 years. Population is growing less than 1 percent per year in developed economies, but 2 percent per year in developing countries excluding China, and 3 percent or more in many Sub-Saharan African countries. These higher growth rates place pressure on available food supplies and on the environment in many low-income countries. Population growth and food production are closely linked, and changing either in a major way takes time, as discussed in Chapter 4. It is clear that continual increases in food production are needed, because regardless of how successful are efforts to control population growth, world population will not stabilize for many years. Rapid urbanization is also occurring as populations continue to grow.

Globalization

Food and economic systems in less-developed countries are affected by the international economic environment far more today than they were in years past. Trade and other economic policies abroad and at home, international capital flows, migration, and oil price shocks have combined to increase the instability of developing countries and reduce opportunities for improving their food and economic security.

International trade in agricultural products (as with other products) has grown rapidly since the 1970s, building on improvements in

9

transportation and information systems. As exports and imports of farm products constitute a higher proportion of agricultural production and consumption, effects of domestic agricultural policies aimed at influencing the agricultural sector are altered. World prices become more important to farmers than they were previously, and possibilities for maintaining a nation's food security at the aggregate level are improved. However, production and policy changes abroad also tend to have a great effect on domestic agriculture as international trade grows. While the need for national food production self-sufficiency has been reduced, the need to be price competitive with other countries has grown, as has the need to participate in international negotiations to alter the policy environment.

International capital (money) markets, through which currencies flow from country to country in response to differences in interest rates and other factors, have become as important as trade to the food and economic systems in developing countries. The volume of international financial transactions far exceeds the international flows of goods and services. Capital flows affect the values of national currencies in foreign exchange markets. The foreign exchange rate, or the value of one country's currency in terms of another country's currency, is an important determinant of the price a nation receives for exports or pays for imports. Speculation in financial markets has led to rapid inflows and outflows of capital in some countries, resulting in sharp changes in asset values and incomes.

Many developing countries also have serious foreign debt problems. Beginning in the 1970s, developing countries increased their rate of borrowing from both public and private sources in developed countries. Latin American countries borrowed particularly heavily. In some cases, the borrowing was necessitated by sharp increases in the cost of oil imports. Banks and governments in developed countries were very willing to lend. When interest rates rose, those loans became difficult to repay. Many countries reduced their rate of government spending in efforts to service this debt, and this decrease in turn lowered the availability of public services, creating further hardships for the poor. The need for foreign exchange to repay external debts also increased the importance of exports for developing countries, forcing some countries to re-examine their trade and exchange rate policies. At the same time, new technologies have been changing the possibilities that countries have for producing and trading particular products.

Slum close to riverbank in Katmandu, Nepal.

Food Prices

In recent years, for most people in the world, the "real" price of food has fallen relative to the prices of other things. The international prices (in nominal or "current" dollars) of maize, rice, and wheat, the world's major food grains, are shown in Figure 1-3. Since historical peaks in 1974 and 1981, each year's average price of all three grains has fluctuated without a trend. The prices of most other things have risen steadily with inflation, so for most people the *relative* or real price of food has fallen. This trend is both good and bad because prices affect economic growth and social welfare in a contradictory fashion. Lower food prices benefit consumers and stimulate industrial growth but can lower agricultural producer incomes and reduce employment of landless workers. To the extent that lower prices reflect lower production costs, impacts on producers may be mitigated. Future food price trends depend on the relative importance of *demand* shifts, resulting primarily from changes in population and income, compared to *supply* shifts, resulting from a variety of forces, particularly new technologies.

Instability in local and world food prices is another problem affecting food security and hunger in developing countries. The three grains shown in Figure 1-3 have exhibited sizable year-to-year price variations. This instability was most severe during the 1970s, but remains a concern today. Food price fluctuations directly affect the well-being of the poor, who spend a high proportion of their income on food. Governments are finding that food price instability increases human suffering and also threatens political stability.

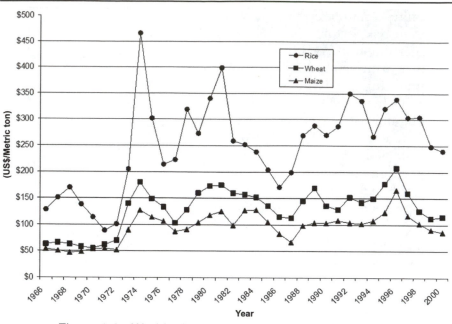

Figure 1-3. World prices of major grains in current dollars
(Source: *FAO Production Yearbook*, various years).

Environmental Degradation

As populations grow, environmental problems become more severe. Deforestation, farming of marginal lands, overgrazing, and misuse of pesticides have contributed to soil erosion, desertification, poisoning of water supplies, and even climate changes. Climates appear to have become less stable, while water has become scarcer. Environmental problems exist in every region of the world. Some degradation is intentional, but most is the unintended result of people and governments seeking means of solving immediate food and economic crises, often at the cost of long-term damage to the environment. Some of this damage may compromise the ability of a country to raise incomes in the long run. When people are hungry, it is hard to tell them to save their resources for the future, and environmental conservation represents a form of savings. However, many potential solutions exist that are consistent with both short-term increases in food production and long-term goals of simultaneously sustaining or improving environmental quality while raising incomes.

Risk and Uncertainty

Most of the factors mentioned above are associated with increased exposure to risk and uncertainty. Fluctuating prices, exchange rate

12

instability, certain crop pests, and rapidly changing weather patterns represent risk factors. Recent research has shown that risks and risk management imply real costs that may compromise long-run improvements in well-being. Risk also lowers welfare in the short run. For example, Hurricane Mitch struck the Central American coastal region during October 1998, causing massive losses in productive capacity, washing out roads, houses, and entire villages. In Honduras alone, it killed more than 8,000 people and injured more than 12,000. Deforestation in hillside areas contributed to the hurricane's damage as landslides and flooding washed out low-lying areas. In December 2004, a large earthquake off the coast of Indonesia caused a tsunami that washed ashore in several countries, especially in Indonesia, Sri Lanka, India, and Thailand, killing more than 228,000.

Risk is not necessarily bad. Innovation and entrepreneurship are risky. It is the way that risks are managed that most influences economic growth. Risk management needs to be conducted in an efficient manner; the proper balance must be found between managing risks and pursuing other goals.

The preceding overview provides brief highlights of some of the dimensions of the food-income-population problem. These and other problems are discussed in more depth in subsequent chapters, and alternative solutions are suggested. First, however, it is important to consider what we mean when we talk about development.

MEANING OF DEVELOPMENT

The term *development* means a change over time, typically involving growth or expansion. *Economic* development involves changes in people's standard of living. For most of human history there was little such change, but over the past 300 years there has been a rapid and (so far) sustained increase in almost every kind of human activity. Growth occurred first and has been sustained the longest in Northwest Europe and North America, but similar kinds of expansion have occurred all around the world.

Development is a process with many economic and social dimensions. For most observers, *successful* economic development requires, as a minimum, rising per capita incomes, eradication of absolute poverty, and reduction in inequality over the long term. The process is a dynamic one, including not only changes in the structure and level of economic activity, but also increased opportunities for individual choice and for improved self-esteem.

Development is often a painful process. Adjusting to new circumstances is always difficult: as Mark Twain famously wrote, "I'm all for progress — it's change I can't stand." There is often dramatic social upheaval: traditional ways of life are displaced, existing social norms are challenged, and pressures for institutional and political reform increase. The physical and cultural landscape of a country can change radically during economic development. And, at the individual level, the standard of living for the poorest people in a society does sometimes decline, even as average real incomes increase. More often, the fruits of improvement are unequally distributed. By any measure, poverty and deprivation remain widespread, despite the astonishing improvements in living standards experienced all across the globe.

As economic activity continues to expand, there is continuous concern with the constraints imposed by natural resources and environmental factors. The World Commission on Environment and Development has defined sustainable development as "development that meets the needs of the present without compromising the ability of future generations to meet their own needs."[3] Thus, the term "development" encompasses not only an economic growth component, but distributional components, both for the current population and for future generations.

Measures of Development

Although development is difficult to measure, it is often necessary to do so in order to assess the impacts of particular programs, to establish criteria for foreign assistance, and for other purposes. Because of its several dimensions, single indicators of development can be misleading. Measures are needed that are consistent with the objective of raising the standard of living broadly across the population. Average per capita income is frequently used as a measure (Fig. 1-4). Is it a good measure?

Average per capita income is not a perfect measure of living standards for several reasons, but finding an alternative indicator that can incorporate each dimension of development is impossible. Because development is multidimensional, collapsing it into a single index measure requires placing weights on different dimensions. Average per capita income is an inadequate measure even of the economic dimensions because it misses the important distributional elements of development and is a crude measure of people's well-being.

Alternative multidimensional development indicators have been suggested. One of the oldest is a level-of-living index proposed by M.K.

---

[3] World Commission on Environment and Development, *Our Common Future* (New York: Oxford University Press, 1987), p. 43.

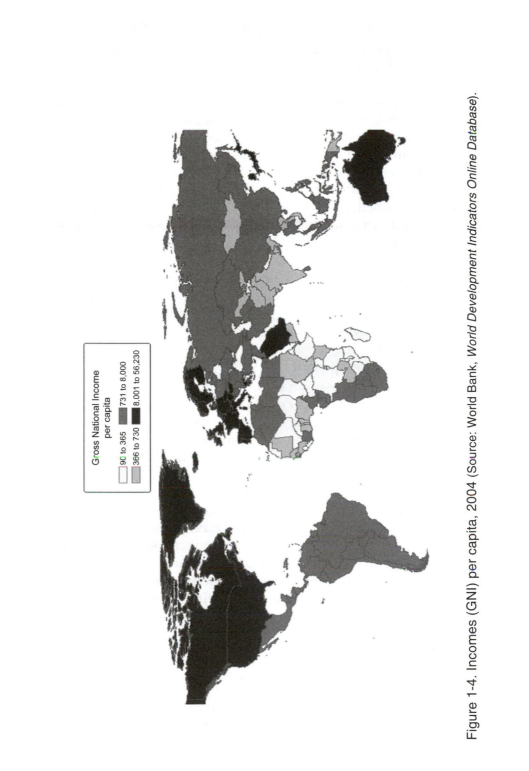

Figure 1-4. Incomes (GNI) per capita, 2004 (Source: World Bank, *World Development Indicators Online Database*).

Gross National Income
per capita

90 to 365
366 to 730
731 to 8,000
8,001 to 56,230

Bennett that weights 19 indicators for which data were available in 1951.[4] Examples of indicators include calorie intake per capita, infant mortality rates, number of physicians per 1000 of total population, and years of schooling. A more recent index is the Human Development Index[5] (HDI), which weights life expectancy, education, and income. Weighting schemes are subjective, however, and average per capita income is highly correlated with many of the indicators. Consequently, average per capita income, measured as gross national product or income (GNP or GNI) or gross domestic product (GDP) per capita is often employed as a first approximation; measures such as income distribution, literacy rates, life expectancy, and child mortality are examined separately or as part of an index. Even these supplementary indicators can be misleading, due to regional disparities within countries.

Several studies have called for the GNP income measure itself to be modified to account for depreciation or appreciation of natural resource-based assets, particularly forests.[6] This modification may be possible once natural resource accounting procedures are further refined.

Incomes and Development

Poverty and low incomes are most frequently associated with underdevelopment, while growing per capita incomes should indicate increasing levels of development. As discussed above, increasing average incomes may not necessarily mean more development, because the distribution of this income often determines whether poverty and inequality are diminished as the mean grows. Some of the relationships between poverty and inequality are discussed in Box 1-1.

Numerous measures of inequality and the extent of poverty exist. For example, the Human Poverty Index (HPI) measures the extent of deprivation with respect to life expectancy, education, and income.[7] If, as is argued above, the meaning of development contains some element of poverty reduction or increased equality of income distribution, then clearly the incomes of the poor and destitute should be raised during the development process.

---

[4] See M. K. Bennett, "International Disposition in Consumption Levels," *American Economic Review*, vol. 41 (September 1951), pp. 632-49.

[5] United Nations Development Programme, *Human Development Report 2005* (New York: Oxford University Press, 2005), p. 341.

[6] See Salah El Serafy and Ernest Lutz, "Environmental and Natural Resources Accounting," ch. 3, in *Environmental Management and Economic Development*, ed. Gunter Schramm and Jeremy Warford (Baltimore, Md.: Johns Hopkins University Press, 1989), for a summary of the concept known as natural resource accounting.

[7] United Nations Development Program, *Human Development Report 2005* (New York: Oxford University Press, 2005), p. 342.

BOX 1-1
POVERTY AND INEQUALITY

Poverty is generally defined as the failure to achieve certain minimum standards of living. By its very nature, poverty refers not just to *averages*, but to *distributions*. Poverty is not, however, synonymous with inequality; countries with perfect equality could contain all rich or all poor people. Measurement of poverty requires three steps: determining an appropriate measure or indicator, deciding on its minimum level, and counting the number or percentage of people falling below it. Alternatively, a measure of degree or intensity of poverty would indicate the amount by which people fall below the poverty line.

While poverty refers to some level or position with respect to a measure such as income, inequality refers to the distribution of that measure among a population. For example, evidence from 21 developing countries indicates that, on average, 6 percent of household income is received by the poorest 20 percent of the households, whereas 48 percent of household income is received by the richest 20 percent. In some countries the extremes are even more dramatic. It is possible for poverty to decrease in a country during the development process, but for inequality to increase, at least for a period of time.

Policies undertaken to promote development have diverse effects on the incomes of the poor. Some people benefit, but often some do not, and, at times, incomes fall for certain population groups. It is important to consider the winners and losers in the development process. Income distributions and changes in them are indicators of the impact of development policies on different groups in society.

Values and Development

Value judgments or premises about what is or is not desirable are inextricably related to development economics. Concerns for economic and social equality, poverty *eradication*, and the *need* to improve education all derive from subjective beliefs about what is good and what is not. Solutions to specific problems often involve tradeoffs, and decisions about public resource allocations always involve tradeoffs. Governments make such tradeoffs every day, as most government actions are costly to some people even as they benefit others. Economics can be a powerful tool for evaluating these tradeoffs, providing insights into the costs and benefits of different actions, winners and losers, and longer run consequences of savings, investment, and consumption decisions. Economics is, however, less well suited for making value decisions.

Even if people share the same set of beliefs and values, they may attach different weights to the individual beliefs and values within that set. Because there is no correct set of weights, people may not agree about appropriate solutions to development problems, even if the suggested solutions appear conceptually sound in terms of leading to their intended impacts.

Most policy suggestions would result in both gainers and losers. In some cases, the gainers could compensate the losers, but sometimes they could not, and often they do not. Because affected groups have differing political strengths within society, economic and social development policies cannot be separated from the political process. These realities must be considered if development policies are to succeed.

ROLE OF AGRICULTURE

Many alternative development paths or strategies exist. The strategy followed by an individual country at a particular point in time is, or at least should be, influenced in part by its resource endowments and stage of development. Some countries with vast oil and mineral resources have generated capital for development by exporting those resources. Others have emphasized cash-crop exports such as coffee, cocoa, and tea. Some have focused on industrial exports, while others have stressed increases in basic food production. The optimal development path will vary from country to country, but the choice of an inappropriate path, given the existing resource endowments and stage of development, can result in long-term stagnation of the economy.

Numerous examples can be found of countries choosing the wrong development path and paying the price. Argentina, a country well-endowed with land resources, pursued government policies in the 1940s and 1950s that stressed industrialization and virtually ignored agriculture. The result was that agricultural exports, previously an important component of economic growth, stagnated in the 1950s, and foreign exchange shortages prevented the import of capital goods needed for industrialization. Economic growth slowed dramatically as a result. India is another country whose potential for agriculture-driven growth was subverted by a disproportionate emphasis on industrialization in the 1950s and 1960s.

Agriculture is not very productive in most low-income countries. Early in the development process, much of the population is employed in agriculture, and a high percentage of the national income is derived from that sector[8] (see Table 1-1). As development proceeds, population grows and per capita income increases. As incomes grow, more food is demanded; either agricultural production or imports must increase.

Table 1-1. Relationship among Per Capita National Income, the Proportion of National Income in Agriculture, and the Proportion of the Labor Force in Agricuture, Selected Countries, 2002

Country	Per capita national income (dollars) [1]	Agriculture GDP as a percentage of total GDP[1]	Percentage of active labor force in agriculture[2]
Ethiopia	110	52	80
Niger	170	39	90
Uganda	250	43	82
Tanzania	280	48	80
Kenya	360	24	75-80
Bangladesh	360	35	63
India	480	25	60
Indonesia	710	17	45
Ukraine	770	23	24
Honduras	920	14	34
Ecuador	1,450	11	30
Colombia	1,830	13	30
Thailand	1,980	11	54
Brazil	2,850	8	23
Mexico	5,910	5	20
Greece	11,660	8	20
Argentina	11,690	5	NA
New Zealand	13,710	8	10
South Korea	13,780	4	9
Italy	18,960	2	5
France	22,010	3	4
Canada	22,300	2	3
Japan	33,550	1	5
United States	35,060	2	2

[1] Source: World Bank, *World Development Indicators*, 2002.
[2] Source: CIA, *World Fact Book*, 2002.

[8] A warning about measurement is appropriate: in most countries it is difficult to measure the number of people employed in agriculture. Multiple job holdings, seasonal labor use in agriculture, and unpaid household labor all complicate the measurement problem. Often, data on the number employed in agriculture are obtained by (generally high-quality) census estimates of the rural population. Even in rural areas, many people are employed outside of agriculture.

Because agriculture commands so many of the resources in most low-income countries, few funds are available for importing food or anything else unless agricultural output grows.

The capacity of the agricultural sector to employ an expanding labor force is limited. As incomes continue to rise, the demand for non-food commodities grows as well. Therefore, economic development requires a structural transformation of the economy involving relative expansion of nonagricultural sectors. The agricultural sector must contribute food, labor, and capital to that expansion. It also provides a market for nonagricultural goods.

This economic transformation is illustrated in Table 1-1. Agriculture accounts for a large percentage of total income, and an even larger percentage of total employment for the lower income countries. A steep decline in agricultural employment relative to the rest of the economy seems to occur at about $1000 per capita income. The contribution of agriculture to national incomes declines from 40 to 50 percent for the lower-income countries, to 20 to 30 percent for the middle-income range, and down to 15 percent or below for the highest income countries.

The initial size and low productivity of agriculture in most developing countries suggests an opportunity for raising national income through agricultural development. Because of the initial size of, and low per capita income in, the agricultural sector, there is real scope for improving the distribution of income and enhancing the welfare of a major segment of the population through agricultural development.

One of the keys to agricultural development is to improve information flows. In primitive societies, economic activities are local and information is basically available to all. Inappropriate activities are constrained by social and cultural norms. As development begins to proceed and economies become more complex, information needs increase but traditional forms of information transmission are incapable of meeting these needs. Modern information systems are slow to develop, creating inequalities in access to new information. Those with greater access than others can take advantage of this situation to further their own welfare, often at the expense of overall agricultural and economic development.

Some changes required to foster broad-based and sustainable development require institutional changes and capital investments. Capital investments necessitate savings. Such savings are channeled into private and public investment, the latter to build the infrastructure needed for development. Saving requires striking a balance between present and future levels of living because it requires abstention from current consumption. Means must be sought to reduce this potential short-run versus long-run conflict during the development process. However, certain types of investments necessary for development, such as education, provide both short- and long-run benefits, as do investments in technologies and employment-intensive industries.

Improving Agriculture

How can agriculture be improved to facilitate its role in providing food and contributing to overall development? There are still areas of the world, particularly in parts of Latin America and Africa, where land suited for agricultural production is not being farmed. Most increases in agricultural production will have to come, however, from more intensive use of land currently being farmed. Such intensive use will require improved technologies generated through research as well as improved irrigation systems, roads, market infrastructure, and other investments. It will require education and incentives created through changes in institutions such as land tenure systems, input and credit policies, and pricing policies (Box 1-2).

Agriculture and Employment Interactions

Agricultural development can provide food, labor, and capital to support increased employment in industry and can stimulate demand in rural areas for employment-intensive consumer goods. Because of their comparative advantage in labor-intensive production, many developing countries will need to import capital-intensive goods, such as steel and fertilizer, and export labor-intensive consumer goods and certain types of agricultural goods. Countries that do not match an employment-oriented industrial policy with their agricultural development policy will fail to realize the potential income and employment benefits of agricultural development.

SUMMARY

Some of the basic dimensions of the world food-poverty-population problem were examined. The aggregate world food situation was reviewed, and questions such as who the hungry are, and why they are hungry even though the world produces a surplus of food, were addressed. The significance of population growth and a series of forces in the global economy that influence developing countries were stressed.

The meaning and measures of development were discussed, as was the importance of development problems and the desirability of suggested solutions that depend on value judgments. While alternative development strategies can be followed, agriculture has an impor-

BOX 1-2
HISTORICAL PERSPECTIVE
ON AGRICULTURAL DEVELOPMENT

The historical progression of agricultural development can be broadly broken into four distinct periods, marked by three "revolutions" in production technology and social institutions.

First, from the time that we first appeared on earth, human beings hunted and gathered their food. Hunter-gatherer societies typically lived in small groups, and experienced little population growth.

Then, more than 10,000 years ago, a combination of climate changes and other factors created conditions for the development of settled agriculture. In the Middle East and elsewhere, people began to collect and cultivate the seeds of plants that eventually became modern barley, wheat, and rye. This development is known as the *first agricultural revolution*, and permitted a slow but significant increase in human population density.

More recently, a few hundred years ago, rising population density and opportunities for trade led to a *second agricultural revolution*. In North-western Europe and elsewhere, farmers developed crop rotations and livestock management systems that permitted rapid growth in output per person, fueling the *industrial revolution* and the eventual mechanization of many important tasks.

Finally, in the late nineteenth and early twentieth centuries, scientific breeding, chemical fertilizer and other innovations allowed rapid increases in output per unit of area. The spread of these biological technologies to developing countries, known as the *green revolution*, has been a powerful engine of economic growth and poverty alleviation, allowing low-income people to produce more food at lower cost than ever before.

These historical trends played out at different speeds and in different ways across the globe. A few people in the poorest countries still devote substantial energy to hunter-gatherer activities, and many millions of farmers still cultivate the same seeds in the same ways as their ancestors. Because of population growth, these techniques and institutional arrangements yield less and less output per capita over time. The development and spread of higher productivity systems to suit these people's needs is among the major humanitarian challenges of our time.

tant role to play in overall development in most developing countries. Development will require a complex set of improved technologies, education, and institutions, and an employment-oriented industrial policy.

IMPORTANT TERMS AND CONCEPTS

Agricultural productivity
Development
Enhanced information flows
Environmental degradation
Food-poverty-population problem
Food price instability
Foreign exchange rates
Globalization
Health problems

Institutions
International capital markets
International trade
Measures of development
Population growth
Structural transformation of
    the economy
Sustainability
Technology

Looking Ahead

In order to visualize more clearly the relationships among food supplies, food demand, population growth, and nutrition, it is important to examine facts, scientific opinion, and economic theory. We make this examination in the remaining chapters of Part One in this book. We turn first in Chapter 2 to the causes and potential solutions to hunger and malnutrition problems.

QUESTIONS FOR DISCUSSION

1. Are people hungry because the world does not produce enough food?
2. Has food production in developing countries kept pace with population growth there?
3. Is malnutrition more widespread today than in the past?
4. What are some factors that will influence the price of food over the next 10 to 20 years?
5. Is there any hope of bringing more land into production to help increase food production?
6. Why is agricultural development particularly important in less-developed countries?
7. Approximately what proportion of the world's population lives on per capita incomes of less than $2 per day?
8. What is development? To what extent are values important when discussing development issues?
9. Is average per capita income a good measure of level of living?
10. Why is most of the labor force engaged in agriculture in many less-developed countries?

11. Does economic development require expansion of the nonagricultural sector in low-income countries?
12. What is the conflict between increasing near- versus long-term levels of living in developing countries?
13. What are the major health problems in developing countries, and what are their primary causes?
14. How fast is population growing in developing countries?
15. Why has international agricultural trade become more important over the past 30 years?
16. Why have international capital markets become more important to developing countries over the past 30 years?
17. Why might low food prices be both good and bad?
18. Why has environmental degradation become an increasing problem in developing countries?

RECOMMENDED READING

Eicher, Carl K., and John M. Staatz, *International Agricultural Development* (Baltimore, Md.: Johns Hopkins University Press, 1998), especially pp. 3–53.

Food and Agriculture Organization of the United Nations, *State of Food and Agriculture 2005* (Rome: FAO, 2005, and various other years).

Pinstrup-Andersen, Per, and Rajul Pandya-Lorch, *The Unfinished Agenda: Perspectives on Overcoming Hunger, Poverty, and Environmental Degradation* (Washington, D.C.: International Food Policy Research Institute, 2001).

Runge, C. Ford, Benjamin Senauer, Philip G. Pardey, and Mark W. Rosegrant, *Ending Hunger in Our Lifetime: Food Security and Globalization* (Baltimore, Md.: Johns Hopkins University Press, 2003).

Todaro, Michael P., *Economic Development* (New York: Addison Wesley, 1999), especially chs 1, 2, and 3.

United Nations Development Programme, *Human Development Report 2005* (New York: Oxford University Press, 2005).

World Bank, *World Development Report 2006* (New York: Oxford University Press, 2006; see earlier volumes as well).

CHAPTER 2

Hunger and Malnutrition

For hunger is a curious thing: at first it is with you all the time, waking and sleeping and in your dreams, and your belly cries out insistently, and there is a gnawing and a pain as if your very vitals were being devoured, and you must stop it at any cost, and you buy a moment's respite even while you know and fear the sequel. Then the pain is no longer sharp but dull, and this too is with you always, so that you think of food many times a day and each time a terrible sickness assails you, and because you know this you try to avoid the thought, but you cannot, it is with you. Then that too is gone, all pain, all desire, only a great emptiness is left, like the sky, like a well in drought, and it is now that the strength drains from your limbs, and you try to rise and find that you cannot, or to swallow water and your throat is powerless, and both the swallow and the effort of retaining the liquid taxes you to the uttermost.

— Kamala Markandaya[1]

This Chapter
1. Describes the world food situation
2. Examines different forms of hunger and malnutrition: their magnitudes, consequences, and measurement
3. Identifies principal causes of and potential solutions to hunger and malnutrition problems.

THE WORLD FOOD SITUATION

World Food Demand and Supply

World food consumption and production have each grown about 2.2 percent per annum since 1970, while in developing countries consumption has grown about 3.7 percent and production 3.5 percent. The second set of numbers implies a small increase in net food imports by developing countries. Cereals are the most important sources of food and, since the mid-1960s, world cereal production has risen by roughly

[1] Kamala Markandaya, *Nectar in a Sieve* (New York: New American Library, 1954), p. 91.

one billion tons per year. It is likely that an additional billion tons in production per year will be needed by 2030 to meet the needs of a world population expanding in numbers and in income. It is also likely that cereal imports by developing countries will increase from about 10 percent of consumption to about 15 percent.

While the overall numbers and projections suggest gradual improvement in reducing malnutrition in the world, there are still several countries in which per capita food consumption has declined and is not likely to increase enough to significantly reduce the number of undernourished people. In addition, the rate of growth in agricultural output for the world as a whole has slowed since the 1980s. A best-case scenario for the world would seem to point to a reduction in severely malnourished to about 400 to 500 million people from the current roughly 800 million people, but for populations in many countries in Sub-Saharan Africa and parts of Asia, the struggle for food will continue. Therefore we turn now to how food shortages manifest themselves in terms of hunger, malnutrition, and, in some cases, famine.

HUNGER, MALNUTRITION, AND FAMINE

Hunger is a silent crisis in the world. In times of famine, it can tear at the heartstrings as media attention focuses on its dramatic effects. But to most people reading this book it is an invisible and abstract problem somewhere out there. We seldom think about it, and when we do we often don't know what to think or how to take action against it. Hunger has many faces, many dimensions, and many causes. The most extreme type of hunger is severe calorie and protein undernutrition during a famine. However, more pervasive is chronic undernutrition and malnutrition associated with poverty, illness, ignorance, maldistribution of food within the family, and seasonal fluctuations in access to food. We begin our discussion of hunger with the contrast between famines and chronic malnutrition.

Famines

Famine is marked by an acute decline in access to food that occurs in a definable area and has a finite duration. This lack of access to food usually results from crop failures, often in successive years, due to drought, flood, insect infestation, or war. During a famine, food may actually be present in the affected area, but its price is so high that only the wealthy can afford it. Food distribution systems may break down so that food cannot reach those who need it.

Famines have occurred throughout history. In recent years, their prevalence has been greatest in Sub-Saharan Africa, but famines also have occurred in North Korea periodically since 1995, in Kampuchea (formerly Cambodia) in 1979, Bangladesh in 1974, India in 1966 and 1967, and China in 1959 to 1961. The latter, the worst famine of the twentieth century, resulted in an estimated mortality of at least 16 million people.

Famine is the extreme on the hunger scale because it causes extreme loss of life and concurrent social and economic chaos over a relatively short period of time. As access to food falls, people begin by borrowing money and then selling their assets to acquire money to purchase foods. Subsistence farmers sell their seed stocks, livestock, plows, and even land. Landless laborers and other poor groups lose their jobs, or face steeply higher prices for food at constant wages. As the famine intensifies, whole families and villages migrate in search of relief. The telltale signs of acute malnutrition and, eventually, sickness and death appear (see Box 2-1).

Fortunately, progress is being made against famine. Although there are large variations in annual food production in individual countries and world population continues to grow, the frequency and intensity of famines has decreased due to improved information and transportation networks, increased food production and reserves, and dedicated relief organizations. Much of the starvation we see during famines today occurs in areas where transportation systems are particularly poor and where political conflict thwarts relief efforts. The recent famine in North Korea was due to a combination of natural disasters, economic collapse, and lack of political will by its government to address the problem.

Chronic Hunger and Malnutrition

As devastating as famines are, they account for only a small fraction of hunger-related deaths. Famines can be attacked by relief agencies in a relatively short period of time if political conflict in the affected country does not hamper relief efforts. Chronic hunger and malnutrition affect a much greater number of people and are more difficult to combat.

Although no accurate figures on the prevalence of malnutrition exist, the World Health Organization (WHO) estimates that a half-billion people suffer from protein and calorie deficiencies and perhaps an equal number suffer from malnutrition caused by inadequate intakes of micronutrients, principally iron, vitamin A, and iodine. Thus, roughly 15 to 20 percent of the world's population suffers from some form of malnutrition. Malnutrition does not affect all segments of the population equally. Preschool children and pregnant and nursing women are particularly vulnerable to the dangers of malnutrition.

BOX 2-1
NATURAL DISASTER AND FAMINE IN BANGLADESH[1]

From June to September 1974, severe flooding in the Brahmaputra River in Bangladesh led to large-scale losses of the dry-season rice crop and created pessimism about the prospects for the transplanted spring crop. The price of rice doubled in fewer than three months during and after the floods. Two months after this sudden upturn in rice prices, unclaimed dead bodies began to be collected in increasing numbers from the streets of Dhaka, the capital city. Similar collections were reported throughout the countryside. The government of Bangladesh officially declared a famine in September 1974. Estimates of the final death toll vary widely, but most agree that more than one million people died of starvation or related causes during and after the famine.

Insufficient food stocks clearly hindered the government's efforts to provide relief. Inadequate relief stocks should not, however, be confused as a cause of the famine; the evidence clearly shows that in 1974 adequate food grains were available in Bangladesh to avoid famine. This same evidence shows that the districts most affected by the famine even had increased availability of food per person compared to prior years.

What, then, caused the famine? Landless laborers and farmers with less than half an acre of land were most severely affected by the famine. These groups, whose only true asset was their labor power, found that the value of their labor declined greatly relative to the price of rice. Despite available food in local markets, they were unable to purchase it. The flood did not immediately affect food supply since the lost crop would not have been harvested until the next year anyway. It did, however, greatly lower employment opportunities. Lower wages combined with higher rice prices were the root causes of the 1974 Bangladesh famine.

[1]Most of this material is drawn from Amartya K. Sen, *Poverty and Famines: An Essay on Entitlement and Deprivation* (New York: Oxford University Press, 1981).

Table 2-1. Estimated Worldwide Number of People Affected by Preventable Malnutrition

Deficiency	Morbidity due to Malnutrition	Estimated Prevalence of Morbidity	Group most affected
Protein and energy	Underweight	150,000,000	Children
Protein and energy	Stunted growth	182,000,000	Children
Iron	Anemia	2,000,000,000	Every age & sex
Vitamin A	Blindness	250,000–500,000	Every age & sex
Iodine	Brain damage	50,000,000	Every age & sex

Source: World Health Organization, 2003.

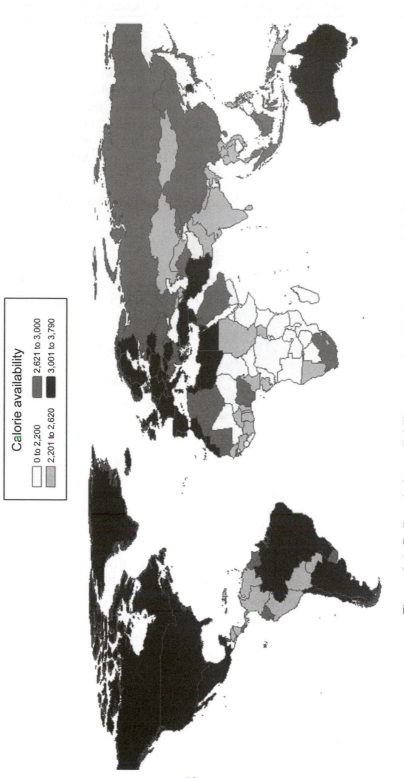

Figure 2-1. Daily calorie availability per capita, 2002 (Source: FAOSTAT data).

Calorie availability

- 0 to 2,200
- 2,201 to 2,620
- 2,621 to 3,000
- 3,001 to 3,790

Woman and child in Ethiopia
(photo: Mesfin Bezuneh).

Serious macronutrient (calories and protein) malnutrition in developing countries reflects primarily undernourishment — a shortage of food — not an imbalance between calories and protein. The availability of calories per capita by country is illustrated in Figure 2-1. Many of the countries with very low per capita calorie availability are found in Sub-Saharan Africa. A close, but not perfect, correspondence exists between low calorie availability and the low-income countries identified in the previous chapter. The major nutritional problem was once believed to be the shortage of protein. Although dietary protein is important, many nutritionists now believe that when commonly consumed cereal-based diets meet energy (calorie) requirements, it is likely that most protein needs will also be satisfied, for most people older than about two years of age. Thus, for everyone except infants, the greatest concern is the total quantity of food available to eat, and this quantity can most readily be measured by total dietary energy in terms of calories per day. In settings where overall energy intake meets minimum needs, remaining protein or micronutrient deficiencies can often be improved with rather small investments to improve the quality of the diet, which is not to say that those investments are necessarily made.

There are areas where calorie intake is adequate but protein or micronutrient intake is seriously deficient. Regions where diets are based on staples such as cassava or sugar rather than cereals are more likely to be deficient in protein even if calories are adequate. Iodine deficiency is common in regions far from the sea, for example, parts of the Andes in South America. Iron deficiency is a particularly serious problem among women of childbearing age all over the world, and vitamin A deficiency is a severe problem in many countries.

Consequences of Hunger and Malnutrition

Stunted growth, reduced physical and mental activity, muscle wasting, increased vulnerability to infections and other diseases, and, in severe cases, death are the most common consequences of calorie deficiencies. Death most frequently results from dehydration caused by diarrhea, whose severity is closely linked to malnutrition. Chronic protein malnutrition results in stunted growth, skin rash, edema, and change of hair color. A diet relatively high in calories but low in protein can result in an illness known as *kwashiorkor*, while a diet low in both calories and protein can result in an illness known as *marasmus*. People can live about a month with kwashiorkor, 3 months with marasmus; 7 to 10 million people die each year from the two diseases.[2]

Iron-deficiency anemia affects muscle function and worker productivity. Vitamin A deficiency is a leading cause of childhood blindness and often results in death due to reduced disease resistance. Iodine deficiencies cause goiter and cretinism.

There is little doubt that hunger and malnutrition result in severe physical and mental distress even for those who survive the infections and diseases. Malnutrition can affect the ability of a person to work and earn a decent livelihood, as mental development, educational achievement, and physical productivity are reduced. People with smaller bodies because of inadequate childhood nutrition are paid less in agricultural jobs in many countries. Lower earnings perpetuate the problem across generations.

Measuring Hunger and Malnutrition

Measuring the extent of hunger and malnutrition in the world is difficult. Disagreement surrounds definitions of adequate caloric and protein requirements, and data on morbidity and mortality combine the effects of sickness and malnutrition.

[2] Michael C. Latham, "Strategies for the Control of Malnutrition and the Influence of the Nutritional Sciences," in *Food Policy: Integrating Supply, Distribution, and Consumption*, ed. J. Price Gittinger, Joanne Leslie, and Caroline Hoisington (Baltimore, Md.: Johns Hopkins University Press, 1987), p. 331.

Nutritional assessments are usually attempted through food balance sheets, dietary surveys, anthropometric surveys, clinical examinations, and administrative records. Food balance sheets place agricultural output, stocks, and imports on the supply side and seed for next year's crops, exports, animal feed, and wastage on the demand side. Demand is subtracted from supply to derive an estimate of the balance of food left for human consumption. That amount left can be balanced against the Food and Agricultural Organization of the United Nations' (FAO) tables of nutritional requirements to estimate the adequacy of the diet. This method provides rough estimates at best, due to difficulties in estimating agricultural production and wastage in developing countries.

Food balance sheets provide only a picture of average food availability. Malnutrition, like poverty, is better measured if the distribution of food intake or of other indicators is also taken into account. Average national food availability can be adequate, while at the same time malnutrition is common in certain areas, or among particular population groups. Even within families, some members may be malnourished while others are not. To measure malnutrition accurately, information on households or individuals is required.

Household and individual information can be obtained from dietary or expenditure surveys and from clinical or field measurements of height, weight, body fat, and blood tests. These methods are expensive and seldom administered on a consistent and widespread basis for an entire country. They can be effective, however, in estimating malnutrition among population subgroups. Since preschool children are most vulnerable to nutritional deficiencies, random surveys to measure either their food intakes or anthropometry (body measurements) can provide a good picture of the extent of malnutrition. Another procedure for estimating the extent of malnutrition is to utilize existing data in hospital, health service, and school records. Anthropometric information from these records can be compared with standards. Unfortunately, these statistics can be biased because the records for rural areas are scarce, the poor are the least likely to have sought medical attention, and the quality of the information in the records is uneven. For example, many countries in Latin America record the heights, weights, and ages of first-year elementary school children. Unfortunately, many of the poorest children do not attend school. Estimates of the percentage malnourished among school-aged children generally understate the true problem. In summary, one reason that malnutrition is misunderstood is that its measurement is so difficult.

CAUSES OF HUNGER AND MALNUTRITION

A variety of factors contribute to hunger and malnutrition, but inadequate income is certainly the most important underlying cause. The World Bank estimates that redistributing just 2 percent of the world's output would eliminate malnutrition. But such redistribution would be feasible only if those who now go hungry had some way to obtain that food, or something to offer in exchange. If people, for whatever reason, produce too few goods and services, they lack income to buy food and they go hungry. Even in times of famine, decreased purchasing power rather than absolute food shortages is often the major problem, as food may be available in nearby regions. Incomes in the affected area have declined so that people cannot afford to buy food from the unaffected areas.

Figure 2-2 contains a schematic diagram of the determinants of nutritional status. Factors affecting income and productive assets determine how much food can be purchased or consumed by the family. Total food purchases and consumption do not, however, tell the entire story. Health status and family food preparation, along with how food is distributed among members of the family, help determine how food available to a family is related to individual nutritional status.

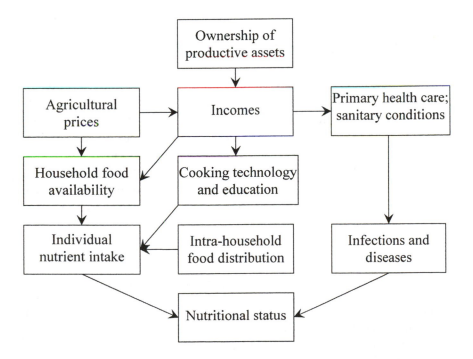

Figure 2-2. Determinants of individual nutritional status.

Health and Malnutrition

Poverty's interaction with malnutrition is often compounded by infectious diseases and parasites that reduce appetites, cause malabsorption of food, or result in nutrient wastage due to fever and other metabolic processes. Health problems and malnutrition exhibit a synergistic relationship: infections and parasites lead to malnutrition while malnutrition can impair the immune system, thus increasing the risk of infection and the severity of the illness.[3] Measles, parasites, intestinal infections, and numerous other health problems are prevalent in developing countries. Many of these health and sanitation problems lead to diarrhea, which in turn can lead to dehydration and death. Health is determined by, among other things, the sanitary facilities of the household. These in turn are affected both by family assets and income, and by government programs. There is room for optimism related to many childhood diseases. The World Health Organization reports that, because of sustained efforts to vaccinate children, the majority of the world's children under one year old are now vaccinated against six common childhood diseases. However, the last twenty years has seen HIV-AIDS become an escalating problem, first in Africa and increasingly in Asia. Malaria also remains a serious problem in many countries, especially in Africa where 13 countries report over 10 percent of their populations infected.[4]

Poor Nutritional Practices

Ignorance of good nutritional practices, maldistribution of food within the family, and excessive demands on women's time can all contribute to malnutrition. The results of studies that have examined each of these factors provide conflicting evidence as to their importance. Each factor is undoubtedly significant in some areas of the world but not in others. For example, in parts of Northern India and Bangladesh, evidence indicates that adult males receive a disproportionate share of food in the family compared to young females, but such unequal distribution of food does not occur everywhere.[5] Problems that appear to be related to

[3] See Joanne Leslie, "Interactions of Malnutrition and Diarrhea: A Review of Research," in *Food Policy: Interacting Supply, Distribution, and Consumption*, ed. J. Price Gittinger, Joanne Leslie, and Caroline Hoisington (Baltimore, Md.: Johns Hopkins University Press, 1987), pp. 355–70 for additional discussion.

[4] United Nations Development Programme, *Human Development Report 2005* (New York: Oxford University Press, 2005).

[5] See Michael Lipton, "Variable Access to Food," in *Food Policy: Interacting Supply, Distribution, and Consumption*, ed. J. Price Gittinger, Joanne Leslie, and Caroline Hoisington (Baltimore, Md.: Johns Hopkins University Press, 1987), pp. 385–92.

ignorance, and are in fact discriminatory, are sometimes related to culture and often to poverty.

There is evidence that whether the male or the female controls income and assets within a family helps determine how food is distributed. Likewise, there is strong evidence that increased educational opportunities for women are linked both to improved nutritional practices and more equitable distribution in the family.

Seasonal and Cyclical Hunger

Many people in developing countries move in and out of a state of malnutrition. There are hungry seasons, hungry years, and hungry parts of the life cycle. A given individual may or may not survive these periods and frequently experiences lasting physical, mental, and emotional impacts even if he or she does survive.

Hungry seasons occur because of agricultural cycles. In the weeks or months preceding a harvest, food can be in short supply. This normal seasonality can be exacerbated if crops in a particular year are short. In certain seasons of the year, particularly the rainy seasons, disease and infection are more common. Likewise, droughts, floods, and insect infestations happen in some years but not in others. Young children are vulnerable, in part due to dangers associated with diarrhea. Pregnant and lactating women experience extra nutritional demands on their bodies while the elderly suffer disproportionately as well, particularly if they lack the support of their children.

SOLUTIONS TO HUNGER AND MALNUTRITION PROBLEMS

Solutions to hunger and malnutrition depend on the types and causes of the problem. Famine relief strategies differ from solutions to chronic hunger and malnutrition. Unfortunately, there is no magic bullet. A concentrated effort on a variety of fronts is required to eliminate hunger.

Raising Incomes

Lifting vulnerable people out of poverty is central to any long-term strategy to alleviate malnutrition in the world. For subsistence farmers, this strategy involves raising productivity levels, increasing the access to land, or creating opportunities to migrate to off-farm employment. For the population in general, it implies a need for increased employment opportunities combined with higher productivity per person. The latter requires rapid growth in jobs and in capital per job in the non-farm sector. Enhanced education, an investment in human capital, will also increase productivity and incomes. Equal access to jobs and expanded economic opportunities in impoverished regions within countries can also help reduce poverty. Economic growth without increased employment for the poorest segments of the population will do little to reduce hunger. Programs to increase employment and earning opportunities for women are particularly important, partly because these opportunities help accelerate the transition to lower birth rates (for reasons discussed in Chapter 4).

Agricultural Production

Agricultural productivity is particularly important for the incomes and nutritional status of the poor, because in most developing countries many of the poorest people have no choice but to be farmers, to try to feed themselves and their families using their own labor and the available land. Increased productivity for those farmers not only raises their incomes and purchasing power, but can also lower the price of food for those who must buy it to feed their families, making it possible for the poor to purchase larger quantities. Hence, methods for increasing food production are a major focus of this book. Increased use of purchased inputs, improved marketing and credit institutions, improved agricultural policies, enhanced education, effective agricultural research, and development of infrastructure such as roads, storage, and irrigation systems are particularly important.

Food Intervention Programs

Food price subsidies, supplementary feeding programs, and food fortification can all help reduce nutritional deficiencies. Few developing countries have come close to eliminating malnutrition without some combination of these practices. At the same time, these programs alone cannot solve problems of chronic malnutrition.

General food-price subsidies were used in Sri Lanka for several years and helped relieve malnutrition and extend life expectancy to a remarkable degree. However, food-price subsidies are expensive, and even Sri Lanka decided to cut back its general subsidy, and instead to target specific groups. A study by the International Food Policy Research Institute (IFPRI) of the Sri Lankan food stamp scheme indicated that the targeted subsidies did reduce program costs substantially, but had mixed results in reaching the poor.[6] Food-price subsidy schemes sometimes lower prices, thereby reducing incentives for domestic food production.

---

[6] Neville Edirisinghe, *The Food Stamp Scheme in Sri Lanka: Costs, Benefits, and Options for Modification*, International Food Policy Research Institute, Research Report No. 58, (Washington, D.C., March 1987), pp. 1–85.

Rural Health Center in Colombia.

Several countries have instituted supplementary feeding programs for vulnerable groups such as children and pregnant and nursing mothers. In some cases these programs provide food to be consumed in a specific location such as in schools or health centers, while in others food may be consumed at home. In either case, while total family food consumption rises, that of the food recipient usually grows by less than the total donation. Some food is shared with family members. The evidence on supplementary feeding programs indicates that they often are associated with measurable improvements in nutritional status, but they tend to be expensive for the benefits received. Administration of these projects can be very difficult. In some cases, these programs have been assisted with food aid from other countries, as discussed below.

Another major food intervention program is the fortification of food by adding specific nutrients to food during processing. The most successful example of fortification is programs that add iodine to salt to prevent goiter. Vitamin A also has proven feasible to add to foods such as tea, sugar, margarine, monosodium glutamate, and cereal products. Attempts have been made to fortify food with iron to prevent anemia, but reducing iron-deficiency anemia has proven to be a complex problem. In general, the effectiveness of adding nutrients to food is reduced by the fact that the poor buy few processed foods, there is often cultural resistance to the fortified product, and the cost of fortification is prohibitive. In many cases, the "fortified" food has been shown to have no more nutrients than unfortified foods; quality control can be prohibitively expensive in developing countries. However, recent success in incorporating vitamin A and iron into rice, through genetic modification, provides a potential avenue for reducing these micro-nutrient problems.

Health Improvements

Programs to improve sanitation, reduce parasitic infections, and prevent dehydration caused by diarrhea can reduce malnutrition and mortality substantially. For example, oral rehydration therapy, which involves the use of water, salt, and sugar in specified proportions to replace fluid lost during diarrhea, can significantly reduce diarrhea-related deaths. Investments in sanitation services, such as potable water and latrine construction, help improve water quality, and, when combined with effective education programs, can significantly improve nutritional status. Improvements in health services including immunization programs can reduce the incidence and intensity of diseases that contribute to malnutrition.

Political, Social, and Educational Changes

Political stability can help alleviate both famine conditions and chronic hunger. The famine in Ethiopia in 1983 and 1984 was exacerbated substantially by political upheaval that hampered relief efforts. The recent famine in North Korea also has political roots. Because programs to curb chronic hunger and malnutrition require long-term commitments, they are necessarily rendered less effective by political instability or malfeasance. Responsible political action can improve income distribution in a country, thereby reducing poverty and malnutrition.

Social, cultural, and educational factors also come into play. For example, declining rates of breastfeeding in some countries have contributed to malnutrition as substitutes often are less nutritionally complete, are often watered down, and in some cases are even unsanitary. In other cases, breastfeeding may continue too long without the addition of needed solid foods. While social and cultural factors change slowly, and economic factors influence decisions, education can help. In fact, few consumption practices are totally unaffected by education. Nutrition education programs, especially when combined with income-generating projects or efforts to increase a family's access to nutrients, such as home gardening, have been shown to lead to improved nutritional status.

International Actions

A variety of international actions can help alleviate both famine and chronic malnutrition. Because increased incomes are so important to improved nutrition, opening of markets in more developed countries and debt relief are actions that can help, especially in the long run. Foreign assistance can provide short-run relief and also facilitate long-run development.

Reduced barriers by developed countries to imports from developing countries will enable low-income nations to gain greater access to world markets. The foreign exchange earned can be used for development efforts and food imports when needed.

Debt relief is a dire need in many countries, particularly where past governments were not held accountable for how loans were spent, so that the funds were not invested productively. When bad debts arise, it is usually in the long-run best interests of both lender and borrower to share some of the burden of adjustment, to reduce expectation of loan repayment in line with the actual productivity of the loan. For more details on this important topic, see Chapter 18.

Foreign assistance includes food aid as well as technical and financial assistance. Gifts and loans of food at low interest rates can help solve part of the hunger problem if the food assistance is properly administered. Food aid can relieve short-term famines and be used in supplementary feeding programs and in other activities, such as food-for-work programs, to help generate wealth in developing countries. Much more important for the long run, financial and technical assistance can help developing countries expand their capital bases and improve methods for producing food and other products, allowing them to import or develop the new technologies they need to break out of poverty.

SUMMARY

In this chapter, the types and consequences of hunger and malnutrition were examined. Even though it is difficult to measure accurately the extent of hunger and malnutrition in the world, it is known that chronic malnutrition affects more people than do famines. Malnutrition results in reduced physical and mental activity, stunted growth, blindness, anemia, goiter, cretinism, mental anguish, and death.

The causes of hunger are many, but virtually all these causes are related to poverty. Infections, diseases and parasites, poor nutritional practices, and seasonal variability in food supplies all contribute to the severity of malnutrition. Solutions to hunger and malnutrition include raising incomes; increasing agricultural production in developing countries; food intervention programs; improving health systems; political, social, and educational changes; and a series of international activities such as food aid and other foreign assistance, debt relief, opening of foreign markets, and price stabilization.

IMPORTANT TERMS AND CONCEPTS

Anthropometry
Chronic malnutrition
Debt relief
Dietary surveys
Famine
Food aid
Food balance sheets
Food fortification
Food price subsidies
Foreign assistance

Kwashiorkor, marasmus, goiter,
   anemia, and cretinism
Maldistribution of food
Oral rehydration therapy
Political upheaval
Price stabilization
Protein and calorie deficiency
Seasonal and cyclical hunger
Supplementary feeding programs
Vitamin and mineral deficiency

Looking Ahead

Hunger and malnutrition imply a need for food but not necessarily a demand for food unless that need is backed by purchasing power. Food demand is influenced by income, prices, population, and tastes and preferences. In the next chapter, we will examine tools that can help measure or project the extent to which various demand factors affect food consumption. We will explore how demand interacts with supply to determine prices. The tools discussed are the first of a set of theories and methods presented in this book that can improve your ability to analyze and not just observe food and development problems and policies.

QUESTIONS FOR DISCUSSION

1. Is famine more widespread today than in the past?
2. Is protein deficiency a more severe problem in developing countries today than is calorie deficiency? Why or why not?
3. If people in the United States moved to a diet in which they consumed more grain and less meat, would there be more food for people in poor countries of the world? Why or why not?
4. What are the principal causes and consequences of hunger?
5. How do we measure the adequacy of food availability in a country?
6. What are some solutions to hunger and malnutrition problems?
7. Why and how does political upheaval contribute to famine?
8. What are the major interactions between health and nutritional problems?

RECOMMENDED READING

Flores, Rafael, and Stuart Gillespie, *Health and Nutrition: Emerging and Reemerging Issues in Developing Countries*, IFPRI 2020 Vision, Focus 5, February 2001 (available at website: http://www.ifpri.org/index1.htm).

Foster, Phillips, and Howard D. Leathers, *The World Food Problem* (Boulder, Colo.: Lynne Reinner Publishers, 1999).

Gittinger, J. Price, Joanne Leslie, and Caroline Hoisington, eds., *Food Policy: Integrating Supply, Distribution, and Consumption* (Baltimore, Md.: Johns Hopkins University Press, 1987), chs 24–34.

Mellor, John W., and Sarah Gavian, "Famine, Causes, Prevention, and Relief," *Science*, vol. 235 (January 1987), pp. 539–45.

Pinstrup-Andersen, Per, and Rajul Pandya-Lorch, *The Unfinished Agenda: Perspectives on Overcoming Hunger, Poverty, and Environmental Degradation*, International Food Policy Research Institute, Washington, D.C., 2001, especially Parts 1 and 2 (available at website: http://www.ifpri.org/index1.htm).

Sanchez, Pedro, M.S. Swaminathan, Philip Dobie, and Nalan Yuksel, *Halving Hunger: It Can Be Done*, U.N. Millennium Project Task Force on Hunger (London: Earthscan, 2005).

Sen, Amartya K., *Poverty and Famines: An Essay on Entitlement and Deprivation* (New York: Oxford University Press, 1981).

United Nations Development Programme, *Human Development Report 2005* (New York: Oxford University Press, 2005).

United Nations Standing Committee on Nutrition, *Fifth Report on the World Food Situation: Nutrition for Improved Development Outcomes* (New York: United Nations, 2004).

World Health Organization, *World Health Report 2005, Make Every Mother and Child Count* (Geneva: WHO Press, 2005).

Economics of Food Demand

Rather than a race between food and population, the food equation should be viewed as a dynamic balance in individual countries between food supply and demand.
— J. W. Mellor and B. F. Johnston[1]

This Chapter

1. Discusses the concept of effective demand and the relative importance of income, population, preferences, and prices in determining the demand for food as development occurs
2. Explains the importance of income elasticities and price elasticities of demand for projecting consumption patterns and for development planning
3. Describes how food supply interacts with demand over time to determine price levels and trends.

EFFECTIVE DEMAND FOR FOOD

The need for food and the effective demand for food are related but distinct concepts. Food needs correspond to the nutrient consumption required to maintain normal physical and mental growth in children, and to sustain healthy bodies and normal levels of activity in adults. The effective demand (often just called demand) for food is the amount of food people are willing to buy at different prices and income levels, given their needs and preferences.

In this chapter, we consider the means for analyzing food demand changes resulting from both income and price changes. The goal is to help you predict the likely impacts on consumption of a change in either factor. Later, we will see how these demand pressures interact with supply conditions to determine changes in economic well-being over time.

[1] John W. Mellor and Bruce F. Johnston, "The World Food Equation: Interrelations among Development, Employment, and Food Consumption," *Journal of Economic Literature*, vol. 22 (June 1984), p. 533.

Determinants of Food Demand

The quantity demanded of food, or of any commodity, is influenced by two major factors: its price, relative to all other goods; and consumers' incomes, relative to all prices. In order to isolate each effect, economists use a thought-experiment in which we imagine a change in only one variable at a time, and trace out the resulting change in another.

When considering the effect of a change in price on quantity consumed, we expect a higher price to cause a lower quantity consumed, and vice versa. This inverse relationship between price and quantity consumed is often called the *law of demand*, and is illustrated on a graph using a market *demand curve* (Fig. 3-1). The slope and the location of the market demand curve are determined primarily by income per person, the number of people and the distribution of income among those people, prices of other goods, and other factors such as consumer

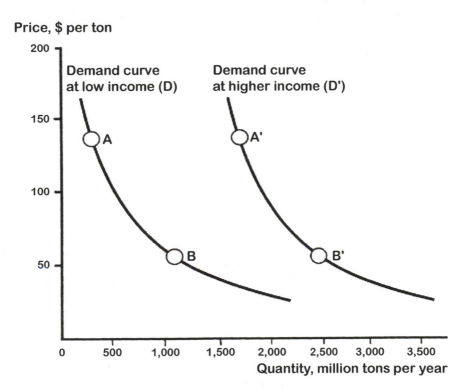

Figure 3-1. Hypothetical demand curves for a commodity. A reduction in the price of the commodity, all other things being equal, will cause a movement along a demand curve, say, from point A to point B, and an increase in quantity demanded. Changes in the determinants of demand—population, income, prices of other goods, and preferences—can cause a shift in demand, say, from point A on demand curve D to point A' on demand curve D'.

preferences and consumption technologies. Changes in any of these factors cause the demand curve to shift, as shown by the shift from curve D to curve D' in Figure 3-1. Such a shift might be caused by a rise in income, which increases the quantity demanded at a given price. Alternatively, the shift might be caused by population growth at a constant per capita income. This income effect on demand varies by commodity. Because the influence of income on food demand is not constant across countries, within countries, or by commodity, it is important to have a measure of the sensitivity of demand for food and for particular goods to changes in income. The measure used is called the *income elasticity of demand*.

Income Elasticities of Demand

The income elasticity of demand is defined as the percentage by which the quantity demanded of a commodity will change for a 1 percent change in income, other things remaining constant.[2] For example, when per capita income increases by 1 percent, if quantity demanded of a commodity increases by 0.3 percent, its income elasticity of demand is 0.3. Typically, for a very low-income country, the elasticity of demand for food as a whole is around 0.8, while for a very high-income country it is around 0.1. This difference in income elasticities means that changes to income have a much larger relative impact on food demand in low-income countries than in high-income countries.

By necessity, poor people have no choice but to spend the bulk of their income on food — at times as much as 80 percent — and when their incomes rise they spend a high proportion of that increase on more food. Eventually, however, further increases in income tend to be spent on other things. This change in the proportion of the family's budget spent on food, or *Engel's law*, says that as income increases, people spend a smaller proportion of their total income on food. This process is reflected in Figure 3-2, which shows the percentage of total income spent on food for a number of countries with different levels of per capita income. The distinct downward slope associated with Engel's law would be similar if the graph were constructed for individuals within a country, where richer people spend a smaller fraction of their income on food.

Engel's law reflects, in part, the limited capacity of the human stomach, but note that total expenditures on food generally continue to rise

---

[2] If we define n to be the income elasticity of demand for a good, ΔQ to be the change in quantity demanded for that good, and ΔI to be a change in income, then:

$$n = \frac{\%\Delta Q}{\%\Delta I} = \left(\frac{\Delta Q/Q}{\Delta I/I} \right) = \left(\frac{\Delta Q}{\Delta I} \right)\left(\frac{I}{Q} \right).$$

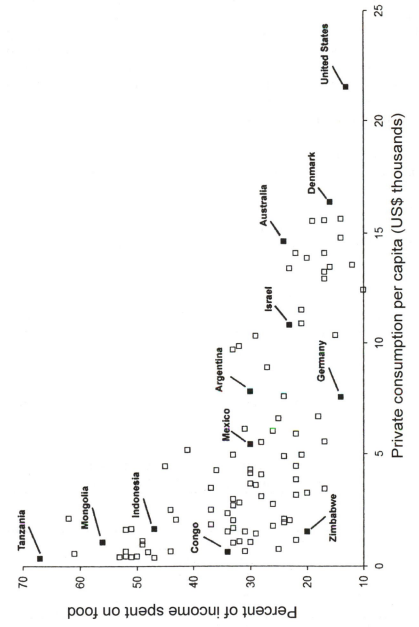

Figure 3-2. Relationship between per capita income and percentage of income spent on food, most countries (Source: World Bank, *World Development Indicators*, 2000).

with income, even as the proportion of the budget spent on food declines. Rising incomes lead people to consume more total calories, and also to consume more expensive foods. These foods are often more highly processed (for example, as people switch to bread instead of porridge) and include more animal products (meat, dairy, eggs, and fish) as well as more fruits and vegetables. The transition in consumption from a few inexpensive starchy staples such as cassava, rice, or corn to this greater variety of more expensive foods is known as *Bennett's law*, named after the same M.K. Bennett mentioned in Chapter 1. But note that when consumers switch from starchy staples to animal products, demand for animal feed can rise very fast: consumers may reduce their direct consumption of cereal grains as food, while increasing their total usage of cereal grains as animal feed.

Diversification and improvement of the diet with rising incomes implies that income elasticities vary by commodity, and by income level. To show patterns of demand among some of the poorest people in the world, Table 3-1 provides estimated income elasticities in various regions of Sub-Saharan Africa for a range of commodities. Estimated income elasticities of demand for other countries and commodities are presented in Table 3-2. Note that income elasticities for animal products are higher than for food grains and root crops. Wheat and rice income elasticities tend to be higher than those of coarse grains, while roots and tubers have consistently small elasticities. The substantial variations in income elasticities across countries reflect differences in income and in preferences for foods. For example, the income elasticity of demand for beef is low in Latin America compared to Africa, partly because initial levels of beef consumption are high in Latin America.

Table 3-1. Selected Income Elasticities of Demand for Agricultural Commodities in Sub-Saharan Africa

Region	Wheat	Rice	Maize	Millet	Roots and tubers	Pulses
The Sahel	0.92	0.93	0.46	0.15	– 0.04	– 0.14
West	0.87	0.65	0.15	0.09	0.12	0.42
Central	0.55	0.93	0.66	0.28	– 0.21	0.14
Eastern	0.51	0.58	0.28	0.01	0.29	0.02
Southern	1.46	0.56	0.35	0.17	– 0.15	– 0.002

Source: Cheryl Christensen *et al.*, *Food Problems and Prospects in Sub-Saharan Africa: The Decade of the 1980s*, U.S. Department of Agriculture, Economic Research Service, Foreign Agricultural Research Report No. 186 (Washington, D.C., August 1981).

Table 3-2. Selected Income Elasticities of Demand for Cereals and Livestock Products in Various Countries

Country	Cereals	Beef	Pork	Poultry	Cow's milk	Eggs
Brazil	0.15	0.58	0.29	0.64	0.45	0.55
Egypt	0.04	0.80	0.70	1.30	1.00	0.70
India	0.25	1.20	0.80	1.50	0.80	1.00
Indonesia	0.29	1.50	0.80	1.50	0.20	1.20
Kenya	0.35	1.00	0.70	1.20	0.59	1.30
South Korea	0.09	0.80	0.73	1.00	0.49	0.80
Malaysia	0.14	0.49	0.41	0.87	0.57	0.73
Mexico	− 0.10	0.59	0.49	0.93	0.68	0.59
Nigeria	0.17	1.20	1.00	1.00	1.20	1.20
Philippines	0.22	1.20	0.93	1.00	1.50	1.00
Thailand	0.06	0.56	0.47	0.50	0.80	0.50
Turkey	− 0.05	0.80	0.50	1.20	0.80	0.80

Source: J. S. Sarma, *Cereal Feed Use in the Third World: Past Trends and Projections to 2000*, International Food Policy Research Institute, Research Report No. 57 (Washington, D.C., December 1986), p. 64.

Most of the estimated income elasticities in Tables 3-1 and 3-2 range between 0 and 1. These goods are called *normal* goods. Goods with income elasticities greater than 1 are called *superior* and represent foods that can be thought of as luxuries in the diet in a particular country. If the income elasticity is less than 0, the goods are called *inferior*, as consumption of them actually declines as income increases.

The fact that income elasticities vary by commodity means that increases in income will result in an asymmetrical expansion in demand for different commodities. Demand for some commodities will expand by a greater percentage than that for others. Depending on the nature of supply, asymmetric expansion of demand can cause different pressures on commodity prices. These changes in commodity prices can influence which crops producers grow and can help determine the direction of development.

Price Elasticities of Demand

So far we have focused on per capita income as the major determinant of food consumption per person, but quantity demanded also responds to price changes. That price response was represented by movements along the demand curve in Figure 3-1, such as movement from point A at a high price to point B with a relatively low price and a higher quantity demanded. The degree of response in demand from a change in price is measured by the (own) *price elasticity of demand*, defined as the

Potatoes in Ecuador.

percentage change in quantity demanded of a commodity given a 1 percent change in its price, other things remaining unchanged.[3] For example, an own-price elasticity of –0.5 means that with a 1 percent change in price, the quantity demanded will change in the opposite direction by 0.5 percent. Own-price elasticities are typically negative, reflecting the negative slope of the demand curve. If the own-price elasticity of demand is greater (in absolute value) than one, the demand is said to be *elastic*. If it is equal to one, it is said to be *unit-elastic*. If it is less than one, it is said to be *inelastic*. In a demand curve such as that shown in Figure 3-1, an elastic demand has a relatively flat slope, as small price changes lead to large quantity changes.

Price elasticities of demand are useful for projecting demand changes that might result from policies that manipulate prices or from supply shifts. *Cross-price elasticities,* which represent the percentage change in quantity consumed of one commodity for a 1 percent change in the price of another commodity, holding all else equal, also are important.[4] If the cross-price elasticity of demand is greater than zero, the

---

[3] If we define E to be the price elasticity of demand for a good, ΔQ to be the change in quantity consumed, and ΔP to be the change in its price, then:

$$E = \frac{\%\Delta Q}{\%\Delta P} = \left(\frac{\Delta Q/Q}{\Delta P/P}\right) = \left(\frac{\Delta Q}{\Delta P}\right)\left(\frac{P}{Q}\right).$$

[4] If we let E_{12} = the cross-price elasticity for commodity 1 as the price of commodity 2 changes, ΔQ = the change in the quantity demanded of commodity 1, and ΔP_2 = the change in price of commodity 2, then:

$$E_{12} = \frac{\%\Delta Q_1}{\%\Delta P_2} = \left(\frac{\Delta Q_1}{\Delta P_2}\right)\left(\frac{P_2}{Q_1}\right).$$

two commodities are said to be *substitutes*. If the cross-price elasticity is zero, the commodities are unrelated, and if it is less than zero, they are called *complements*.

When the price of a commodity changes, the change in relative prices causes most consumers to adjust the composition of the commodity bundle they purchase so that they buy less of the good that increased in price. This substitution is known as the *substitution effect*. Also, if the price of a commodity increases, the real purchasing power of a given amount of income is reduced, causing demand to change because of an *income effect*. In most cases, this income effect is a second factor that reduces demand for the commodity experiencing the price increase.[5] For inferior goods, however — commodities such as potatoes and cassava — the income effect may work in the opposite direction and partially offset the reduced consumption induced by the relative price increase.

A price increase for a good will increase consumption of substitutes, and decrease consumption of complements. Part of these consumption changes are caused by changes in relative prices and part of them are due to income effects. Because the income elasticity of demand for food is large for low-income consumers and because they spend a high proportion of their income on food, low-income consumers often make larger adjustments in their commodity purchases than do high-income consumers when prices change.

Obtaining Elasticity Estimates

The effects of changes in consumer behavior discussed above have important implications for food policies and nutrition in less-developed countries, so food-policy analysts often need updated local estimates of the sizes of the income elasticities, own-price elasticities, and cross-price elasticities of demand for various commodities. For example, if a policymaker wants to project domestic food demand and the increased production or imports needed to meet that demand, the income elasticity of demand for food is one of the pieces of information needed. If an estimate of the effect on the calorie and protein intakes of the poor resulting from a decrease in the price of rice is needed, it is important to have the own-price elasticity of demand for rice and the cross-price

---

[5] If the consumer is also a producer of the good, which is often the case in rural areas of developing countries, this income effect can be positive. Commodity price increases can actually raise disposable income by increasing farm profits. This profit effect can be important when examining price responses among agricultural households that both consume and produce goods.

elasticities of demand between rice and other major foods in the country, disaggregated by income group.

How are elasticity estimates obtained? There are several approaches, and the appropriate procedure to use depends on the data available and the questions being asked.[6] One type of estimate uses national aggregate data on consumption, production, trade, and prices. Often these data are published by international sources for several countries. If data are available on the same factors for several countries or for several regions in one country for one period of time, they are called *cross-sectional data*. If data are available for the same factors for one country for several years, they are called *time-series data*. Often we have combined cross-sectional and time-series data, that is, time-series data for the same factors for a number of countries at the aggregate level. These aggregate data are not very useful for studying short-term consumption behavior for commodities within countries because tastes and preferences vary by country. However, the data may be helpful in making long-term projections.

Sometimes, household-level, cross-sectional data are obtained by sampling many households to obtain information on income, expenditures on different commodities, prices paid, and educational levels and other demographic characteristics.[7] Occasionally the data are collected over time as well, although not often because of the cost involved. If one is interested in microeconomic issues associated with consumer behavior for different income groups, these household-level data are preferred.

Data (aggregate or household level) are usually analyzed graphically and then in a statistical or *econometric* (statistical model which incorporates economic theory) model containing a set of demand equations.[8] These equations include variables representing the factors mentioned above. Elasticities are calculated from the estimated coefficients. These elasticities can be used for a variety of policy and planning purposes. Sometimes when data do not exist in one country or at a period

[6] See C. Peter Timmer, Walter P. Falcon, and Scott R. Pearson, *Food Policy Analysis* (Baltimore, Md.: Johns Hopkins University Press, 1983), pp. 48–56, for additional but brief discussion of approaches for obtaining elasticities. More detailed information is provided in several textbooks on consumer demand analysis.

[7] Collecting household data is a difficult and costly undertaking. For an excellent overview of topics in household data collection, see Joachim von Braun and Detlev Puetz, *Data Needs for Food Policy in Developing Countries* (Washington, D.C.: International Food Policy Research Institute, 1993).

[8] See Angus Deaton, *The Analysis of Household Surveys* (Baltimore, Md.: Johns Hopkins University Press, 1997), especially ch. 1, for an advanced treatment of types of data and their uses for policy analysis.

in time, studies from other countries or at a different period of time are used. Elasticities from other studies may not be ideal, but they are frequently used.

Some countries have serious deficiencies in aggregate and household-level data. Often these data are unreliable or even nonexistent. Policy analysts who have little time or money to collect new data and estimate a model sometimes rely on relationships from economic theory to obtain rough approximations of missing elasticities. For example, there is a useful working assumption (called the homogeneity condition) that the sum of the own-price elasticity, the income elasticity, and the cross-price elasticities of demand for a commodity is equal to zero.[9] Typically, the sum of the cross-price elasticities for a commodity is greater than zero and the own-price elasticity is negative. Therefore, the absolute value of the own-price elasticity is usually larger than the income elasticity of demand. One may have an estimate of the income elasticity of demand but not the own-price elasticity. The homogeneity condition can be used to obtain a rough estimate of the size of the price elasticity of demand given that income elasticity and assumptions about cross-price elasticities. The homogeneity condition is just one example of the use of demand theory. The main points are that data availability and quality limit the potential for economic analysis, but a variety of techniques can often be exploited to interpret the available data in useful ways.

USING CONSUMPTION PARAMETERS FOR POLICY AND PLANNING

The purpose of obtaining income and price elasticities is to assist with policy analyses and planning. A variety of questions can be answered with the help of these elasticities. For example, what will happen to the consumption of rice, wheat, sugar, or meat when income rises? What will happen to the aggregate demand for food? How will the demand change for different commodities as absolute and relative prices change? What will be the effects of price and income policies on the poor? The answers to these questions help policymakers anticipate future demand changes and production needs, and provide information for designing price and income policies (see Box 3-1).

---

[9] That is, for the ith commodity out of T commodities, $E_i + \eta_i + \sum_{\substack{j=1 \\ i \neq j}}^{T} E_{ij} = 0.$

BOX 3-1
IMPACTS OF RICE PRICE POLICY
ON THE POOR IN THAILAND

Angus Deaton used household-level data from Thailand to examine how policies affecting the price of rice would affect households in rural and urban areas and at different levels of income. Because rural households are both producers and consumers of rice, increased prices may or may not benefit them. They will gain as producers, but lose as consumers (all urban rice consumers will lose as a result of higher rice prices). The key to the analysis is to determine the "net benefit ratio" or the difference between the value of production and the value of consumption divided by total household expenditures. This ratio varies by total household income, and the analysis shows that middle-income producers will benefit most from rice price increases. High-income rural households benefit very little from high prices (they earn their incomes outside agriculture or do not produce much rice). Very low-income rural households benefit by relatively small amounts, because their marketed surplus is low. Compared to plantation-type products (such as sugar and bananas), where product price increases benefit larger-scale producers, rice price policy has its strongest impact on the middle of the income distribution in rural areas of Thailand. The study shows that the impacts of price policy depend on the commodity in question and the socioeconomic conditions of producer and consumer groups.

Source: Angus Deaton, *The Analysis of Household Surveys* (Baltimore, Md.: Johns Hopkins University Press, 1997), pp. 187-90.

Income-induced Changes in the Mix of Commodities Demanded

For commodities with high income elasticities, demand can grow very rapidly when income rises. Anticipating income growth, policymakers may want to support research or use other means for encouraging increased production of those commodities. Otherwise, prices will rise or imports increase in response to demand growth.

Many highly income-elastic commodities such as milk and vegetables have high nutritional value. However, some goods with relatively high nutritional value have low income elasticities.[10] If a government wants to increase consumption of a good with a low income

---

[10] Elasticities reflect people's preferences for different attributes of the good, including taste, convenience, and nutritional value. A low value for an elasticity is not necessarily "bad"; it reflects consumer choices given income, preferences, prices, and information about the good.

Cattle in Colombia.

elasticity, it may have to resort to educational or subsidy programs. Educational programs help change people's perceptions about physical (nutrient) needs and the amount of these needs the food provides. These programs essentially lower the costs associated with acquiring information about nutrient needs and food nutrient content.

At the world level, differences in income elasticities by commodity imply that as per capita income grows over time, a relative shift will occur in demand toward agricultural commodities with high income elasticities. Many of these are high-protein foods such as livestock products. One can also expect the grains fed to livestock, such as corn, to increase in demand relative to food grains such as rice. These types of changes have already been occurring over the past several years.

Another impact of these patterns of income elasticities is that the average income elasticity of demand for food grains will decrease as development occurs. Small income elasticities are associated with small price elasticities of demand. With lower price elasticities, increased production of food grains would put sharp downward pressure on their prices. Lower prices should help poor consumers, who continue to spend large shares of their budget on grains, but may force many of the farmers producing these grains to switch to other commodities or leave agriculture.

Changes in Aggregate Food Demand as Development Proceeds[11]

The demand for food is influenced by population, per capita income, prices, and preferences. As development proceeds, the two primary factors shifting the demand for food outward are increases in population and in per capita income. These two major forces are captured by the simple relation $D = p + ng$, where D = rate of growth in the demand for food, p = rate of population growth, n = income elasticity of demand for food, and g = rate of increase in per capita income.

In the above equation, population influences food demand in two ways. First, as presented by the term p, it causes a proportional increase in demand. However, per capita income equals total income divided by population. Therefore, the net effect of population growth will not be a proportional increase in demand, because population growth may slow the rate of per capita income growth.

At the extreme, if income does not expand at all with increased population, the drop in per capita income will almost completely nullify the direct effect of population growth. For example, developing countries often experience a population growth rate of 3 percent per year during the early stages of development. The income elasticity of demand for food may be as high as 0.9. If total income remains constant, then per capita income will decline by 3 percent and the rate of growth of demand will be $D = 3 + 0.9(-3) = 0.3$.[12]

On the other hand, if per capita income is growing at 3 percent per year while population is also growing, at 2.5 percent (rates that are not uncommon in middle-income developing countries), even if the income elasticity of demand for food drops to 0.7, the rate of growth in demand for food would be 4.6 percent per year. Few countries have been able to maintain such a rate of growth in agricultural production over time. Thus, food imports may be needed to meet growing demands.

These examples ignore the fact that income growth in most less-developed countries is heavily dependent on agricultural output. If agricultural output fails to grow, per capita income will grow very slowly. As development proceeds, the proportion of employment and of total national income derived from agriculture shrinks. Even so, total per capita income still may be affected by the rate of growth of agricultural production because agriculture provides food, capital, and a

---

[11] Material in this section draws on John W. Mellor, *Economics of Agricultural Development* (Ithaca, N.Y.: Cornell University Press, 1966), pp. 73–9.

[12] The negative consequences of such a scenario should be obvious: total demand will increase by 0.3 percent but *per capita* demand will decline by 2.7 percent.

market for non-agricultural products. These issues will be more fully discussed in subsequent chapters.

The determinants of food demand are interrelated, but as development proceeds, certain patterns tend to hold for some of these factors (see Table 3-3). As incomes increase, population growth rates generally increase slightly at first, as death rates decline. For a number of reasons discussed in the next chapter, population growth rates eventually fall as income continues to grow. The rate of per capita income growth is frequently highest in the middle-income countries, and the income elasticity of demand for food declines continually as income grows. The result is that the rate of growth in food demand is highest for middle-income countries. These are the countries that are most likely to need food imports. Data indicate that middle-income countries frequently exhibit the largest increase in per capita income and food imports even though they also experience the largest increases in agricultural production.

INTERACTIONS OF DEMAND WITH SUPPLY

If markets operate freely with numerous buyers and sellers, supply interacts with demand to determine the quantity supplied and demanded as well as the price. Market supply is defined as the amounts of a product offered for sale in a market at each specified price during a specified period of time (Fig. 3-3).

A given supply curve assumes that the following factors are held constant: (1) technology of production (the way the good is produced), (2) prices of inputs used in production, (3) prices of products that may be substituted in production, and (4) number of sellers in the market. Changes in these factors can cause the supply curve to shift inward or outward. For food as a whole, changes in technology are a major factor

Table 3-3. Comparison of Growth of Demand for Agricultural Goods, Hypothetical Cases

Level of development	Rate of population growth	Rate of per capita income growth	Income elasticity of demand	Rate of growth of demand
Very low income	2.5	0.0	1.0	2.5
Low income	3.0	1.0	0.9	3.9
Medium income	2.5	4.0	0.7	5.3
High income	2.0	4.0	0.5	4.0
Very high income	1.0	3.0	0.2	1.3

Source: Adapted from John W. Mellor, *Economics of Agricultural Development* (Ithaca, N.Y.: Cornell University Press, 1966), p. 78.

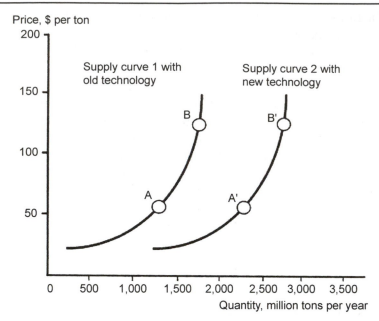

Figure 3-3. Hypothetical supply curve for a commodity. An increase in the price of the commodity, all other things being equal, will cause a movement along a supply curve, say, from point A to point B, and an increase in quantity supplied. Changes in the determinants of supply—technology, input prices, other output prices, number of sellers—can cause a shift in the supply curve, say, from A along supply curve I to A' along supply curve 2, or vice versa if there is a worsening of productivity.

causing shifts in supply over time. A new technology that lowers the cost of production will shift the supply curve downward to the right (such as from supply curve 1 to supply curve 2 in Figure 3-3).

Price and Policy Implications

The rate of growth or decline in agricultural prices over time depends on the net effects of supply and demand shifts (Fig. 3-4). Because of outward shifts of the demand curve caused by population and income growth, it is unlikely that agricultural prices will experience major declines resulting from supply growth in a country during the early stages of development.[13] If the supply curve for food shifts out very little, population- and income-driven demand growth could lead to price increases. However, these increases are likely to be small because of the close relationship between agricultural production growth and income growth during early stages of development. As noted earlier, it is difficult to

---

[13] However, there may be substantial local or regional variation (see Box 3-2).

get large increases in income, and therefore effective demand, without corresponding increases in agricultural production.

Other important determinants of the effect of supply and demand shifts on agricultural prices are the elasticities of supply and demand. The more elastic the supply curve (roughly the flatter it is in Figure 3-4), the less prices will change as demands grow. Open economies (those where imports and exports are common) tend to be characterized by more elastic commodity supplies. One means of minimizing demand-induced price increases is to permit food imports. Another is to increase the responsiveness of the food production sector.

The expected relative stability in food prices during the early stages of development (except as prices are affected by short-run phenomena such as weather) implies a need to place emphasis on policies to shift out the agricultural supply curve and to raise incomes, rather than on pricing policies. Thus, the focus of public investment needs to be where the return is highest, whether it is inside or outside agriculture. Because it is difficult to increase incomes of the poor without increasing employment, the country may need to focus investments on labor-intensive commodities and industries.

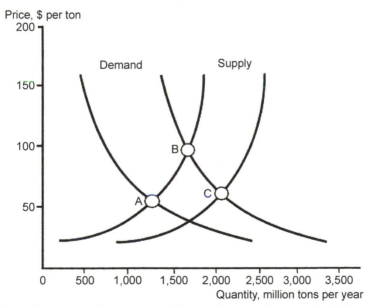

Figure 3-4. Hypothetical supply and demand curves for a commodity. Changes in determinants of demand—for example, income and population—can cause a shift in demand while changes in the determinants of supply—for example, technology—can cause a shift in supply. When both are shifting out, whether the net effect is a price increase or decrease (whether A is higher or lower than C) depends on the relative size of shifts of supply and demand and the slopes of the curves.

BOX 3-2
MARKETS AND REGIONAL PRICE VARIATION

Developing countries are often characterized by poor transportation systems, sparsely populated areas, or isolated pockets of high population densities and limited means of knowing what economic conditions exist in these isolated regions. Because of these factors, regional food markets tend to be isolated and independent. Prices can vary widely from region to region, with little relationship to average national prices, or to those prices prevailing in markets in large cities. In addition, local prices tend to be more variable than national prices since, with few market participants, changes in behavior by small numbers of participants can affect prices.

The consequences of these market problems can be high regional food prices and less ability to meet consumption needs for given incomes. High price variability causes uncertainty to producers and consumers of the products. These factors worsen national welfare, and can cause isolated pockets of poverty. Increases in national supply will do little to improve such situations.

Regional price differences caused by high marketing costs due to poor transportation systems can only be lowered by improvements in infrastructure and market information. Poor information causes these differences when costs associated with gathering price and demand information impairs the effectiveness of the marketing system. Measures to enhance information flows include collection and dissemination of market-related information and telecommunications systems to transmit the information.

As development proceeds, incomes grow, and demand shifts outward, the possibilities for rapid increases in food prices arise even in countries experiencing rapid growth in agricultural production. The reasons for this effect were discussed earlier and illustrated in Table 3-2. Middle-income countries experiencing rapid rates of income growth are likely to need increased agricultural imports.

Eventually, when high income levels are reached, income elasticities of demand for food and population growth rates become smaller. These small income elasticities relieve the upward pressure on food prices but create the potential for food surpluses and low farm prices. Policies at this stage tend to be concerned with easing the cost of adjusting large portions of the labor force out of agriculture, directing producers into those commodities for which the country has a relative advantage in world markets, and stabilizing domestic farm prices, which are now more heavily influenced by swings in world prices.

The existence of structural changes in the market for agricultural goods over time suggests a strong need to tailor development policies to each country's stage of development. It also suggests a need for each country to consider the stages of development of other countries in the world when making projections about future demands for agricultural products.

SUMMARY

The effective demand for food is determined by the physical and psychological need for food combined with the ability to pay for it. Demand is influenced by prices, population, income, and preferences. The level of per capita income is a major determinant of food demand in low-income countries. The income elasticity of demand for food varies systematically by income level, by commodity, and by places and socioeconomic groups within a country. The income elasticity of demand for food declines as development proceeds, and shifts in consumption occur away from starchy staples toward higher protein foods. Own- and cross-price elasticities of demand are useful for projecting demand changes. Several procedures are available for obtaining income and price elasticities. Middle-income developing countries generally experience the most rapid rates of growth in demand for food.

IMPORTANT TERMS AND CONCEPTS

Aggregate versus household data
Bennett's law and why it holds
Contradictory role of agricultural
    prices
Cross-price elasticity of demand
Cross-sectional versus time-series
    data
Econometric model
Effective demand
Elastic versus inelastic demands
Engel's law and why it holds
Factors that shift the demand
    curve
Factors that shift the supply curve
Homogeneity condition and its use

Income effect
Income elasticity of demand
Law of demand
Major determinants of long-run
    price trends
Normal, superior, and inferior
    goods
One-price elasticity of demand
State of development
Substitutes or complements
Substitution effect
Supply
Use of aggregate versus
    household-level data

Looking Ahead

Rapid population growth over the past few years has dramatically increased the world's population and made the task of raising per capita income and reducing hunger in some countries more difficult. Population growth is influenced by many factors, and several policies have been tried or suggested for controlling it. In the next chapter, you will learn about population growth, including implications for food consumption and natural resource use. You will examine population projections and policies for the future.

QUESTIONS FOR DISCUSSION

1. As incomes increase, do people spend greater, smaller, or the same proportion of their income on food?
2. Distinguish between an income elasticity of demand and a cross-price elasticity of demand.
3. What tends to happen to the income elasticity of demand for food as the per capita income of a nation increases? Why?
4. To estimate the effect on the calorie and protein intake of a population resulting from a decrease in the price of rice, why is it important to know something about the cross-price elasticities of demand between rice and other major foods in the country?
5. Assume the price elasticity of demand for eggs in India is –0.75. By what percentage would the price of eggs have to change to increase egg consumption by 15 percent?
6. Do you expect the price of food in the world to be higher or lower 10 years from now? To answer this question. draw a graph with supply and demand curves and show how you expect the curves to change over time and why.
7. If population is growing at 2.6 percent per year, the income elasticity of demand for food is 0.6, and per capita income is growing at 4 percent per year, what would be the growth in demand for food per year, assuming prices remain constant?
8. What tends to happen to the mix of foods consumed as per capita income in a country increases? Why?
9. If agricultural development is successful at increasing the level of per capita food production in several less-developed countries over the next 10 years, why might these same countries become less self-sufficient in food (have to import more food than before) during that period of time?
10. Assume you have the following cross-price elasticities for a particular country:

Commodity	Cross-price elasticity
Rice and beans	– 0.35
Rice and wheat	0.40
Rice and chicken	– 0.10
Rice and milk	– 0.05
Rice and other goods	0

a. You are a planner for the country represented above and you want to raise the consumption of rice by 6 percent to improve calorie intake of the population. The income elasticity of demand for rice is 0.4. Use the information above and the homogeneity condition to determine the necessary percentage change in the price of rice.

b. If rice consumption increases by 6 percent, what else besides the calories obtained from rice would you need to consider when assessing the impact on calorie consumption?

11. What distinguishes the need for food from the effective demand for food?

12. Which of the following factors shift primarily the demand curve and which factors shift primarily the supply curve: per capita income changes; new technologies; population growth; tastes and preferences; prices of inputs used in production; prices of other goods consumed; prices of substitute goods in production?

13. Why is there a close relationship between agricultural production growth and a nation's income growth during the early stages of development?

14. Even if agricultural production increases rapidly, why is it unlikely that countries in early stages of development will experience major price decreases as a result?

15. Why do middle-income countries experiencing rapid rates of growth in food production often need food imports while very poor countries that are experiencing slower rates of food production growth do not?

RECOMMENDED READING

Foster, Phillips and Howard Leathers, *The World Food Problem* (Boulder, Colo.: Lynne Rienner, 1999), ch. 8.

Mellor, John W., *Economics of Agricultural Development* (Ithaca, N.Y.: Cornell University Press, 1966), ch.4.

Mellor, John W. and Bruce F. Johnston, "The World Food Equation: Interrelations Among Development, Employment, and Food Consumption," *Journal of Economic Literature*, vol. 22 (June 1984), pp. 531–74.

Population

> When poverty is tied to rapid population growth rates (as it gener-
> ally is), the risk of widespread hunger is ever present.
> — C. Ford Runge, Benjamin Senauer,
> Philip G. Pardey, and Mark W. Rosegrant[1]

This Chapter

1. Presents basic facts about the distribution of the world's population, the rate of population growth, and the consequences of rapid
   population growth
2. Explains the determinants of population growth and policies that
   can affect that growth
3. Examines causes and implications of migration from rural to urban areas.

BASIC FACTS ABOUT POPULATION GROWTH

The human race dates back about 3 million years. During more than 99
percent of this time there was virtually zero population growth. Average life expectancy was 20 to 25 years, and world population probably
never exceeded 10 million people. After agriculture replaced hunting
and gathering of food, around 6000 to 8000 B.C., population began to
grow more quickly because larger numbers of people could be supported and fed. By the year A.D. 1, there were about 300 million people
and by 1650, 500 million.

Population began to grow more rapidly during the industrial revolution in the eighteenth century and really accelerated after World War
II when populations in developing countries began to grow dramatically. World population reached 1 billion around 1800, 2 billion in 1930,
and 3 billion in 1960. It grew to 4 billion in 1975, 5 billion in 1986, 6

---

[1] C. Ford Runge, Benjamin Senauer, Philip G. Pardey, and Mark W. Rosegrant, *Ending Hunger in Our Lifetime: Food Security and Globalization* (Baltimore, Md.: Johns Hopkins University Press, 2003), p. 21.

billion in 1999, 6.5 billion in 2006, and will exceed 7 billion before 2015, based on projected future growth rates (Fig. 4-1).

The rate of population growth in the world peaked at 2.0 percent per year in 1965 and has declined since then to its current (2005) rate of about 1.2 percent. However, population itself will continue to grow for many years since the future number of parents will be much larger than the current number because of the rapid population growth in the recent past.

Distribution of the World Population

The world's population is distributed unevenly across the globe, reflecting the degree to which each location attracted migrants and was able to sustain growth in its local population over time. The earliest human ancestors lived in Sub-Saharan Africa, and migrated from there to other regions. By far the greatest accumulation of population has occurred in Asia, which holds over 60 percent of the world's population and has the highest population densities. Large populations are also found across Europe, along the coasts of North and South America, and within Africa.

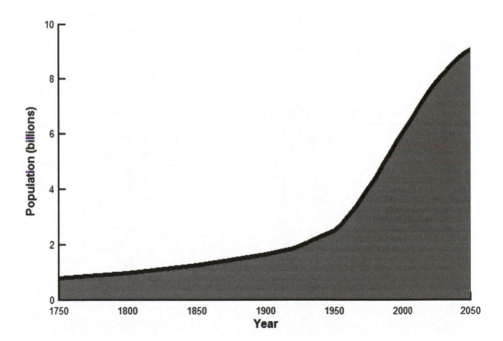

Figure 4-1. Past and projected world population, 1750 to 2050 (Source: Population Division, Department of Economic and Social Affairs, U.N. Secretariat, *World Population Prospects: The 2004 Revision* of *Population Database*).

The current size and density of the ten most populous countries are shown in Table 4-1. The list is dominated by China and India, but several other Asian countries have large populations and also have very high density, with more than 300 people per square mile. These countries account for the bulk of historical population growth. Today, population growth in Asia and elsewhere has slowed, and the fastest growing countries are mainly in Sub-Saharan Africa (Table 4-2). Eight of the ten fastest growing countries are in Africa, with annual rates of population increase at or above 2.8 percent per year. Such rapid growth is almost unprecedented in human history. It is occurring in the world's poorest places, where purchasing power per capita is below a dollar a day, and it is often occurring in places where rapid population growth is a fairly recent phenomenon. At the other end of the spectrum, the slowest growing countries are presented in Table 4-2. Some of these countries are actually losing population. Countries that have negative population growth rates are mainly the former socialist countries of Eastern Europe, but also include some high-income countries in Europe (Germany and Italy).

Consequences of Rapid Population Growth

Rapid population growth is a problem for most developing countries mainly because it changes the age composition of the country, with a larger fraction of the population being children. Population growth

Table 4-1. The World's Most Populous Nations

Nation	Mid-2005 population (millions)	Population density (people/mile²)
China	1,304	353
India	1,104	869
United States	296	80
Indonesia	222	302
Brazil	184	56
Pakistan	162	528
Bangladesh	144	2,594
Russia	143	22
Nigeria	132	375
Japan	128	876
Total (10 nations)	3,819	
Total (world)	6,477	125

Source: Population Reference Bureau, Inc., 2005 World Population Data Sheet.

Table 4-2. Population Growth Rates in the World's Fastest and Slowest Growing Nations (with 7 million or more population)

Fastest growing nations	Annual growth rate (percentage, 2005)	Mid-2005 population (millions)
Niger	3.4	14.0
Yemen	3.3	20.7
Mali	3.2	13.5
Uganda	3.2	26.9
Malawi	3.2	12.3
Dem Rep. of the Congo	3.1	60.8
Benin	2.9	8.4
Somalia	2.9	8.6
Guatemala	2.8	12.7
Burundi	2.8	7.8

Slowest growing nations	Annual growth rate (percentage, 2005)	Mid-2005 population (millions)
Ukraine	− 0.7	47.1
Russia	− 0.6	143.0
Belarus	− 0.6	9.8
Bulgaria	− 0.5	7.7
Hungary	− 0.4	10.1
Romania	− 0.2	21.6
Czech Republic	− 0.1	10.2
Germany	− 0.1	82.5
Italy	0.0	58.7
Poland	0.0	38.2

Source: Population Reference Bureau, Inc., 2005 World Population Data Sheet.

mainly takes the form of a rising number of children and young people, which imposes a strain on the natural resource base, increases pressures for jobs, reduces food production gains per capita, contributes to pollution, and strains the capacity of schools and other social services. While it would be an oversimplification to say that population growth is the root cause of natural resource problems, unemployment, and so forth, it certainly intensifies these problems.

Differences in age structure associated with different rates of population growth are illustrated in Figure 4-3. Those with rapid growth have large numbers of very young children relative to working-aged people. This high dependency causes increased current consumption and reduced savings and investment. The impacts of rapid population

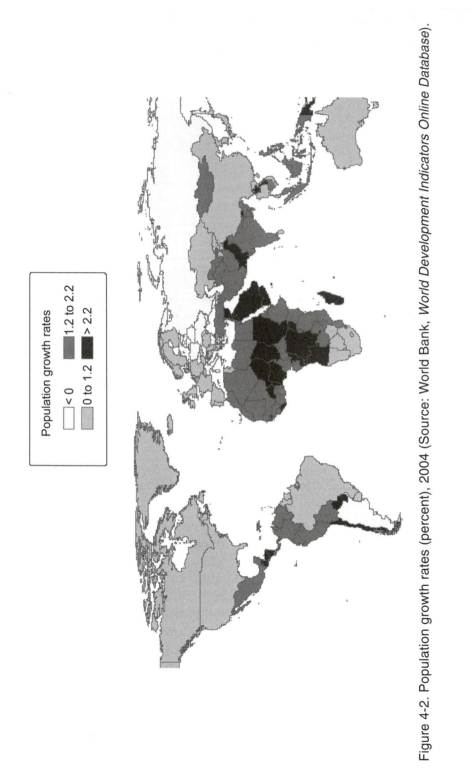

Figure 4-2. Population growth rates (percent), 2004 (Source: World Bank, *World Development Indicators Online Database*).

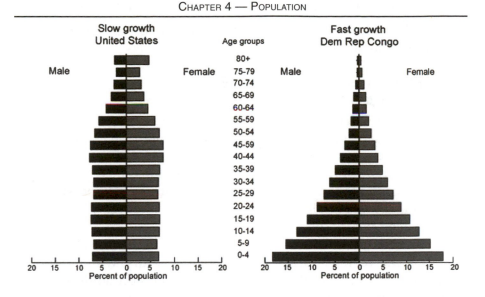

Figure 4-3. Population profiles, growth, and momentum: The age distribution of the people in a country has a major impact on the future rate of growth of its population. The population pyramid is a tool that demographers use to describe this distribution. Shown above are two population pyramids, reflecting differing rates of current and future population growth. The broad base on the Congo pyramid means that there is population growth "momentum" which will cause population to grow, even if fertility, or the number of children that each family has, slows immediately to replacement levels. As the large number of people in the younger age groups in Congo reach childbearing age, the number of births will rise dramatically, even if the number of births per couple falls. The United States has a relatively even age distribution, and is unlikely to experience a large increase in population (Source: Population Reference Bureau, 2001).

growth on schooling can be particularly important. Since about 25 per-cent of the people in developing countries are of school age, compared to 15 percent in typical developed countries, equal amounts of budget outlay for education translate either to low expenditures per pupil or low enrollment rates. Inadequate investments in either physical or hu-man capital will hurt the long-run possibilities for development.

The argument that most countries need more population to pro-vide labor and markets is not very compelling, given the abundance of unskilled labor relative to capital in many countries and the fact that increased consumption of manufactured goods is heavily dependent on per capita income growth.

Hunger, famine, and poverty were serious problems long before population began its rapid rise. However, the population explosion has made it difficult for some countries to invest and has magnified the lack of social justice in others.

CAUSES OF FERTILITY CHANGE AND POPULATION GROWTH

Population growth occurs for the world as a whole when births exceed deaths.[2] Years ago, births and deaths were both high, on the order of 40 to 50 every year per 1000 people in the population. About half of the deaths occurred before age ten, and death rates fluctuated from year to year with contagious diseases and with variations in food supplies. During this time, population fluctuated but did not grow rapidly for any sustained period of time.

Sustained population growth began in Europe and other now-industrialized regions during the eighteenth century, with a slow but steady decline in the death rate. Technological and economic progress resulted in improved nutrition and health, which reduced infant deaths and extended life expectancy well before scientists or medical doctors understood what caused disease or knew how to cure people once they fell ill. Population growth accelerated as death rates fell with no change in the birth rate for about one hundred years, until the late nineteenth

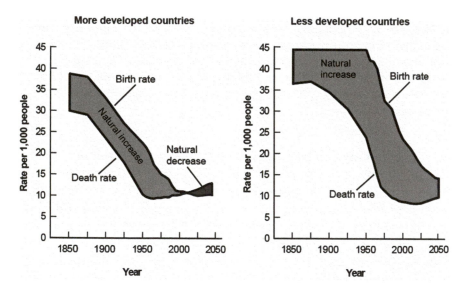

Figure 4-4. Trends in births and deaths, 1850-2050 (Sources: World Bank, *World Development Report 1980* (New York: Oxford University Press), p. 64; and Population Division, Department of Economic and Social Affairs, U.N. Secretariat, *World Population Prospects: The 2004 Revision* of *Population Database*).

[2] Population in individual countries also depends on immigration and emigration.

century, when birth rates began to fall as women delayed marriage and had fewer children (Fig. 4-4). Birth and death rates declined in tandem until the 1950s, when death rates stabilized and the total population growth rate slowed. It took roughly 200 years for the now-industrialized countries to transition from high birth and death rates in the early eighteenth century to low birth and death rates in the late twentieth century. During this period, births exceeded deaths by about 10 per 1,000 people, for a population growth rate on the order of 1 percent per year.

In contrast, today's less-developed countries experienced no significant decline in mortality until the twentieth century, when their death rates declined more rapidly than they ever had in the now-developed countries. This precipitous drop in the death rate was not due to slow improvements in nutrition and wealth, but to the sudden introduction of technological improvements developed through scientific research. Once scientists and doctors understood the causes of disease and the principles of nutrition, especially after World War II, countries rapidly deployed the new antibiotics, immunizations, and insecticides to control disease-bearing insects. They invested heavily in sanitation and maternal and child health programs. After the decline in death rates, it took several decades for birth rates to begin falling, but by then the gap between deaths and births was on the order of 20 per 1,000 people, or 2 percent per year, and in many countries it was over 3 percent per year.

In summary, population growth has been much faster in today's low-income countries than it ever was in today's high-income countries for one reason: the low-income countries' death rates fell faster, due to the sudden introduction of life-saving technologies. It is hard to imagine any serious observer wishing that those techniques had *not* been introduced, since they saved millions of lives and made possible much of the population we have today, but the speed of introduction made it relatively difficult for those countries to raise their per capita incomes, until the transition to lower birth rates could be completed

The historical *demographic transition* shown in Figure 4-4 has repeated itself in country after country. Each has a different timing and speed of transition, but all began with high birth and death rates and a relatively stable population size, then a decline in the death rate that initiates population growth. For those countries which have completed the demographic transition, a decline in the birth rate has followed, closing the gap between birth and death rates and stabilizing the population size at a new higher level.

The fact that this demographic transition has been observed in many countries in the past does not, of course, guarantee that it will be

Child weeding onions in the Philippines.

observed in the future. If population growth outstrips society's resources, death rates could rise again, and indeed in much of Africa they already have due to the ravages of HIV/AIDS as well as continued high levels of child malnutrition and disease. To understand where and when the demographic transition can be completed without rising death rates, we need to examine the causes of fertility (birth rate) changes and consider policies that might influence those changes.

Causes of Fertility Changes

Family size is largely determined by parental motivation, and this motivation reflects rational, and in many cases, economic decisions. Tastes, religion, culture, and social norms all play a role; yet evidence suggests that differences in economic factors as well as family planning education and access to birth control play the major roles. Female education is particularly important in reducing family size.

People receive pleasure and emotional satisfaction from children. Thus there is a consumption benefit from having children, and in poor societies there may be little competition from other consumption goods. It costs time and money to raise children, but these costs (both out of pocket and in terms of earnings forgone while caring for children) may be relatively low, especially in rural areas.

Children are also an investment. This investment value increases the benefits associated with having children. They frequently work during childhood. In rural areas they gather firewood, collect water, work in the field, move livestock, and do other chores. In urban areas, a child's ability to contribute work to the family is more limited; however, income opportunities exist for very young children in urban areas of most less-developed countries (LDCs). An important source of urban employment of children is the "informal sector," often in petty trading and services. When older children leave home, especially if they go to the city, they may send cash back home. Children also provide security during old age. Most developing countries have no social security system. These benefits from additional children raise the number of desired children in less-developed countries, especially among poor families. In many countries child mortality is high, so that extra births may be necessary to ensure that the desired number of children survive. All these factors increase birth rates.

As people obtain more education and earn more money, they delay marriage and have fewer children. Parents have more options, and come to prefer keeping their children in school rather than earning income from children's work. An increase in per capita income is inherently a rise in the value of time. A rise in the value of time, particularly if women have expanded employment opportunities outside the household, creates strong incentives to have fewer children and to invest more in the health and education of each child.

Thus, poverty and high fertility are mutually reinforcing. Social and economic factors such as income, literacy, and life expectancy account for as much as 60 percent of the variation in fertility changes among developing countries. The strength of family planning programs also accounts for a significant share.

Birth rates do not decline immediately when incomes begin to increase. Expectations about desired family size may take years to evolve, and in any case they will change at different rates for different social groups. Within each country, people with fewer opportunities — especially fewer opportunities for women — will often continue to have higher birth rates than other groups, further slowing the transition. And of course the speed of reduction in birth rates depends on the availability of effective family-planning techniques. To reduce fertility, households must both want to reduce their total family size, and be able to control the number and timing of births through effective contraception.

Policies That Influence Population Growth

Virtually everyone favors public and private actions to reduce death rates, but measures to reduce birth rates are more controversial. The controversy arises because some question the cost-effectiveness of family planning programs and others find efforts to control fertility in conflict with their strongly held values and beliefs. Family planning programs in at least one country appear to have been coercive, and some argue that more people are needed to provide labor and domestic markets.

Those who call for public actions to help curb birth rates argue that public costs (schools, hospitals, pollution, etc.) associated with large families exceed social benefits. Therefore society has a right to at least inform its citizens of ways to control births. Evidence from countries that have had strong family planning programs, such as Colombia and Indonesia, shows that these programs can be effective.[3]

China combined educational programs, social pressure, and economic incentives to reduce rates of birth. These were effective, but many people consider China's family planning program too strong; they particularly object to the use of abortion to control family size. These critics can point to less coercive educational programs that appear to have been equally effective in Sri Lanka, in parts of India, and in other countries.

Measures to improve income growth and distribution, develop social insurance and pension programs, and expand education and employment opportunities for women are all likely to help reduce birth rates. These efforts take time, however, which is why the policy debate often centers on family planning issues. Increased populations make all these programs more expensive and difficult to implement, so that current investments in family planning will save money in the long run. Most people in developing countries consider the fertility rates in their countries too high, and only a few consider them too low.[4]

Future Population Projections

The United Nations has projected that by the year 2050 world population will have grown to around 9 billion people.[5] Projections vary, however, from 7.7 billion to 10.6 billion due to the uncertainty in factors affecting the projections (Fig. 4-5). Most of the growth will be concen-

---

[3] World Bank, *World Development Report 1984*, p. 9.

[4] The Hunger Project, *Ending Hunger: An Idea Whose Time Has Come* (New York: Praeger, 1985), p. 30.

[5] United Nations, *World Population Prospects: The 2004 Revision* (New York: United Nations, 2005).

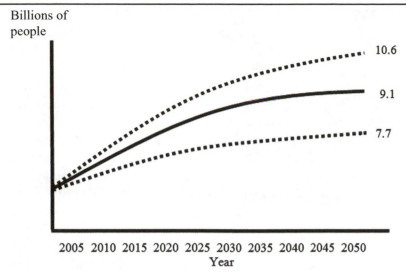

Figure 4-5. Future world population projections (Source: United Nations, *World Population Prospects: The 2004 Revision*).

trated in the developing countries. Future population projections are uncertain because they depend on income increases, educational improvements, family planning programs, and the future progression of the AIDS epidemic that are hard to predict. If present trends in growth rates continue however, the middle estimate appears the most likely.

URBANIZATION
Regardless of the total increase in population, it is clear that urbanization will continue at a rapid pace and that by the year 2050, the world will be substantially more urban. While total population in developing countries grew roughly 2 percent annually from 1990 to 2005, urban population grew at an annual rate of more than 3 percent. Natural population increases in urban areas account for about 60 percent of this growth rate, and another 8 to 15 percent is attributable to reclassification of rural areas to urban areas. At least 25 percent of the rapid growth in urban areas is caused by migration from rural to urban areas. Because a large proportion of the migrants are of childbearing age, a sizable part of the "natural increase" in urban populations can also be attributed to recent migrants. The percentage of urban population growth due to migration is highest in those countries in the early stages of development.

Many of the migrants in Dhaka Bangladesh seek work as
bicycle rickshaw drivers.

Causes of Rural-to-urban Migration

Rural-to-urban migration is, in a broad sense, a natural reflection of the
economic transformation from agriculture to industry which economies
undergo during the development process. As we discuss in Chapter 5,
the process of industrialization increases the demand for labor in the
manufacturing and service sectors. In the early stages of development,
much of this labor must come from the rural areas.

By and large, people move to urban areas because they expect increased economic opportunities in terms of both employment earnings
and access to goods or services produced by others. Landlessness and
rural poverty, natural calamities, lack of educational opportunities,
unequal public services provision, and other factors come into play as
well. Although living costs are higher in urban areas, migrants are
searching for a better level of living; they are pushed out of rural areas
by poverty and desperation, and pulled to the cities by hope and opportunity.

The vast majority of people who migrate to cities perceive that the
benefits of the move exceed its costs (these costs include forgone rural
income and the cost of the move), or they would not make the move.
Migrants tend to be young, disproportionately single, and better educated than the average of those left behind. The first two of these characteristics tend to lower the costs of the move, while the third raises the
benefits. Better-educated people can expect higher returns from their

education (wages) in urban areas. Most migrants to large cities in developing countries have relatives or friends already living there, a fact that tends to lower the cost of the move.

Rural-to-urban migration has been persistent despite rising unemployment rates in urban areas. The likely reasons for this persistence are that workers consider both rural–urban wage differentials and the probability of obtaining a job (which is often much less than 100 percent) and still perceive that they will be made better off by moving. Many of these migrants realize it is unlikely that they will obtain a high-paying or "formal" job immediately, but they are willing to work in low-paying jobs such as selling goods on street corners, "watching over" parked cars, or doing other jobs in the "informal" sector. For some of these migrants, these high-paying jobs may come only to their children, and then only if the children receive a better education than their parents.

The importance of educational opportunities and other public services cannot be overlooked as reasons for rural-to-urban migration. In many countries, an urban political bias has created a large disparity between the levels of services, including quality of public education, in rural and urban areas. Furthermore, and perhaps more important, the political bias extends to economic policies such as pricing policies. Food prices are often kept artificially low (through policies discussed later in this book). This policy helps urban consumers but discourages investment in food production and lowers incomes in rural areas. These distortions help explain some of the attractions of cities.

Consequences of Rural-to-urban Migration

Urbanization per se is not a problem. There are economies of scale resulting from the concentration of suppliers and consumers for industry and public services. Innovative and knowledge-intensive industries are more likely to form and prosper in high population-density areas. The problem arises when cities become "too large, too quickly," often because rural-to-urban migration increases the urban population at a rate faster than industry, schools, sewage systems, and so forth, can expand. The result is substandard housing, poor sanitation, and lack of other services for recent migrants (Box 4-1). While migrants have been shown to be assets to the cities, the shanty towns that surround almost all large cities in less developed countries attest to the growing disparities that occur within cities if urbanization occurs too rapidly. Many people live in absolute squalor, often without sewage systems and sometimes in garbage dumps. The fact that people are willing to live in these areas highlights the poverty and lack of opportunity in rural areas.

BOX 4-1
MEXICO CITY: AN EXAMPLE OF RAPID URBAN GROWTH

The situation in Mexico City, whose population more than tripled over the past 25 years, is an example of some of the strains imposed by rapid urbanization. The growth of Mexico City outstripped the growth in the availability of services. The city opened an ultramodern subway system in 1969, began large-scale construction of housing in the early 1970s, and inaugurated a deep-drainage sewer system that was hailed as an engineering marvel in 1975. Now, however, the subway and other transportation systems are hopelessly overloaded, 30 percent of the families in the city live in single rooms, and fully 40 percent of houses lack sewerage. Congestion and air pollution are severe, water is pumped into the city from as far away as 50 miles; rainwater and sewage are pumped out. The sewer system is so overtaxed that sewers back up and overflow into the streets during downpours. The city's garbage dumps are overflowing, and thousands earn their livelihood by picking garbage at the public dump.

Rural-to-urban migration continues in spite of these problems, with about 400,000 rural Mexicans moving to Mexico City each year. The hope of a better life provides a strong pull. While roughly 23 percent of the country's population lives in the city, 40 percent of the GDP is produced there, and more than one-third of factory and commercial jobs are located in the capital. Rural Mexico is very poor, with high rates of malnutrition, low literacy, and poor services even compared to the capital.

Evidence suggests that farm output has not been affected greatly by the loss of migrants and their labor to urban markets. In most low-income countries, the number of farmers keeps rising despite rural–urban migration, because the total population is growing faster than cities can expand. And migrants help sustain their relatives on the farm, when they remit money back to rural areas. However, some rural areas have suffered because the brightest and most educated workers have migrated.

Governments have employed many approaches to the task of slowing down rural-to-urban migration. Some countries are restricting migration, implementing resettlement schemes, and providing services to smaller towns and cities. It appears, however, that unless the urban bias in economic policies is removed and economic development proceeds to the point where living conditions improve in rural areas, rural-to-urban migration will continue in many countries at a very fast rate.

SUMMARY

The current world population of about 6.5 billion is growing at an annual rate of 1.2 percent, an extremely high rate by historical standards. The developing world is experiencing a population explosion caused by rapid decline in death rates due to improved health and nutrition. While birth rates have begun to decline due to higher incomes, family planning, education, and other factors, world population is likely to continue to grow for more than a century. Effective measures to control population growth should consider the economics of fertility and how different economic and social policies affect childbearing decisions. Rural-to-urban migration is proceeding at a rapid rate in many developing countries as migrants seek to achieve higher standards of living. Rapid urbanization has caused a strain on public services, pollution, and other problems.

IMPORTANT TERMS AND CONCEPTS

Birth rates and death rates
Causes of fertility changes
Causes of rural-to-urban migration
Characteristics of migrants
Consequences of rapid population growth
Demographic transition

Family planning
Population density
Population distribution
Population growth
Rural-to-urban migration
Urban political bias
Why death rates decline

Looking Ahead

This chapter concludes our overview of several dimensions of the world food-income-population problem. Hunger and development problems are both severe and complex. We move now to a set of two chapters, which examine economic theories that have been used in attempts to identify the heart of the development process. We begin in the next chapter with a discussion of important factors related to production growth. Subsequent chapters then incorporate these factors into development theories.

QUESTIONS FOR DISCUSSION[6]

1. Has population increased at a fairly constant rate since prehistoric times?
2. What is the current world population and how fast is it growing? When will it stop growing?

---

[6] Some of these questions are taken from Murphy, *World Population*.

3. At present growth rates, how long will it take to add one billion people to the world population?
4. Why is population increasing more rapidly today in LDCs than it did during early stages of development in Europe and the United States?
5. What are the major determinants of birth rates in LDCs?
6. What are the impacts of rapid population growth?
7. What policies can be used to help reduce population growth?
8. Are population growth rates more likely to increase or decrease over the next 15 years?
9. Which are the fastest and slowest growing countries in the world (in terms of population)?
10. What proportion of the world's population lives in Asia?
11. Why are we seeing rapid rural-to-urban migration in many developing countries?
12. What are the consequences of rapid rural-to-urban migration?
13. Describe the characteristics of the most common type of migrant.
14. How can high fertility be viewed as a consequence of poverty as well as a cause of it?
15. Describe the demographic transition that tends to occur as development takes place and why it occurs.

RECOMMENDED READING

Gelbard, Arlene, Carl Haub, and Mary M. Kent, "World Population Beyond Six Billion," *Population Bulletin*, vol. 54, no.1, Population Reference Bureau, March 1999.

Johnson, D. Gale, and Ronald D. Lee, *Population Growth and Economic Development: Issues and Evidence* (Madison: University of Wisconsin Press, 1987).

Murphy, Elaine M., *World Population Toward the Next Century* (Population Reference Bureau, New York, November 1981).

Population Reference Bureau: http://www.prb.org/content/navigationmenu/other_reports/2005.

PART 2

Development Theories and the Role of Agriculture

Rice in Peru.

Economic Transformation and Growth

Economic growth depends ultimately on the impact of productive resources and the efficiency with which they are used.
— Angus Maddison[1]

This Chapter
1. Describes the economic transformation that occurs with economic development, involving a decline in the size of agriculture relative to non-agricultural activities
2. Introduces the concept of a production function and the law of diminishing returns
3. Identifies potential sources of economic growth.

THE ECONOMIC TRANSFORMATION
Economic growth is almost always accompanied by an *economic transformation* from agriculture into other activities. As the economy expands, the agricultural sector grows more slowly than do manufacturing and services, and agriculture accounts for a declining fraction of employment, output, and consumer expenditures. The transformation from farm to non-farm activities as incomes rise applies to regions, countries, and the world as a whole. It is among the most dependable relationships in the world economy, and has major effects on people's lives. This chapter explores its causes and its consequences, both within agriculture and for society as a whole.

The tendency for richer countries to derive a smaller share of their income from agriculture is shown in Figure 5-1, and their tendency to have a smaller share of total employment in agriculture is shown in Figure 5-2. These two figures show remarkable similarity

---

[1] Angus Maddison, *Economic Progress and Policy in Developing Countries* (New York: W.W. Norton and Co., 1970), p. 34.

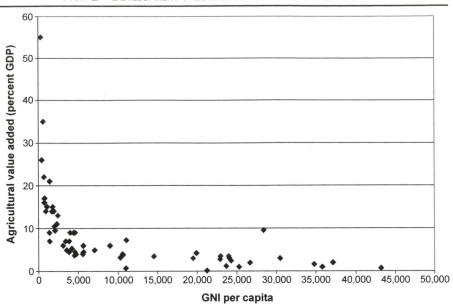

Figure 5-1. Agriculture's share of total output and Gross National Income per capita, 2002 (Source: World Bank, *World Development Indicators*, 2002 and *CIA World Factbook*, 2002).

and an interesting difference. The similarity is the clear downward trend. All poor countries derive a significant share of their income from agriculture, while all rich countries derive only a small fraction from it. Note that agriculture never disappears entirely in the rich countries, and there is wide variation in its share among the poorest countries. A key difference between the two figures is that, in poor countries, agriculture accounts for a larger fraction of employment than of output. Roughly speaking, countries below $1,000 per year in per capita income have 40 to 90 percent of the workforce engaged in agriculture, and these people earn 20 to 50 percent of their country's total income. In other words, within poor countries, on average each farmer earns roughly half of what non-farmers earn.

Causes of the Economic Transformation

In low-income countries, labor productivity is low and people, out of necessity, spend a high proportion of their income on food. Labor and small amounts of land are their primary assets, and many have no choice but to devote at least some of their labor to farming, to feed themselves and their families. Many low-income farmers are actually net food buyers, using small amounts of non-farm income or the sale of high-valued

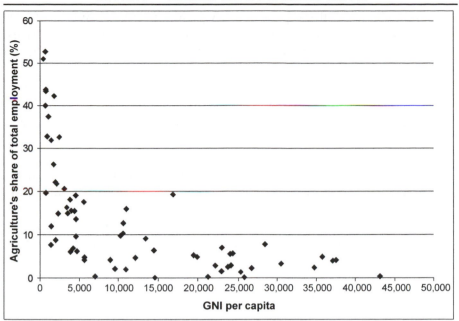

Figure 5-2. Agriculture's share of total employment and Gross National Income (GNI) per capita, 2002 (Source: World Bank, *World Development Indicators*, 2002 and *CIA World Factbook*, 2002).

crops and livestock to supplement the basic foods they grow on the farm. To emerge from poverty, these semi-subsistence farmers must improve their productivity either on the farm or in non-farm activities.

As the productivity of labor and other factors increases, four major factors drive the transformation from farm to non-farm activities. The first factor is that incomes rise due to the productivity increase, causing a gradual shift in demand from food to non-food items. This consumption shift occurs primarily because the income elasticity of demand for food is less than 1.0 and tends to decline as income grows. Declining income elasticities mean that for each percentage increase in income, progressively lower proportions are spent on food (See Engel's law in Chapter 3). These changes in demand for agricultural and nonagricultural products imply that, as development proceeds, relatively more labor inputs and other resources are devoted to nonagricultural activities.

The second factor driving the transformation is that at any given income level, the quantity of food demanded changes relatively little when its price changes. In other words, the price elasticity of demand for food is low, less than 1.0 in absolute value, and it may be even smaller at higher levels of income. This "price-inelastic" aspect of food demand means that, if agricultural productivity grows, prices received by farmers will fall by a higher percentage than the quantity demanded rises, creating incentives to remove resources from farming and transfer them to non-farm activities.

These two "demand-side" drivers cannot explain the transition in settings where farmers are selling their produce at prices that are determined in a world market. In those cases, prices received by farmers depend little on local demand, so there must be "supply-side" explanations for the transformation as well.

A third, supply-side, factor driving transformation is specialization. Even if the mix of activities in the economy remains the same, during economic growth the availability of capital and market opportunities allows people to expand production of what they do best, and then trade with others for the products they want to consume. Thus farmers produce less of their own food, clothes, furniture, and so forth, and an increasing share of these kinds of activities is re-classified from "agriculture" to "industry."

Another supply-side factor that could drive transformation is the fact that land supply is fixed, while other forms of capital can expand. As people accumulate savings from year to year, they find fewer and fewer opportunities to add resources to their farms, and so prefer to invest their savings in non-farm enterprises. For example, the farmer who already has good buildings, fencing, livestock, and equipment will tend to invest her savings in something else, such as a retail trade or services.

Does Agriculture Actually Shrink?

The fact that having higher incomes leads to a smaller fraction of output and employment in agriculture does not mean that the absolute size of the farm sector declines. Indeed, as countries get richer, the level of farm production and consumer expenditure on farm goods usually keeps rising, and in countries with rapid farm productivity growth, output in the sector can grow as fast as non-farm output. As agricultural productivity and incomes grow, labor is gradually transferred from work on farms to work in other enterprises. Some of this work occurs in the same rural areas where the farms themselves are located — people find employment in small-scale manufacturing, in value-added processing of agricultural products, in transport and services, etc. Others, as noted in Chapter 4, migrate to cities and find work in the formal and informal sectors.

In most countries, the land area available for farm use is roughly constant over time, so any change in the number of farm workers translates directly into a change in number of acres available per worker. One might expect economic development to influence the number of people working on each farm, and it does, but in an unexpected way. Across countries and over time, the number of workers on each farm stays close to the number of workers in the family. Family farming dominates the sector, and so the number of workers per farm varies with family size, which tends to decline as the economy grows. Thus poor countries may have five to eight workers per farm while rich countries may have only one or two, but that is mainly because of the declining number of workers per family. Furthermore, at every level of income, many family members work only part-time on the farm, and hire themselves out for off-farm work. A few do hired farm work, but hired workers are less common in agriculture than in other sectors.

Family workers dominate farming for many reasons, but perhaps the primary reason is that many field operations are difficult to supervise and monitor, and are therefore done better by self-motivated workers. For example, a farm owner would have great difficulty ensuring that a hired worker plows, plants, or fertilizes appropriately, because these operations are dispersed across the field and many other factors intervene to determine that field's eventual yield.

Since family farming dominates the sector, any change in the number of farm families translates directly into a change in the average cropped area per farm. Figure 5-3 illustrates this process for the United States. The number of farms peaked in the 1920s, but as farm labor moved into cities, the acreage per farm increased as exiting farmers rented or sold their land to the remaining operators. Note that the decline in the number of farmers cannot go on forever. In the United States, there has been no further decline since 1990, with roughly one-third the number of farms as there were in the 1910 to 1920 period, and farm sizes roughly three times as large.

A great deal of variability in farm sizes over time exists across countries. Several middle-income countries in Asia are now in a period of rapid decline in the number of farmers, much like the United States in the 1960s. The poorest countries, however, have growing rural populations and fixed land bases. Many regions in South Asia and Africa have experienced decades of decline in the available acreage per farmer, sharply reducing their ability to feed themselves or initiate the economic transformation out of agriculture.

Implications of Changes in the Number of Farmers

The key fact about the economic transformation presented above is that, as incomes rise the share of agriculture falls, but the absolute number

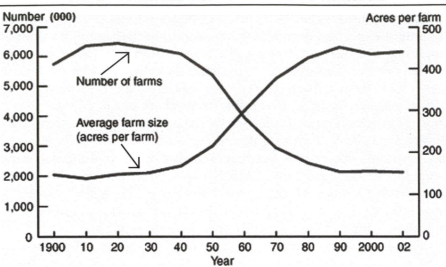

Figure 5-3. Number and average size of farms in the United States, 1900-2002 (Source: Carolyn Dimitri, Anne Effland, and Neilson Conklin, 2005. *The 20th Century Transformation of U.S. Agriculture and Farm Policy.* Washington, D.C.: Economic Research Service, USDA).

of farmers *rises and then falls*. The initially rising number of farmers in low-income countries translates directly into the rising number of workers per acre of available land. If output per acre cannot rise at least as fast as the number of workers, output per worker must fall. This downward pressure on farmers' income accounts for much of the deterioration in social conditions that we observe in the world's poorest regions.

An essential aspect of rural population growth is that it is temporary. If economic development continues, eventually non-farm employment becomes large enough to absorb all new workers, rural population growth slows, and any growth in output per acre translates directly into growing output per worker. Many of the people moving off the farm incur significant adjustment costs during the transition.

The fact that an economic transformation occurs with development does not explain the sources of economic growth and development. Understanding those sources of growth and how they contribute to development requires knowledge of a few basic economic principles related to production economics. In the next section we introduce a set of principles that can be used to help explain the output and economic effects of input and technology choices.

EXPLAINING PRODUCTION CHOICES

Economic growth requires transforming a country's basic production resources into products and doing so in ever more efficient ways. Economists have developed ways to characterize how that transformation occurs, utilizing the concepts of a *production function*, a *marginal product*, and *economic optimality*.

Production Functions

Production requires resources or inputs such as labor, natural resources, and tools or other capital items. These inputs are often called factors of production. Production also requires that these factors be combined by a producing unit that can organize their use to obtain desired goods and services. A description of the way in which factors of production are combined to produce goods and services is commonly called a production function. A production function describes, for a given technology, the different output levels that can be obtained from various combinations of inputs or factors of production.

The relationship between the level of production that can be obtained when only one input is allowed to vary (say, labor) while all other inputs are held fixed may look something like that shown in Figure 5-4. This relationship is also referred to as an *input response curve*, or a *total product curve*. In the case of labor, when no work is done the production level is usually zero, so the input response curve starts at zero. Output may then rise at an increasing rate, showing "increasing

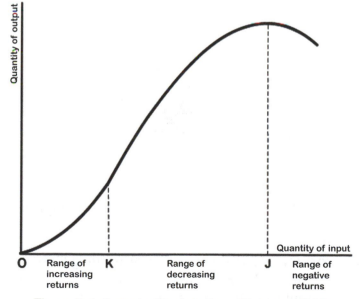

Figure 5-4. A production function with one variable.

returns" to each additional unit of input. In farming, for example, the initial effort of planting is more productive if followed by additional effort spent weeding, so doubling labor time could more than double the resulting output. Eventually, however, all such opportunities will be exhausted and each additional hour of labor or unit of other input begins to offer "decreasing returns": output continues to rise, but at a decreasing rate. Finally, at very high levels of input use, all opportunities to do *anything* productive may be exhausted, and additional inputs might actually reduce output.

On the particular curve drawn in Figure 5-4, the transition from increasing to decreasing returns occurs at the input level marked K. Beyond that point, for each additional unit of labor, the *additions* to output become smaller and smaller, until eventually, at point J, additions to output may stop entirely. Beyond that point, additional units could actually reduce output, so the curve begins to slope down.

The input-response curve in Figure 5-4 shows the productivity of one input, when all the other inputs are held constant. Changing the quantity of this one input results in a movement along the curve. If other inputs were to change, that would be shown as a shift in the curve. We will see an example of such a shift later in this chapter.

If two inputs are allowed to vary simultaneously, the resulting production function can be illustrated as in Figure 5-5, with each curve (called an *isoquant*) representing a different level of output. Curves higher and to the right represent greater output levels than curves lower and to the left. For example, point C represents a higher output level (200 units) than points A or B (100 units).

The isoquant that represents 100 units of output illustrates that the same level of output (100 in this case) can be produced with different combinations of labor and capital (combination A versus combination B). Thus, if a country has abundant labor and little capital, it might produce using the combination of labor and capital represented by A. If it has abundant capital and little labor, it might produce at B. The isoquant through points A and B shows all the different combinations of labor and capital that can be used to produce 100 units of output. It also tells us how easy it is to substitute labor for capital in the production of that output. When isoquants are very curved, inputs are not easily substituted for each other. Straighter isoquants imply easier substitution.

Marginal Product and the Law of Diminishing Returns

The idea illustrated in Figure 5-4 that, after some point, adding additional units of input tends to generate less and less additional output is

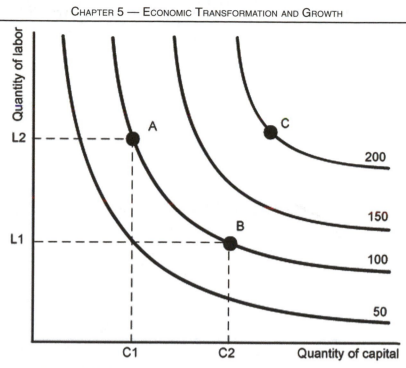

Figure 5-5. Production function with two variable inputs.

known as the *law of diminishing returns*. Specifically the law says: 'In the production of any commodity, as we add more units of one factor of production to a fixed quantity of another factor (or factors), the additions to total output with each subsequent unit of the variable factor will eventually begin to diminish." What is diminishing is the *marginal output gain* or *marginal product* of the factor (labor in Figure 5-4).[2] As discussed below, the law of diminishing returns has important implications for countries experiencing rapid population (and labor) growth with a fixed natural resource base.

A marginal product curve can be obtained (derived) from Figure 5-4 by examining *changes* in total output for each successive unit of input. The marginal product curve corresponding to the production function in Figure 5-4 is shown in Figure 5-6. To the left of K, the slope of the production function is increasing (Fig. 5-4); thus the changes in output are growing and the marginal product curve is rising (Fig. 5-6). To the right of K, the changes are smaller and the marginal product curve falls. If total output eventually ceases to grow at all as more labor is applied,

---

[2] The marginal product of an input is equal to the slope of the total product curve, or $\Delta Y / \Delta X$, where Δ represents a small change. Therefore, anything affecting this slope changes the marginal product.

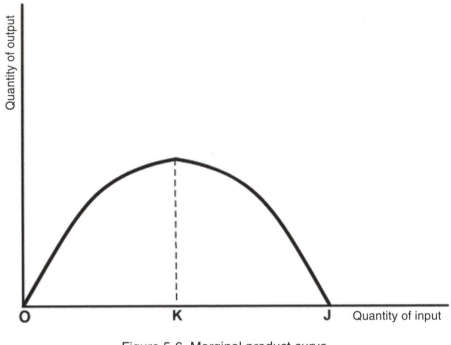

Figure 5-6. Marginal product curve
derived from the total product curve in Figure 5-4.

the marginal product goes to zero; this is point J on the production function and on the marginal product curve. Marginal productivity is important because it helps determine payments to factors of production, such as wages paid to labor. In addition, the marginal productivity of an input, together with prices of outputs and inputs, determines the demand for the input.

Economic Optimality: What Output and Input Levels will People Choose?

All points along a production function are equally possible to achieve. But are they equally likely to be chosen? What factors might motivate a farmer to choose one point as opposed to another? When people are asked what explains their choices, they mention a variety of factors such as input scarcity, the need for output of particular products, traditions or habits, and a desire to minimize risk. Repeated studies have found that actual choices by large numbers of people over several years are best explained by *economic optimality*. Economic optimality means that farmers are rational and choose options that will give them the highest level of well-being attainable given the prices they face, the available resources and technology, and their ability to absorb risk.

Even in very low-income settings and across cultures, farmers generally attempt to optimize. They may consider cultural and risk factors as they optimize, but economic well-being plays an important role. Because farmers optimize, they will generally choose to be somewhere along the production function and not below it. For any given level of input(s), they prefer to obtain as much output as they can attain. In other words, they prefer to be on the production function and not below it. But where along the total product curve would they prefer to produce? Prices help determine the answer. Even for farmers whose production is largely for home consumption, some of their outputs and inputs are sold and purchased at prices set in markets off the farm. When markets set prices, farmers can often reach the highest possible level of well-being by *maximizing profits,* subject to acceptable risk, and then trading those profits for goods they want to consume. This kind of economic optimality typically leads to a single point along the production function, as illustrated in Figure 5-7. In Figure 5-7, each level of profits can be represented by a straight line, whose slope is the price of the input divided by the price of output. In the left-hand panel in Figure 5-7, the highest such line, representing the highest attainable level of profits, occurs where the line touches (is tangent to) the production function, and their slopes are the same. At this point of tangency, the marginal revenue from the output equals the marginal cost of the input

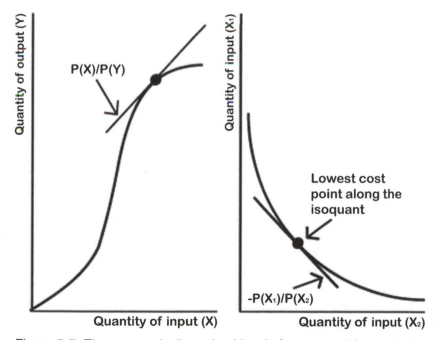

Figure 5-7. The economically optimal level of output and input choice.

(MR=MC). On the right-hand panel in Figure 5-7, the profit line is the ratio of the two input prices and also represents the total cost of production. When farmers are producing on their production functions and employing the correct amount of inputs to equate their marginal revenue to their marginal cost of obtaining the last unit of output, they are said to have achieved *price or allocative efficiency*.

SOURCES OF ECONOMIC GROWTH

We can now use the production economics principles described above to explore the possible sources of growth in an economy over time. One of the major ways that economic growth can occur is through increases in the amounts of inputs used in production. While production functions usually refer to a particular type of output (say, rice), one can think of an aggregate production function relating total inputs to total output or total national product. Additional inputs can move a country out of its aggregate production function to higher isoquants and higher levels of output. Therefore, (1) *population growth* (which affects labor availability and cost), (2) *natural resource availability* (which affects the cost of environmental factors such as land with its associated soils, water, and forests), and (3) *capital accumulation* (which affects the availability of man-made inputs) are three major elements in the development process. These sources of growth cause movement along a given multi-factor production function.

A second means of spurring economic growth is to change the way in which a country uses its factors of production, increasing the amount of output produced by these inputs. These output increases can result from better organization of production or from shifts in the production function. For example, a new technology can shift the total product curve upward so more output is produced per unit of inputs. There are three ways to get increased output per unit of input: (1) increases in scale or specialization, (2) increases in efficiency, and (3) technological change. In many cases markets can change, in turn stimulating changes in these factors. Movements along a given production function versus shifts in the function are illustrated in Box 5-1.

A third means of stimulating economic growth is through increased *human capital* as embodied in people (e.g., improved education and health) and improvements in *social institutions* (the rules of the game). Human capital can make labor more productive, contributing to technological progress and increased efficiency (especially when technologies and markets are rapidly changing). Social institutions help define property rights.

Let us examine more closely each of the sources of economic growth.

BOX 5-1
SOURCES OF GROWTH AND THE PRODUCTION FUNCTION

Growth in output can occur either through a change in market opportunities and relative prices, a change that leads farmers to add inputs using existing technologies, or because of an innovation that allows production of more output at a given level of inputs.

The left-hand panel below illustrates how profit-maximizing farmers would respond to increasing abundance and hence lower relative price of an input. For example, in poor countries when rural labor becomes more abundant over time, there is a decline in wages relative to other prices, leading farmers to apply more labor in land preparation, weeding, etc. in an effort to obtain more output.

The right-hand panel shows how those same farmers might respond to a new invention, such as better-performing seeds or veterinary medicine for their livestock. Now the farmer can obtain more output at each level of input. This particular innovation was drawn so that the new profit-maximizing level of input use happens to be exactly the same as before; thanks to the innovation, the farmer has gotten more output for no change in the input.

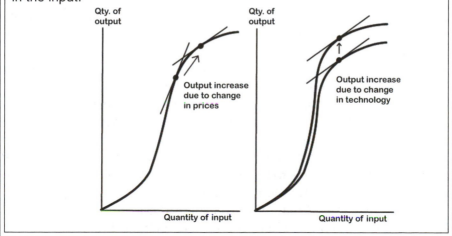

The Demographic Factor: Effects of Population Growth on Agriculture and the Economy

For most of history, population growth was a major source of output growth in the world. People worked with primitive tools and more people meant more labor and output. Crop and pasture areas expanded with the rural workforce, although output per person remained roughly the same. A greater population density also reduced the distance between people and made it easier to develop cost-effective services such as transportation, communications, schooling, and so forth. Population growth, however, is a mixed blessing because, while there are more productive hands, there are more mouths to feed. As long as farm land is plentiful, land frontiers can be pushed back and growth continues in the agricultural sector, but in most areas of the world, the best farm land has been exhausted and rising numbers of farmers have no choice but to invest more time in each field. In this situation, diminishing returns to labor cause farm incomes to fall, unless farmers can turn to an alternative source of growth.

Population growth may also mean an increasing number of children relative to adults. If the number of consumers is growing faster than the number of producers, then the effect of population growth is also more likely to be negative. If population growth results from extending the productive life of workers, the odds of its effect being positive improve.

Natural Resources: Environmental Influences on the Location and Pace of Development

Natural resources — including land and its associated soil, water, forests, and minerals — have played an important role in economic development. The extension of the frontier in the United States brought more land and mineral resources into production and helped create wealth. Similar expansions occurred in other countries. Extensive use of other types of natural resources has been important as well. For example, in the eighteenth and nineteenth centuries one of the most important resources was coal, as countries with large and easily accessible coal deposits such as Britain used it to fuel their local industrial revolution. In the twentieth century, oil became important in some countries. Will natural resources continue to be an important source of economic growth or will they be a limitation to future growth?

Some have argued that the Earth is like a spaceship, that its natural resource capacity is finite. There is only so much land and, indeed, we see increasing problems with soil erosion, deforestation, and overgrazing. Increased combustion of fossil fuels releases carbon into the atmosphere and depletes a finite supply of these resources. Water resources are exploited to their fullest potential (or over-exploited) in many places.

While technologies change, and in essence create new resources, there is no question that land is limited and that the opening of new uninhabited fertile lands will be much less important to future economic growth in most countries than it has been historically. It is also clear that many resources, particularly forests and minerals, are being depleted in many countries and are thus becoming less available than they once were to stimulate growth. The real question for most coun-

A plow and bullock can be a sizable investment
in many developing countries.

tries may not be whether exploitation of natural resources will be a
significant source of growth, but whether natural resources will act as a
constraint to growth, and, what will be the cost involved in transitioning
from one natural resource regime to another. This issue is discussed
further in Chapter 9.

Accumulation of Physical Capital

Physical capital may be defined as a country's stock of human-made
contributions to production consisting of such items as buildings, fac-
tories, bridges, paved roads, dams, machinery, tools, equipment, and
inventory of goods in stock. Physical capital, as we refer to it here, means
human-made physical items and not money, stocks and bonds, etc. It
refers to private physical goods but also public investments in physical
infrastructure.

Capital accumulation is the process of adding to this stock of buildings, machinery, tools, bridges, etc. Another name for capital accumulation is investment. Capital investment is important because it can
increase the amount of machinery and tools per worker, thereby increasing the output or marginal product per worker. A higher marginal
product per worker usually leads to a higher income per worker.

Capital accumulation is also related to the possibilities of making
changes in the scale of technology of production. Furthermore, the process of capital accumulation involves a choice between consumption
today and investing for future economic growth. The choices of how

much to invest and in what types of capital have important implications for the rate and direction of economic development. As will be argued throughout this text, investment should be guided along an appropriate path by signals (prices) that reflect the true scarcity of resources.

Technological Progress

Increases in input levels (land, labor, and capital) accounted for much of economic growth prior to the nineteenth century. However, evidence suggests that changes in the ways goods are produced have been the engine of modern economic growth for many, if not most, countries. The three sources of growth mentioned above involve increasing inputs with a given production technology. Economic growth can occur, but only through exploitation of natural resources and labor, or accumulation of costly resources through savings and investment from year to year. More important, this type of growth is subject to diminishing returns, as movements along the production function generate smaller and smaller increments of output for each additional unit of input. Sustaining economic growth over time requires the constant invention of new technologies, to shift the production function and overcome diminishing returns (Box 5-2).

If technological progress allows the same or fewer resources to provide more output, the value of output per unit of resources rises, and this rise can lead to increases in per capita income. Resources can also be freed up to provide new types of goods. The phenomenon of technological progress is not new and has been occurring for many years. What is new is the rapidity with which new technologies are being developed. Modern technological progress is the result of both *applied science* and *new knowledge* in the basic sciences.

Specialization

As innovation occurs and capital is accumulated, increasing opportunities arise for people to specialize and trade with each other. Such *specialization and trade* can raise productivity and attract savings and investment. Specialization is related to scale as well. As firms increase in size, specialization is facilitated. "Division of labor" can make workers more efficient as they become proficient at just a few tasks. Adam Smith argued that this type of division of labor is at the heart of economic growth. In his famous book, *The Wealth of Nations* (1776), Smith noted that specialization is limited only by "the extent of the market," or the ease with which one person can trade with others, both within and across countries. As markets expand, the possibilities

BOX 5-2
NEW TECHNOLOGIES, INPUT USE, AND THE DEMAND FOR INNOVATION

Technological innovations can have different impacts on a farmer's input use and output levels, and changes in resource availability can lead to different kinds of innovation. The diagrams below illustrate how farmers' profit maximization affects their response to new technologies, and affects the kinds of new techniques that are most needed in various countries.

The left-hand panel shows an innovation that, with no change in relative prices, would lead a farmer to cut back on input use — in this case, it is drawn so that, with no change in relative prices, the farmer keeps the same level of output. Most input-saving innovations are mechanical devices such as bigger, faster implements, which take less capital and labor to do a given task.

The right-hand panel shows another kind of innovation, drawn here so that, with no change in relative prices, the farmer is led to increase input use. The most important examples of such technologies are "green revolution" crop varieties, whose growth habits and stress tolerance make it worthwhile for farmers to apply more labor, fertilizer, and water to the plant.

"Input-using" innovations involve the discovery of new techniques *to the right* of existing input levels, whereas "input-saving" innovation involves discovery of new techniques to the left of them. Price changes, by leading farmers to look for new techniques in one direction or the other, help influence which kind of innovation is more likely to be discovered and adopted. Most notably, in poor countries where the farm labor force is rising, labor-using innovations are demanded. In contrast, once the farm labor force starts falling in richer countries, labor-saving mechanization is the farmers' priority.

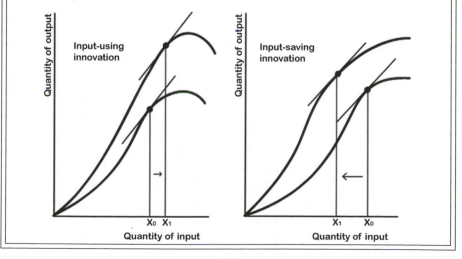

of mass-producing goods enable firms to gain efficiency in both production and marketing. Increased scale and specialization allow more output per unit of input and, hence, growth.

Efficiency Improvement

Another type of organizational change that can lead to economic growth is improved production efficiency. Improved efficiency means getting more for the same inputs.

Efficiency can be divided into different types. *Technical efficiency* relates to whether producers are producing on the production function as opposed to below or inside it. Using the same amount of inputs, some producers obtain higher output levels than others due to differences in management and effort. *Price or allocative efficiency*, mentioned above and illustrated in Figure 5-7, relates to the degree to which producers, operating on their production functions, employ the correct amount of inputs to equate their marginal revenue to their marginal cost of obtaining the last unit of output. By definition, producers who maximize profits are both technically and allocatively efficient.

Market efficiency is related to the type of economic system and the degree of market power within it. Improvements in resource allocation occur through market efficiency when increased competition or new technology lowers the margin between buyers and sellers. A country that has a relatively free market with many buyers and sellers, so that no producer or consumer can affect prices, has greater market efficiency than one with a few producers who are able to control prices. The availability of good information affects the degree of market efficiency, and improved information flows can help create growth due to more efficient allocation of productive resources.

Human Capital

So far in this chapter we have explained economic growth without assuming any change in the people themselves. Much of economic growth is driven by changes in people's capabilities or their *human capital*, as affected particularly by their education and health. The nature of these capabilities is easily misunderstood. Even the most illiterate, impoverished person is often intelligent and skilled, but educated, healthy people can more easily contribute to the generation of new technologies and more readily utilize those technologies. Education is therefore an important source of economic growth, inextricably linked with technological progress, and, of course, with the productivity of labor. Part of the economic benefits of education is derived from improved productivity of workers, part from improved quality of management, and part from education's contribution to producing new or improved technologies.[3] The term *human capital* is used in referring to education because education is an investment, in many ways similar to that of physical capital in requiring an investment of resources that pays off over a long period of time and eventually depreciates.

Education is important, but in the lowest-income countries an equally important form of human capital is a person's health. Undernutrition and preventable diseases remain the world's biggest killers, and they sharply reduce the productivity of those who survive. Improvements in nutrition and disease control raise output directly, and also make it easier and more worthwhile to keep children in school, leading to more education as well.

Education and health are forms of human capital that are embodied in particular individuals. If you were to trade places with a lower-income person, the odds are you would be more educated and healthier than that person, and that might influence what you could do. But if others in your society were *also* healthier and better educated, that would allow you to develop different expectations about their behavior as well. You could rely more on other people, using your mutual education to develop and communicate new ideas about how to work together.

Institutional Change

Historical patterns of economic growth exhibit remarkable differences across countries and over time. Levels and rates of growth differ significantly even among neighboring countries. Many of these differences are not solely attributable to the sources mentioned above, but to institutions as well. Institutions include government policies, legal structures, and market structures. If markets exhibit distortions, efficient price signals will not be received by producers; if financial markets are incomplete or characterized by excessive risks, savings and capital accumulation will be constrained. If people are unsure about their ability to recover investments, due to political instability or ill-defined property rights, they will not undertake investments. The ability of institutions to adapt to new needs and demands can itself be a source of economic growth.

During economic growth there is often explosive growth in many kinds of social institutions. This new *social capital* may displace previous institutions, such as family or village networks, which might have been helpful but are not as well adapted to new circumstances. Some of

---

[3] Education can, of course, have other benefits associated with the capacity to develop new institutions and with many non-economic factors.

these institutional changes are a result of economic growth, but in some they may play a causal role in economic development, so that a transfer of institutions could accelerate growth. For example, many countries benefit from the introduction of quality certification systems to enforce grades and standards, uniform procedures for contract enforcement and commercial law, and well-adapted property rights of various sorts.

SUMMARY

Economic growth involves a transition from low-income agricultural societies to higher-income non-farm employment. The process is driven by capital accumulation, technological innovation, and specialization in either sector. An economic transformation occurs for several reasons. First, demand for food is relatively fixed. It is "income-inelastic," so when incomes grow, demand for other things grows faster. Second, productivity increases in agriculture free up resources for non-agricultural production. Third, as people specialize and trade with each other, many tasks that were previously done on the farm are now classified as non-agricultural.

Although agriculture declines as a share of the economy, the sector does not shrink. Typically, total farm output continues to rise during economic growth. Furthermore, when the total population is growing, the number of farmers tends to rise for many years, until the absolute size of the non-farm sector is large enough to absorb all those entering the workforce each year. The resulting change in land area per farmer will often place downward pressure on rural living standards during the early stages of economic development, even as the rest of the economy grows.

To explain the causes and consequences of economic growth we use production functions that describe, for a given technology, the different amounts of product that can be obtained from different levels and combinations of inputs. An isoquant shows different combinations of two inputs that can be used to produce the same level of output, given a particular technology. The law of diminishing returns has important implications as population or capital increases against a fixed land base. To overcome diminishing returns and sustain growth over time, people need technological change, increased specialization and trade, and improvements in efficiency that may be related to improvements in human capital and institutions.

IMPORTANT TERMS AND CONCEPTS

Capital accumulation
Economic efficiency
Economic transformation
Education
Input demands
Isoquant
Law of diminishing returns
Marginal product
Human capital

Natural resources
Non-farm job opportunities
Population growth
Production function
Scale and specialization
Sources of economic growth
Technological progress
Input response curve
Institutional change

Looking Ahead

The sources of growth discussed above relate to whole economies, to sectors within economies, and to individual firms, including farms. Various theories have been proposed to explain how the sources of growth have been or could be combined to transform economies from low to higher standards of living. We examine these theories in the next chapter. In subsequent chapters we consider how these growth factors can affect firms within the agricultural sector.

QUESTIONS FOR DISCUSSION

1. What is meant by the term *factors of production*?
2. What are the three major factors of production and how do they relate to the major sources of economic growth?
3. What is the law of diminishing returns and what might be its significance in relation to population growth?
4. Will natural resource limitations be a serious restriction to future economic growth or growth in food production?
5. What is capital accumulation and why is it important to development?
6. Why are specialization, efficiency, and technological progress important to agricultural and economic development?
7. Why is an economic transformation inevitably associated with economic development?
8. What factors determine the rate at which an economy becomes transformed from an agricultural to a mixed economy with significant nonagricultural as well as agricultural activities?
9. If the total labor force were growing 2 percent per year and 50 percent of the labor force were in agriculture, how fast would nonagricultural employment need to expand in order to hold the number of people employed in agriculture constant? Why is this important?

10. What are the implications of the economic transformation for the agricultural sector?
11. What is meant by the terms *human capital* and *institutional change*?

RECOMMENDED READING

Anderson, Kym, "On Why Agriculture Declines with Economic Growth," *Agricultural Economics*, vol. 3, no. 1 (October 1987), pp. 195–207.

Thirlwall, A.P., *Growth and Development* (Boulder, Colo.: Lynne Reinner, 1995), chs 3–4.

CHAPTER 6

Development Theory and Growth Strategies

We can realistically envision a world without extreme poverty by the year 2025, because technological progress enables us to meet basic human needs on a global scale. — Jeffrey Sachs[1]

People respond to incentives; all the rest is commentary. — Steven Landsburg[2]

This Chapter
1. Reviews how economic development and growth theories have evolved over time, including the role of institutions
2. Considers the interaction of technology and institutions
3. Considers the distinctive characteristics of agriculture as opposed to other sectors as the economy develops.

THE HISTORICAL EVOLUTION OF DEVELOPMENT THEORY
In the previous chapter, we identified potential sources of economic growth and the inevitable structural transformation that accompanies economic development. We turn now to ideas and theories that attempt to explain how these sources of growth can be integrated into transformation processes that produce higher living standards. The search for appropriate theories of economic development has received economists' attention for two centuries. Different theories have led to different implications for what governments, private firms, or individuals might do to achieve their goals. One especially important contrast concerns the relative roles attributed to technology and productivity (reflected in the words of Jeffrey Sachs, above), as opposed to institutions and

---

[1] Jeffrey Sachs, *The End of Poverty: Economic Possibilities for Our Time* (New York: Penguin, 2005), p. 347.
[2] Steven Landsburg, *The Armchair Economist: Economics and Everyday Life* (New York: Free Press, 1995).

incentives (reflected in Steven Landsburg's words). Emphasis has shifted over time, partly because of changes in constraints that limit economic growth, partly because of changing technological possibilities, and partly because of experiences with what has or has not worked. We consider in this chapter the historical progression of thinking among economists. Over time, a synthesis of ideas has emerged, with increased focus on the interaction between technology and institutions.

The Classical Period

The late eighteenth century is known as the *classical* period in economic thought, and the books written then remain widely debated today. One of the most enduring debates concerns the role of international trade. At the time, conventional wisdom held that a country's wealth, like the wealth of an individual, could be measured by the amount of its gold and other monetary assets. Exports were believed to be better than imports, and this *mercantilist view* provided an important argument for trade restrictions in Britain and elsewhere. **Adam Smith** challenged the mercantilist idea, arguing that freer trade in both directions would produce higher standards of living, especially if combined with a more competitive, equal-opportunity environment at home. Adam Smith's arguments were extended by **John Stuart Mill** and **David Ricardo,** and their ideas about the division of labor and specialization, comparative advantage, and trade remain key concepts in modern economics. Their theories about the value of freer trade were not easily accepted at the time, however, and many mercantilist ideas remain widespread today.

The eighteenth century was a period of both economic expansion and population growth. Many political leaders argued that having more people would help make each country richer. In the early nineteenth century this idea was challenged by **Thomas Malthus**, who argued that population was limited mainly by the food supply, and by a fixed supply of high-quality land. Ricardo agreed with Malthus and was pessimistic that growth could be sustained in the long run in a country because of the implications of population growth, given the law of diminishing returns. Their classical theory in its simplest form proceeds as follows: (1) There are two broad types of people: workers, whose only asset is their labor, and capitalists, who own land and capital. With a certain amount of labor, just enough wages are paid to cover workers' subsistence. (2) If a new invention or some other favorable event creates an increase in production, a surplus above that necessary to pay the subsistence wage is generated, which is accumulated by capitalists. (3) Such accumulation increases the demand for labor, and, with a given population, in the short run wages tend to rise. (4) As wages exceed the level of subsistence, population grows, generating an increased demand for food. (5) But, if high-quality land is essentially fixed, the rise in food demand is met by bringing lower-quality land into production. The price of food rises to cover the higher cost of production on lower-quality land. (6) The effects of increased population (supply of labor) and higher-priced food drive the real wage, or the wage paid divided by food prices, back to the subsistence level, and the rate of population growth declines.

Thus, in the classical model, diminishing returns to increments of labor applied to a relatively fixed supply of high-quality land, and higher costs of production on lower-quality land, represent constraints to growth, so that living standards remain at subsistence levels. If technological progress occurs, the situation may change temporarily but not permanently. Ricardo's policy prescription was for Great Britain to remove its corn laws, which would free up trade, and allow food imports to keep the price of food from rising and choking off industrial growth. History has shown that the classical model underestimates the role of technological progress. It also fails to consider factors that tend to lower birth rates as economic growth occurs. It oversimplifies the forces influencing wages and the complexity of the sharing or distribution objective found in many societies. Nevertheless, as we will see below, certain aspects of the classical model had a significant influence on subsequent theories of economic development, especially its emphasis on diminishing returns and its implications for trade.

Growth Stages: From Marx to Rostow

By the late nineteenth century, there had been enough economic development in Europe and North America for observers to notice a clear shift in the mix of activities. Many economists focused on patterns of such change, arguing that economies moved through sequential *growth stages*. While the suggested sets of stages were based on different principles, most growth stage theories attempted to emphasize that economic development involves a structural (economic and/or social) transformation of a country.

In the late nineteenth century, **Frederick List**, a German economic historian, developed a set of stages based on shifts in occupational distribution. His five stages were savage, pastoralism, agriculture, agriculture-manufacturing, and agriculture-manufacturing-commerce. Concurrently, another German, **Karl Marx**, visualized five stages of development based on changes in technology, property rights, and ideology. His steps were primitive communism, ancient slavery, medieval feudalism, industrial capitalism, and socialism and communism. He felt that class struggles drive countries through these stages. One class possesses the land, capital, and authority over labor while the other possesses only labor. Class struggles occur because economic institutions allow the exploitation of labor. Prior to reaching the final stage, labor is never paid its full value. For example, if wages rise in the fourth stage (industrial capitalism), labor is replaced by machines, thereby creating a "reserve army of the unemployed" that brings wages back down. Because capitalists derive their profits from labor, more machines and fewer laborers mean lower profit rates. The pressure of lower profits leads to more exploitation, more unemployment, mass misery, and eventually revolution. Labor then gains control over all means of production under communism.

A different kind of thinking about growth stages emerged in the early twentieth century, when **Alan Fisher** and later **Colin Clark** developed a theory in which the transition from agriculture to manufacturing and services occurs not because of government intervention, but because of increases in output per worker, and advances in science and technology. Another growth stage theorist, **Walt W. Rostow,** argued in the 1950s that these changes were closely related to the rate of growth in per capita incomes, which would experience a "take-off" into sustained growth once enough capital had been accumulated. Rostow believed, however, that an eventual slowdown in the rate of growth would be the normal path for any sub-sector in an economy, due to declining price and income elasticities of demand for the goods produced by a sector. In this view, the secret to growth is to find and support emerging or "leading sectors".

Thinking of the economy in terms of distinct sectors has some advantages, but the idea of growth stages fell out of favor in the 1950s. Countries experienced a wide variety of growth paths during the 1950s and 1960s, and some experienced sharp reversals of fortune. Most economists no longer thought of economic growth as a predetermined sequence of stages, which had relatively little prescriptive power, but instead focused on the gradual accumulation of productive resources, particularly capital.

Capital Accumulation: From the "Financing Gap" to Technology-driven Growth

The first widely used theory of growth based on capital accumulation was developed by **Roy Harrod** and **Evsey Domar.** They used mathematical formulas to show how the rate of output growth would be limited by the level of investment and hence the national savings rate, multiplied by the productivity of those investments. The Harrod-Domar model was simple and elegant, and yet could still be fitted to real data using the observed capital/output ratio of the economy to project the productivity of additional investment.

In the 1960s, when the Harrod-Domar approach was applied to low-income countries, it was recognized that national savings was not the only possible source of capital. Borrowing from abroad could add to national savings, permitting an even faster growth of the capital stock. Such "two-gap" models, popularized by **Hollis Chenery** and others, implied that foreign aid to fill a "financing gap" could accelerate growth significantly, as each dollar of aid would have the same productivity as a dollar of savings.

The Harrod-Domar-Chenery approach focused primarily on the rate of national savings or borrowing from abroad, with less attention to the efficiency with which additional funds were spent. In the mid-1950s, **Robert Solow** worked out the mathematics of a model in which additional capital experiences diminishing returns. In that case, the long-run rate of growth of per capita income is driven by the rate of technological progress, not savings as such. Solow did not explain how technological progress is generated: he treated new technology (and hence the growth rate of the economy) as exogenous to (outside of) his model. Much later, a new generation of economists would make growth models in which people choose how much to invest in new technologies, so that technical change and hence the growth rate is endogenous, explained by property rights and government policies. Those models are described in the final section of this chapter.

Dual-economy Models: "Surplus Labor" and Unemployment

The first mathematical models of growth used a single sector to describe the whole economy, and focused on capital accumulation. Soon thereafter, economists produced models with two sectors, in which growth and poverty alleviation depend crucially on the allocation of labor. The most influential *dual-economy* (or two-sector) model was developed by **W. Arthur Lewis.** His model was subsequently modified by **John Fei** and **Gustav Ranis, Dale Jorgenson,** and others.

A simplified version of the dual-economy model can be illustrated using the total and marginal product curves shown in Figure 6-1. This version of the model is designed to relate most closely to the situation in large labor-surplus but relatively natural-resource-poor countries in which domestic (as opposed to international) characteristics of the economy dominate. The model could potentially represent (albeit roughly) the situation in a country such as India or China.

The model includes several sources of growth discussed in Chapter 5, and illustrates the potential for using "surplus" labor and technological progress in agriculture to achieve economic growth. It assumes the existence of a large population in the traditional agricultural sector, for which the marginal product of labor is below the wage rate, which is determined by society's rules about sharing output. There is disguised unemployment in the sense that if the people who appear to be working were removed, production would not drop or would drop very little. In other words, labor is applied in the agricultural sector up to the point where it is redundant in the upper left-hand graph in Figure 6-1; or to the right of N_3 or N_2 in the lower left-hand graph.

The wage rate in agriculture (W) is assumed to initially approximate the average productivity of labor in that sector (and eventually be determined in an inter-sector labor market). Land is fixed. Wages in the modern industrial sector are assumed to be higher than in the agricultural sector in order to attract labor from the agricultural sector. Firms in the modern sector hire labor up to the point at which the marginal product of labor equals the wage rate. Initially this is the point P_0 in the lower right-hand graph of Figure 6-1. Labor in industry is hired up to L_0 at the wage P.

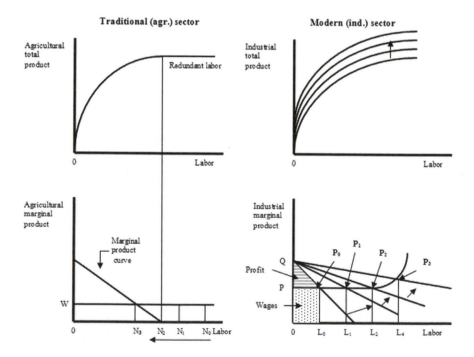

Figure 6-1. Graphical representation of labor-surplus dual-economy model.

In a "labor surplus" economy, the development process can be driven by transfer of labor from agriculture to the industrial sector, where it creates a profit that can be used for further economic growth. In the lower right-hand graph in Figure 6-1, total wages initially paid to labor in the industrial sector equal the area PP_0L_0O while profits equal the triangular area QP_0P. This profit, or part of it, is reinvested in capital items such as equipment, machinery, and buildings — items that make labor more productive. This greater productivity shifts the total product of labor in industry upward (see the upper right-hand graph of Figure 6-1) and the corresponding marginal product of labor (demand for labor) out to the right (see the lower right-hand graph of Figure 6-1). This demand for labor is met by drawing more labor out of agriculture.

In the model, a shift of labor from agriculture to industry continues to drive economic growth as long as the marginal cost of labor remains constant (represented by the horizontal line between P_0 and P_2 in the lower right-hand graph in Figure 6-1). Once the supply of "surplus" labor from the traditional farm sector has been absorbed, the marginal cost of labor supplied to the modern sector turns upward (as it does to the right of L_2), and the growth in demand for labor by industry slows, because fewer profits are available for reinvestment.

Why might the wage rate in industry increase and the demand for labor stop shifting out? First, surplus labor in agriculture might be used up so industry would have to offer higher wages to compete with agriculture for labor. Second, food production will start to decline if fewer than N_2 workers are employed in the agricultural sector. If population is increasing and incomes in the industrial sector are rising, then the demand for food will rise. Unless an increase in agricultural production occurs, agricultural prices eventually rise relative to industrial prices. This rise, in turn, raises the wage at which employers are able to obtain workers from agriculture for industry. The major implication is that economic growth becomes constrained unless there is technological improvement in both sectors.

The labor-surplus dual-economy model is a highly simplified view of the situation in countries with underemployed people. It has several limitations. First, evidence indicates that few if any situations exist where the marginal product of labor in agriculture is close to zero. Few countries have excess labor in agriculture. However, Jorgenson and others have pointed out that the presence of an active labor market in which the two sectors compete for labor can generate the same implication of the need for technological improvement in both sectors. Second, the model ignores the possibility of international trade, although it could be added without much difficulty. Third, and more important, the model

Philippine farm workers.

fails to recognize the cost of resources used in conducting research and educating farmers to produce more and facilitate adoption of new technologies. The issue of how to endogenize (build in the process for self-generating) the development of new technologies in a model of economic development was not addressed. Despite these limitations, it is a useful means of thinking about linkages between multiple economic sectors in a developing-country context.

Dependency Theory and Trade Protectionism

In the 1950s and 1960s, a number of theorists saw international trade and investment as a cause rather than a remedy for poverty in low-income regions, arguing that trade made the poor increasingly dependent and weak. **Immanuel Wallerstein**, for example, popularized the idea that prosperity of the "center" was linked to the impoverishment of the "periphery". *Dependency theory* encompassed a range of arguments, generally leading to the conclusion that the governments of low-income countries should protect their local economies from foreign trade and investment, pursuing self-sufficiency as a form of political and economic independence.

A few dependency theorists, notably **Andre Gunder Frank**, adopted a *Marxist* perspective, arguing that the income of wealthy countries was derived from the output of poor countries. In this view, wealthy countries use military and political power to limit poor countries' options, and thereby extract income that would otherwise belong to the poor. Some expropriation of this type has clearly occurred, in the colonial period and through other kinds of intervention, but most economic historians believe the output of poor countries can explain only a very small fraction of the wealth we see in industrialized countries.

A more widely accepted set of ideas comes from *structuralists* such as **Raul Prebish** and **Hans Singer,** who argue that market forces limit the degree to which poor countries can develop through trade with richer countries. In this view, the terms of trade (the ratio of prices of exports to prices of imports) tend to turn against developing countries over time, because they produce mainly primary products (agricultural and mineral) for which prices decline over time relative to the manufactured products they import. This deterioration in the *terms of trade* is believed to be generated by (1) low price and income elasticities of demand for primary products compared to manufactured products, (2) slow productivity growth in primary product production, and (3) monopolistic elements in the production of products imported by developing countries while primary products are produced competitively. To the extent that demand for poor countries' exports is price- and income-inelastic, then output expansion in the poor countries or in the world as a whole does indeed worsen poor countries' terms of trade, although again this influence can explain only a fraction of the income gap between rich and poor countries.

The trade restrictions favored by dependency theorists could also be justified by much older arguments in favor of government intervention to protect domestic markets from foreign competitors, notably the idea that *infant industries* can get started only if they are temporarily protected from foreign competition, and the idea that a *big push* to expand many industries simultaneously could help countries take advantage of synergies between them. During the 1970s and 1980s, however, it became increasingly clear that industrialization aimed at replacing imports for the domestic market could generate only a temporary burst of economic growth. Export-oriented industrialization proved to be more successful.

Contemporary Growth Theory: Technological Innovation and Public Institutions

By the mid-1980s, enough statistics on national income across countries were available for researchers to test the basic predictions of the standard growth model, posited thirty years earlier by Robert Solow. Results were surprising, and sparked a burst of academic research on economic growth and poverty reduction that continues today.

The Solow model predicted that poor countries would eventually catch up with rich ones, because of diminishing returns to capital. Statistical tests showed that this type of "convergence" did indeed occur, but only among subgroups of countries. The highest-income group of countries continued to grow with no sign of diminishing returns, while some poorer countries grew even faster to catch up, and other poor countries just stayed poor.

Economic theorists attempted to explain these results. **Robert Lucas, Paul Romer**, and others showed how rich countries' growth could be explained by a flow of new technologies, which help overcome diminishing returns. Their models hinge on the idea that new knowledge is a public good: once discovered, it can be used repeatedly in new technologies without being used up, and so technological innovations can accumulate without limit. But not all countries are able to generate or use these innovations.

What determines whether a country develops and applies appropriate new techniques? Knowledge itself is a public good, whose development and dissemination depends on public education and government-funded research. Individuals and private firms will never have enough incentive to invest as much in these resources as they are worth to society as a whole. But knowledge is economically valuable only when embodied in goods and services that meet consumer needs. Successful countries promote both public knowledge and also private enterprise, encouraging new enterprises with new technologies.

A key question is the degree to which innovators should be given monopoly rights over the sale of new products, through patents and other forms of intellectual property rights. Government-enforced protection from imitators is both good and bad: it makes each invention more profitable than it otherwise would be, but it does so by restricting its use! The patent policies that are most economically successful limit the scope and duration of protection, to be just enough to reward past innovators, while encouraging others to make use of the innovation. The British and U.S. patent systems were early pioneers in this regard, offering protection only to a specific product (to allow the entry of other, somewhat similar products), and limiting the time period of protection (to hasten the entry of other firms), while allowing competitors to challenge others' patents in a free and fair judicial system.

The interplay among technology, natural resources, human capital, and institutions remains an active area of research today. It is clear that other sources of growth are only effective if they operate in an institutional environment conducive to growth. The importance of the rule of law, enforceable property rights and contracts, absence of serious government distortions to markets, and relatively low levels of corruption are all important to economic development. The high costs of transacting also seem to prevent many countries from realizing improved levels of living. Improved information flows may help reduce the cost of transacting and make it more difficult for inefficient institutional and political structures to survive. We return to this issue of how to reduce transactions costs in Chapter 11.

FROM THEORY TO ALTERNATIVE STRATEGIES

The concept of a *development strategy* implies a long-term road map that encompasses a series of fundamental decisions with respect to sector emphasis (agriculture versus industry), factor use (capital-led versus employment-led growth), international market orientation (inward versus outward), concern for growth versus distribution, and the roles of the private versus the public sector. Many of these decisions present conflicting choices that countries must make when designing their development strategies. The appropriate path for a particular country depends on its starting characteristics and global economic conditions.

Industry versus Agriculture

The question of whether to channel public and private investments into the agricultural or industrial sectors has been asked by policymakers for many decades. In most countries, agriculture is initially the dominant sector containing most of society's resources, but it contains the poorest and least politically influential people and so is often relatively neglected by government. Investments in agriculture are slowed by this weak political base, but other factors inhibit such investments. Impacts of agricultural productivity growth can be difficult to observe. As seen in Chapter 5, an increase in farm output generally leads to an increase in *other* activity, as farmers invest their resources in non-farm enterprises, and a lower cost of food helps non-farmers buy more of other things. So agriculture appears to be a slow-growth sector, even as it drives the expansion of other sectors. Politicians generally want to please urban constituents and often adopt policies to lower food prices. Lower food prices, in turn, reduce the profitability of investments in agriculture. There is usually much stronger political pressure for urban investments, and for policies that produce immediate, highly visible results.

The degree to which governments support agriculture as opposed to industry also depends on world market conditions: in the late 1960s and early 1970s, the threat of food scarcity associated with Asian population growth led many countries to invest heavily in irrigation and crop breeding to raise agricultural productivity, especially within Asia.

During the 1980s and 1990s, the pay-off from those investments produced a relative abundance of food on world markets, which reduced demand for further investment, even in regions such as Africa where food was increasingly scarce.

Inward- versus Outward-led Growth

A persistent debate in the development literature has centered on the merits of an inward (import substitution, self-sufficiency) oriented strategy versus an outward (international trade, export promotion) oriented strategy. Some observers have argued that developing countries are hurt by trade because they produce mainly primary products for which prices decline over time relative to the manufactured products they import. In addition, the colonial heritage in several developing countries included the export of certain primary products to developed countries with the profits going to foreign companies or to small groups of elites in the developing countries. Proponents of an inward strategy have also argued that countries following an inward-oriented path suffer less from debt crises, protectionist policies in the developed countries, and the adverse effects of high wages due to labor market imperfections.

The impact of inward-directed strategies depends largely on the policies used to implement the strategy. Policies such as overvalued exchange rates, import restrictions, and explicit export taxes, which discourage exports and stimulate substitution of domestically produced goods for imports, have generally been shown to be counterproductive. They lead to distortions in resource prices, create monopoly profits, high government budget deficits and, usually, inflationary pressures. Policies supporting production of foods for internal consumption via research, infrastructure, and other public investments can be called inward oriented, yet are not associated with some of the distortions caused by measures typically used to promote import substitution.

Proponents of outward strategies argue that by removing the bias against exports, countries can achieve significant economic benefits from specialization and comparative advantage, from the import of products manufactured by highly capital-intensive industries abroad, and from the stimulus to employment provided by reduced pressures to concentrate capital in a limited number of capital-intensive industries. Economies of scale can be achieved due to enlargement of the effective market size. Some countries that have been successful at promoting export-led growth have, in fact, also relied on government interventions in exporting industries.

Theoretical arguments support either position. However, over the past 30 years, empirical evidence is weighted heavily in favor of an outward-looking strategy that biases the economy neither for nor against exports. Evidence shows that policies often used to create an inward-looking strategy can lead to inefficiency. The economic efficiencies sacrificed in attempts to insulate a country from world market forces can be significant. Open markets expose a country to the effects of protectionist policies and interest rate fluctuations abroad. However, they also offer insurance against risks originating at home.

Outward-looking strategies will be most successful if international markets are truly competitive and if access to markets is unrestricted. International trade agreements, covered later in this book, have moved the world markets toward more transparency and fewer trade restrictions. Some restrictions, however, still exist.

Growth versus Equity

The persistence of abject poverty even in countries experiencing rapid rates of economic growth has spurred a debate over the appropriate focus of development efforts. Most of us accept the goal of lifting as many people as possible out of extreme poverty, but there are many competing ideas on how to do it. Essentially three general approaches have been suggested, sometimes in combination. The first is to make direct transfer payments (money, goods, services) from the more well-to-do to the poor. The second is for the country to concentrate entirely on growth as a goal, no matter who receives the income, in the expectation that part of the benefits will trickle down to the poor. A third

Many developing countries have a comparative advantage in exporting sugar, but face protectionist sugar policies in developed countries.

approach is to direct specific efforts toward raising the productivity of the poorest segments of society during the growth process.

Direct transfer payments are difficult for developing countries to afford unless obtained as grants from international sources. The most important role of direct transfers can occur (1) during short-run weather-induced famines or other emergency situations, and (2) among the perpetually disadvantaged elderly, orphaned, and handicapped.

The majority of the poor in most developing countries, however, are the unemployed and underemployed rural landless. Even unskilled urban workers are usually better-off than the rural landless. The landless live close to the margin and may fall below it during bad crop years. Therefore, the important question is whether the benefits of growth will trickle down to the poor or whether development efforts must be directed at the poor.

During rapid growth some benefits are captured by the poor. However, the distribution of income will often worsen (become more unequal) during initial stages of growth unless specific efforts are directed toward incorporating the poor into productive activities. The poor can be bypassed by growth-oriented investments, especially when possession of assets, particularly land and education, is skewed. Countries that begin with a more equal distribution of assets tend to experience growth with equity more than others. Growth can actually stagnate under conditions of extremely inequitable asset distribution. Growth itself can be affected by the wider spread of assets, institutional changes, and employment-creating activities.

The mere widening of the income distribution as development occurs is not as much a concern as what happens to income *levels* of the poor. Neither the level nor the distribution of income will be improved for the poor in most countries unless they have improved access to assets such as land and education, which can make their primary asset, labor, more productive during the growth process. Development strategies that increase employment opportunities and promote the supply of wage goods (mainly food) will have the best chances for reducing poverty under virtually all circumstances.

Private versus Public

The appropriate mix of public and private activity varies by country, and by sector. Some services are almost always best funded through the public sector, such as an independent judicial system and roads. These are *public goods*, whose provision is limited by *free rider* problems: people can benefit without paying, so government intervention is needed to force everyone to pay a share of their costs. Other activities can be funded voluntarily through private activity, but must be regulated by the public sector or they will be provided inefficiently.

Activities that are typically regulated by government, if not provided directly in the public sector, include *natural monopolies* such as water supplies, or services with *positive externalities* such as sanitation and health. Too little of these services would be provided by private firms if they were not regulated in some way by government. On the other hand, unregulated firms would provide too many goods that generate *negative externalities* such as pollution.

The outcome of interactions between the public and private sectors is often determined not by who does what, but by the degree of transparency and accountability in what they do. Private firms that can be held accountable to their investors and customers tend to work efficiently, as do public institutions that are accountable to voters and taxpayers. Either kind of institution can become corrupt and inefficient, in the absence of appropriate checks and balances, within and between each sector.

A useful way to explain the degree of accountability in the economy, over both public and private institutions, is through the relative size of *transactions costs* in the market or political system. Lower transactions costs typically make either system more accountable to a larger number of people. Easier transactions between customers and suppliers make the market more efficient, and easier transactions between citizens and their government usually make the public sector more efficient.

A range of institutional arrangements can keep transactions costs low and sustain checks and balances over time. Private markets must be regulated by public institutions, and the public sector must be kept accountable to private individuals. Otherwise, even if new technologies are available, growth can be hindered by an inefficient or inequitable institutional structure.

SUMMARY

The classical model of economic growth stressed the importance of diminishing returns to labor as a constraint to growth, and the mid-twentieth-century Solow model stressed diminishing returns to capital. Contemporary experience, however, shows how countries with institutions that reward innovation can sustain rapid economic growth far beyond these constraints.

Growth-stage theories attempted to categorize the growth process into successive stages through which countries must pass as they develop. Dual-economy models focused on movement of labor out of agriculture and how the agricultural transformation can be smoothed by balanced growth in both sectors. Dependency theorists argued that developing countries became increasingly exploited as they become more integrated into world markets, and so should withdraw into self-sufficiency. Each of these classes of theories provides some insights into the development process, but does not provide a comprehensive theory of growth and development.

Contemporary development strategies recognize the role of agriculture as an engine of economic growth. Agricultural growth frees up labor and other resources that may be used in other sectors. It helps alleviate poverty by improving food availability and stimulating broad-based employment growth. Most economists agree that international trade should be kept relatively open, and that governments should provide public goods, promote innovation, regulate monopolies, and make markets more efficient. The exact development strategy for each country depends on its resource mix, stage of development, and institutional structure. New institutional arrangements will have to be designed in many countries to enhance information flows and lower transactions costs, to make markets more efficient, and to promote accountability in the public and private sectors.

IMPORTANT TERMS AND CONCEPTS

Capital-led growth	Income distribution
Center and periphery	Institutional arrangements
Classical model	Integrated rural development
Comparative advantage	Labor-surplus dual economy
Dependency theory	Open versus closed economy
Employment-led growth	Public good
Enclave dualism	Resource mix
Export-led growth	Sociological dualism
Growth stage theory	Stage of development
Growth versus equity	Terms of trade
Harrod-Domar model	Transactions costs
Import substitution	

Looking Ahead

In this chapter, the roles of agriculture in economic development were mentioned along with the need for countries to tailor their development strategies to their resource bases and stages of development. In much of the rest of the book we will be examining how to develop the agricultural sector itself. Before we do that, however, it is important to discuss the nature of existing agricultural systems in developing countries. In the next chapter, we discuss the characteristics of traditional agriculture, including the particular roles of livestock.

QUESTIONS FOR DISCUSSION

1. What is the major factor that is hypothesized to constrain economic growth in the classical model?
2. What are the major features of the labor-surplus dual-economy model and what are its primary weaknesses?
3. Why might the wage rate eventually increase in the industrial sector in the labor-surplus dual-economy model?
4. What implications does technological change in the agricultural sector have in the labor-surplus dual-economy model?
5. What is the distinguishing feature of dependency theories? What are the policy implications of dependency theories?
6. Why is agricultural development important in most developing countries?
7. What is employment-led growth and why is employment important to development?
8. What are the arguments for and against inward- versus outward-oriented development strategies?
9. What are the three general approaches that have been suggested for alleviating abject poverty?
10. Why might both the private and public sectors have important roles to play in development?

RECOMMENDED READING

Hayami, Yujiro, and Vernon W. Ruttan, *Agricultural Development: An International Perspective* (Baltimore, Md.: Johns Hopkins University Press, 1985), ch. 2.

Lewis, W. Arthur, "Economic Development with Unlimited Supplies of Labor," *Manchester School of Economics and Social Studies*, vol. 22 (May 1954), pp. 139–91.

North, Douglas, "Institutions, Transactions Costs, and Economic Growth. *Economic Inquiry*, vol. 25 (1987), pp. 415–18.

Olson, Mancur, Jr., "Big Bills Left on the Sidewalk: Why Some Nations Are Rich and Others Are Poor," *Journal of Economic Perspectives*, vol. 10 (Spring, 1996), pp. 3–24.

PART **3**

Agricultural Systems and Resource Use

Traditional farm in Nepal.

CHAPTER 7

Agriculture
in Traditional Societies

In low-income countries, peasant agriculture tends to be characterized by low levels of utilization of certain resources, low levels of productivity, and relatively high levels of efficiency in combining resources and enterprises.
— John W. Mellor[1]

This Chapter
1. Describes the common characteristics of traditional agriculture
2. Identifies the multiple roles of livestock in traditional farming systems
3. Discusses implications of characteristics of traditional farming systems for agricultural development.

CHARACTERISTICS OF TRADITIONAL AGRICULTURE

The world food-income-poverty problem is serious, and solutions depend in part on agricultural development. Before considering how to foster development, one needs knowledge of the nature of agriculture in developing countries. Without this knowledge, it is difficult to understand the changes required to promote agricultural development and how these changes will affect the people involved. In this chapter, we examine several general characteristics of traditional agriculture. Then in Chapter 8 we compare specific types of agricultural systems in various stages of development.[2]

The term *traditional agriculture* conveys part of its own meaning. The word "traditional" means "to do things the way they have usually been done." Because natural resources, culture, history, and other factors vary from place to place, the way things have usually been done

---

[1] John W. Mellor, *Economics of Agricultural Development* (Ithaca, N.Y.: Cornell University Press, 1966), p. 134.
[2] Agricultural systems include production practices, or *how* things are produced, as well as the types of enterprises, or *what* things are produced.

also differs greatly from one location to another. And, because conditions change, no type of farming system, no matter how traditional, is ever completely stable. Nevertheless, farms in traditional agricultural systems do have several common characteristics.

Intermixing of Farm and Family Decisions

Traditional agriculture takes several forms, but small peasant farms predominate in most developing countries. Business decisions on these farms are generally intermixed with family or household decisions. The importance of the family and the close relationship between production and consumption decisions occur because much of the labor, management, and capital come from the same household. A sizable proportion of the production is consumed on the farm or at least in the community where it is produced. Success in the farm enterprise can enhance nutritional status, which, in turn, leads to higher productivity on the farm.

This intermixing of production and consumption decisions, along with the low levels of income common among peasant farms, adds an

---

BOX 7-1
KEFA VILLAGE IN EASTERN ZAMBIA

The anthropologist Else Skjonsberg visited Kefa Village first in 1977 and several times since. Her book, *Change in an African Village: Kefa Speaks*, portrays a traditional agricultural system in Eastern Africa. Villagers in Kefa depend on their land, which is controlled and allocated by the local chief. Some inherit cultivation rights from their parents, others request unused land from the chief, and others borrow land from relatives and neighbors. When land shortages arise, groups of villagers break away and search other areas for unused lands.

Typical households cultivate 1-4 hectares, with maize, groundnuts, sweet potatoes, and pumpkins produced for their own consumption, and tobacco and cotton produced in small quantities for sales. More fortunate farmers have access to wetland dambos, where they grow vegetables year-round. The agricultural year starts prior to the first rains in October, when the ground is broken by hand hoes. Maize, the most important food crop, is planted first following the rains; it is weeded first and harvested before other crops. Most villagers plant open-pollinated "local" maize varieties, which have been used for generations. They store surplus maize in household granaries, and in years of abundance use it to brew beer or barter or sell. The primary objective of most Kefa farmers is to fill their maize bins with enough to feed family and visitors until the next harvest. Groundnuts are rotated with maize to maintain soil fertility and provide dietary protein. Hybrid maize varieties, with higher yields and shorter grow-

---

BOX 7-1, continued

ing seasons, have been introduced, but most Kefa villagers are suspicious of their quality and only produce them for sale. Hybrids require purchased fertilizer inputs and, in the world of dryland farming, exposure to risk invites trouble. Many believe that use of fertilizers will breed dependency and bring ruin to adventurous farmers.

Work is shared by family members. Hoeing, although physically demanding, is considered women's work. Women prepare meals, carry them to the fields, hand-cultivate all day, then return home with pots and pans, loads of firewood and water. During December and January, the women take responsibility for weeding. It takes as long as three weeks to weed a hectare of maize, so family time is fully occupied. On rainy days, men make repairs around their huts, while women manage household affairs. When labor is scarce, some mobilize village workers by throwing home-brewed beer parties; others trade labor and work together. During April through June, labor is in short supply and entire villages participate in harvests. Women are chiefly responsible for harvest, but men and older children assist. Women headload food crops in 50 kilogram bags from the fields to storage bins. In rare cases where oxen or motorized transport is used, men take responsibility for the task.

Although agriculture is the main source of well-being in Kefa, all households are engaged in non-agricultural activities. Some brew and sell beer, others practice crafts such as weaving or woodworking, many engage in petty trading; others are healers or scribes, or have specialized skills. Cattle raising and off-farm incomes supplement farm incomes and help families buy farm and household equipment, clothes, blankets, and services such as school fees. Off-farm activities are divided by gender; women brew and sell beer, trade, and weave, while men more often have specialized skills, work with wood, or do repairs. Incomes earned in these activities are held separately by men and women, and women are eager to engage in such activities because the money they earn provides them a degree of autonomy in decision-making.

Families in Kefa are structured in different ways. Only about half of the households are nuclear in the sense of two parents and children. About a third of households are headed by women, some divorced or widowed, some whose husband is absent. Some polygamous households with one husband and several wives are found; in most cases, polygamy is associated with higher wealth and social status. Children participate actively in household economic life; by age 5 most contribute to household tasks, and after 8 years, farmwork increases. Boys are responsible for tending cattle, while girls assist their mothers in the house and care for younger siblings. Most children attend schools, but are excused during periods of peak agricultural labor. The elderly live with their families or are cared for by family members. The poorest of the poor have few relatives and depend on handouts from other villagers.

element of conservatism to family farming (Box 7-1). A farm disaster usually means a family disaster. Consequently, traditional farms use crop varieties and breeds of livestock that have proven dependable under adverse conditions, such as low fertility or rough terrain, even if yields or productivities are modest.

For example, in many regions of the world cassava is grown on a portion of a traditional farmer's land. Cassava grows slowly and has low value, but it grows on relatively poor soils, and under a variety of weather conditions. It is a root crop that can be pulled from the ground at different times of the year when other foods may be short, and represents a form of informal insurance or savings.

Traditional farm households consume most of their products at home. However, the surplus they trade or sell connects them to the local market. Consequently, traditional farms are influenced by market-price relationships in their decisions to allocate family resources. Sometimes, traditional farms are called *subsistence* farms. The surplus traded or sold by traditional farms varies from country to country, region to region, and farm to farm, but few of these farms are entirely subsistence farms in the sense that they consume all they produce. However, many are *semi-subsistence*, consuming part and selling part. The percentage of the total output that is marketed may be planned or unplanned and can vary significantly from year to year as production itself fluctuates according to climatic conditions.

Labor and Land Use

Traditional farms are generally very small, usually only 1 to 3 hectares (about 2.5 to 7.5 acres). Labor applied per hectare planted, however, tends to be high. In many areas, land is a limiting factor and is becoming more limiting over time as populations continue to grow. Labor is often underemployed at certain times of the year, while capital assets are fully exploited. Much sharing of work and income occurs on traditional farms, so there is little open unemployment during slack times. This sharing means that the individual's wage may be determined by the average rather than the marginal productivity of labor, as mentioned in Chapter 6. Family members also supplement their incomes by working off the farm part-time, often on larger farms, and sometimes outside agriculture.

Although family labor is important, traditional farms may hire some labor, at least during the busy times of the year. Low wages caused by high underemployment in peasant agriculture create incentives to hire laborers. That is, traditional farmers can hire labor or buy a small amount of leisure and enhance their social status at relatively low cost.

Therefore the people with the lowest economic and social status are usually not the owners of small traditional farms, but the landless workers hired by those farmers.

Seasonality

Labor use in traditional agriculture exhibits marked seasonal variation corresponding to agricultural cycles. During slack seasons, those immediately following planting or preceding harvest, labor may be abundant. However, during peak seasons, especially during weeding and harvest, labor can be in short supply. Wages often exhibit similar seasonal fluctuations.

The seasonal nature of agricultural production causes variations in consumption and nutritional status, particularly in African settings. Because storage facilities may be lacking and mechanisms for saving and borrowing incomplete, consumption patterns can follow agricultural cycles. It is common to find "lean seasons," when consumption is low and short-run malnutrition high, especially immediately prior to harvest (see Box 7-2). Similarly, traditional feasts often follow harvests, when foods are abundant and labor use is less intensive.

Productivity and Efficiency

Traditional farms are characterized by low use of purchased inputs other than labor. Yield per hectare, production per person, and other measures of productivity tend to be low. These factors do not mean, however, that traditional farms are inefficient. As T. W. Schultz points out, traditional farms tend to be *poor but efficient*.[3] Why?

The crop varieties, power sources, methods for altering soil fertility, and certain other factors available to traditional farms constrain productivity growth, and hence reduce returns to labor and traditional types of capital. Efficiency, as measured by equating marginal returns to resources in alternative uses, is often high. In other words, given the technologies available to traditional farmers, they tend to do a good job of allocating labor, land, and other resources. The implication is that just reallocating the resources they currently have will not have a major impact on output.

It makes sense that with static levels of technology, physical conditions, and factor costs, farmers would gradually become very efficient at what they do. When conditions change rapidly, many of the mistakes in resource allocation occur. Also, one must be careful not to

---

[3] Theodore W. Schultz, *Transforming Traditional Agriculture* (Chicago, Ill.: University of Chicago Press, 1964), p. 38.

BOX 7-2
SEASONAL MIGRATION: A RATIONAL RESPONSE

Seasonal weather patterns cause traditional farmers to adopt production and consumption patterns to help smooth variations. Seasonality also induces migration as people search for employment opportunities and food. Other seasonal causes of migration are trade and marketing, cultivation of secondary landholding, and pasturing cattle. Seasonal migration is a worldwide phenomenon. In some rain-fed areas of Africa, 30 to 40 percent of the economically active population migrates, while in rural Nepal as much as 30 percent of the households have at least one member who migrates.

Why does seasonal migration occur? During the lean season, labor demands on the farm are low, incomes are stretched, and food can be in short supply. Other rural regions may have crop conditions (due to environmental factors, technologies, or irrigation) that alter the agricultural calendar and create counter-cyclical demands for workers. Large plantations commonly producing many export crops also demand labor on a seasonal basis. Seasonal rural-to-urban migration involves workers migrating to towns, cities, and mines in search of work. These reasons combine to push migrants out of regions where their labor is temporarily in surplus and pull them into areas with high demands for labor.

Seasonal migration is not inefficient nor is it caused by factors such as imperfect labor markets. It is a natural adaptation to highly seasonal agricultural cycles and can smooth family incomes and consumption. Seasonal migration also provides insurance; in the event of a crop failure family income can be maintained in the short run by migration.

Seasonal labor flows have benefited countries by minimizing labor shortages in harvest times. Exports of cocoa and coffee from forest regions of Western Africa are largely made possible by seasonal migrants who provide labor during harvest. Other regions of the world have seen their total production possibilities shift outward as labor moves to fill seasonal gaps.

Source: Material was drawn from David E. Sahn, ed., *Causes and Implications of Seasonal Variability in Household Food Security* (Baltimore, Md.: Johns Hopkins University Press, 1987).

---

equate limited education (another common characteristic in traditional agriculture) with lack of intelligence.

A situation with low use of certain inputs, low productivity, but high economic efficiency under static conditions has important implications if productivity is to be increased. First, new technologies can help to change the production possibilities available to farmers. Second, investments to improve the quantity and quality of productive assets such as land can stimulate income growth. Third, education may be needed to help farmers learn to adjust resource use to changing conditions so as to maintain their high levels of efficiency. However, under the static conditions of traditional agriculture, education will do little to improve productivity, since peasant producers are already relatively efficient.

Rationality and Risk

Traditional farmers are economically rational. They are motivated to raise their standard of living while, of necessity, they are cautious. Traditional farmers are not averse to change, but proposed changes must fit into their farming systems without altering too abruptly the methods they have developed over time to reduce risk and spread out labor use (Box 7-3). Traditional farmers face many risks, including weather-related uncertainty, agricultural pests and diseases, price and market-related risks, and human health risks. Decisions made by traditional farmers often reflect attempts to manage this risk. Because formal risk management mechanisms such as insurance are often not available, traditional households turn to informal mechanisms in response to a risky environment.[4]

One mechanism by which traditional farmers spread risk is by exchanging labor and other resources through joint and extended families. By joint and extended families, we mean relatives (and sometimes friends) beyond parents and their children. In many countries, a substantial degree of sharing labor and goods occurs among friends and neighbors, which not only adds to social status but spreads risk. Reciprocal agreements to assist others in times of need can spread risk across space, through agreements with people facing other agro-ecological conditions or in different regions, across economic sectors, through migration and work choice, and across time, through inter-generational sharing. Some of these informal arrangements may deteriorate as development proceeds, creating a need for new institutional arrangements to manage risk.

Another risk-spreading mechanism is reliance on multiple sources of income on and off the farm. Traditional farmers frequently plant multiple crops on a single plot of land in a single season. For example, maize and beans are planted together throughout Latin America; in Africa, maize is intercropped with sweet potatoes, groundnuts, and other foods, depending on the location. Intercropping reduces reliance

---

[4] See Paul B. Siegel and Jeffrey Alwang, *An Asset Based Approach to Social Risk Management*. SP Discussion Series 9926, Human Development Network, Social Protection Unit (Washington, D.C.: World Bank, October 1999).

BOX 7-3
THE NATURAL ENVIRONMENT CAN BE CRUEL
TO TRADITIONAL FARMERS

"That year the rains failed. A week went by, two. We stared at the cruel sky, calm, blue, indifferent to our need. We threw ourselves on the earth and we prayed. I took a pumpkin and a few grains of rice to my Goddess, and I wept at her feet. I thought she looked at me with compassion, and I went away comforted, but no rain came.

"Perhaps tomorrow," my husband said. "It is not too late."

We went out and scanned the heavens, clear and beautiful, deadly beautiful, not one cloud to mar its serenity. Others did so too, coming out, as we did, to gaze at the sky and murmur, "Perhaps tomorrow."

Tomorrows came and went and there was no rain. Nathan no longer said perhaps; only a faint spark of hope, obstinately refusing to die, brought him out each dawn to scour the heavens for a sign.

Each day the level of the water dropped and the heads of the paddy hung lower. The river had shrunk to a trickle, the well was as dry as a bone. Before long the shoots of the paddy were tipped with brown; even as we watched, the stain spread like some terrible disease, choking out the green that meant life to us.

Harvest time, and nothing to reap. The paddy had taken all our labor and lay now before us in faded, useless heaps....

Then, after the heat had endured for days and days, and our hopes had shriveled with the paddy — too late to do any good — then we saw the storm clouds gathering, and before long the rain came lashing down, making up in fury for the long drought and giving the grateful land as much as it could suck and more. But in us there was nothing left — no joy, no call for joy. It had come too late."

Source: Kamala Markandaya, *Nectar in a Sieve* (New York: New American Library, 1954), pp. 76, 81–2.

---

on success in a single crop and helps manage risk. Off-farm employment further diversifies income sources.

Off-farm Employment

Because agriculture is so visible in developing countries, it is easy to assume that rural dwellers are only farmers. In reality, in most countries, off-farm income is an important source of earnings especially for the rural poor. Many landless and near-landless families provide labor to other farmers; these agricultural labor markets are described in more detail in Chapter 13. Others work in non-agricultural enterprises; some are self-employed, producing goods and services for sale. Non-farm employment involves small-scale rural manufacturing, transport, services, and petty trading. Income from these enterprises helps offset fluctuations in earnings from agriculture, representing a risk management strategy. It can smooth intra-year variations in on-farm labor demands. Rural non-farm employment accounts for about 30 to 35 percent of income across the developing world.[5] Non-farm income is particularly important for women who can combine their household obligations including child care with work. The percentage of rural workers in the non-farm sectors varies from country to country, but generally is in the range of 20 to 50 percent.[6] Between 1960 and 1990 in Asia, the proportion ranged from 67 percent in Taiwan to 20 percent in China. Off-farm employment represents a higher proportion of total employment in Asia and Latin America than in Africa, but even in Africa it exceeds 60 percent in countries such as Botswana and Swaziland.

THE ROLES OF LIVESTOCK

Livestock play many vitally important roles in traditional farming systems, roles that are sometimes misunderstood by outsiders. Some have even questioned the need to improve animal productivity in developing countries. There is little doubt that when crops and livestock directly compete for the same resources, it is usually more efficient for humans to consume grain than it is to feed the grain to livestock and consume meat. However, in most traditional farming systems, livestock consume little grain, and meat production is often one of its least important roles. Let us consider several roles of livestock.

Buffers and Extenders of the Food Supply

Farm animals provide a special protection to farm families, acting as a buffer between the family and a precarious food supply. Animals are like a savings bank. Farmers can invest surpluses in them, they grow, and they can be consumed or sold during crop failures. Investments in livestock, thus, represent an informal means of savings and insurance.

In most traditional agriculture, livestock do not directly compete with crops because they eat crop residues, feed off steep slopes and poor soils, and generally consume materials that "extend" the food supply. Many types of animals are ruminants (e.g., cattle, goats, sheep, and

[5] See Steven Haggblade, Peter Hazell, and Thomas Reardon, *Strategies for Stimulating Poverty-Alleviating Growth in the Rural Nonfarm Economy in Developing Countries*, EPTD Discussion Paper No. 93 (International Food Policy Research Institute, 2002).

[6] See Nurul Islam, *The Nonfarm Sector and Rural Development: Review of Issues and Evidence*, 2020 Discussion Paper No. 22 (International Food Policy Research Institute, 1997).

buffalo) that eat grass and other forages that humans cannot and then convert the forages to products for human consumption.

Of course livestock make an important contribution to the quality of the diet as well, by providing meat, milk, and eggs. Small amounts of these high-protein, nutrient-rich foods can have a significant impact on human health.

Sources of Fertilizer, Fuel, Hides, and Hair

Animal manure is vitally important as a source of fertilizer and fuel in many countries. For example, in the remote hills of Nepal, it is difficult to obtain chemical fertilizer. Animal manure increases soil fertility and adds organic matter. In countries where wood is scarce, animal dung is dried and burned for fuel. Often, these two uses of animal manure compete. Dung that is burned cannot be used to increase soil fertility. In India and other countries, methane digesters have been developed; the gas produced is used for cooking, and the residual nitrogen applied to crops.

Few livestock products are wasted in traditional society. Clothing and blankets are made from animal hides and hair of not only cattle and sheep, but buffalo, goats, and other livestock.

Providers of Power and Transport

In many countries, livestock are the principal source of power. They plow the fields, transport products to market, and are used in processing tasks like grinding sugar cane. In some remote areas, animals help to market crops by eating grain and other plant products and then walking to market.

Tractors are still relatively rare in many developing countries. The large investments needed to purchase tractors make them prohibitively expensive for traditional farmers. In most regions there is a rising number of farm workers, and therefore limited demand for purely labor-saving devices. And, on the steep slopes and rough terrain in parts of some developing countries, it will be many years, if ever, before mechanical power replaces animal power.

Social and Cultural Symbols

Livestock, particularly cattle and goats, are highly valued in some societies for social and cultural reasons. A family's social status may be measured by the number of animals it owns.[7] Cattle are given as gifts during ceremonial occasions. While livestock serve major economic

[7] In nomadic societies where no individual family owns the land, animal ownership is almost the only criterion available for measuring social status.

A cow is a type of savings bank in Kenya.

functions, they serve these other social and cultural functions as well. Of course, it is possible that the social and cultural values placed on livestock have evolved over the years because of their importance as capital and income-earning assets.

IMPLICATION OF TRADITIONAL FARMS FOR AGRICULTURAL DEVELOPMENT

Despite the common features described above, one of the striking characteristics of farms in developing countries is their diversity. How land is organized and controlled within farms, gender roles, ties to formal markets, use of mechanical or animal traction, institutional relationships with respect to water rights and access to irrigation, and many other factors differ markedly across regions and sometimes within countries. Farms in much of Sub-Saharan Africa are still quite traditional, whereas farms in many parts of Asia and the Pacific have begun to intensify and modernize. In the next chapter, we discuss factors that cause farming systems to change over time.

Traditional farms are efficient but poor. As population grows and less land is available per farmer, poverty increases unless agriculture changes; as noted in Chapter 5, unless agricultural productivity growth outstrips population growth, rural poverty will increase over time. But change brings additional risks and the danger of increasing income disparities. The distribution of income generated through new plant varieties or power tillers can be affected by asset distribution patterns and

133

Farmer plowing with bullock in Thailand.

institutions that govern the rules of behavior in society. Risks must be managed, and institutions that substitute for historical sharing arrangements must be created. Improved transportation systems are needed to improve information flows and build market linkages.

Several Asian countries face a need to alter their farming systems and diversify out of rice. While rice will remain the dominant agricultural commodity, certain types of vegetable and livestock production become increasingly attractive because of changing consumer demands as incomes grow and the need to intensify labor use increases. Additional education and non-farm employment opportunities become important elements in an overall development strategy. Otherwise the law of diminishing returns will doom traditional farmers to poverty for the foreseeable future. African farmers face problems of low soil fertility, lack of access to markets, and low opportunity costs of time.

As incomes grow in many developing regions, consumer demands change and the global economy will respond to these changes in demand. For instance, estimates show that the demand for meat products in developing countries will nearly double between 2000 and 2020.[8] Growth in meat and milk demand will put pressure on traditional livestock grazing systems, and policies may be needed to smooth the transition to more commercially oriented confinement and open-access

[8] C. Delgado, M. Rosegrant, H. Steinfeld, S. Ehui, and C. Courbois, *Livestock to 2020: The Next Food Revolution.* 2020 Vision Initiative Food, Agriculture, and the Environment Discussion Paper No. 28. IFPRI, FAO and ILRI (Washington D.C.: International Food Policy Research Institute, May 1999).

Traditional farmers in Bangladesh.

grazing systems. Without such policies, market-based pressures may lead to social dislocation and environmental degradation in livestock-producing areas.

SUMMARY

Traditional agriculture is diverse, but traditional farms have some common characteristics. Traditional agriculture is generally characterized by small farms, with intertwined farm and family decisions. Traditional farm families consume, sell, or trade most of their products locally. Their labor use and land area per farm are small, but labor input per hectare is high. Hired labor is often important. These product and labor sales and purchases mean that farmers are, in general, closely linked to the local economy and respond to market signals. Productivity and use of purchased inputs are low but efficiency is relatively high. Traditional farmers are rational but risk averse. They often live in extended or joint families. Livestock play many roles, including extending the food supply; providing a buffer against poor harvests; improving the quality of the diet; generating fertilizer, fuel, hides, and hair. They also provide power and transport and meet social and cultural needs. Traditional farms differ by region; as farms change, some people, particularly the landless, may be left behind unless new technologies are accompanied by improved institutions and education.

IMPORTANT TERMS AND CONCEPTS

Asset distribution pattern
Biological technologies
Buffers and extenders
Diversification
Institutional arrangements
Intermixing of farm and family decisions
Joint and extended families
Landless labor
Mechanical technologies

Mixed cropping
Off-farm employment
Poor but efficient
Rational but cautious
Role of livestock
Seasonality
Semi-subsistence farms
Traditional agriculture

LOOKING AHEAD

A wide variety of agricultural systems is found in the world. These systems evolve over time. In the next chapter we examine the factors that influence the types of farming systems found in a particular country at a point in time. The importance of technical, human, institutional, and political factors is discussed. Several common types of agricultural systems are described, and the significant roles of women and children are highlighted.

QUESTIONS FOR DISCUSSION

1. Why might traditional farms be fairly conservative or slow to change current practices?
2. Are traditional farms subsistence farms? What is meant by "subsistence?"
3. Why are livestock important in many traditional farming systems?
4. Distinguish between productivity and efficiency. Why do traditional farms tend to have high levels of efficiency? Why do they tend to have low or high levels of productivity?
5. What factors influence resource allocation on traditional farms? If a farmer fails to adopt a new, apparently more profitable, farming practice, is he or she irrational?
6. If traditional farmers use resources efficiently, why should we be concerned with raising productivity by increasing the use of new technologies?
7. Are the farmers who own 1 to 3 hectares the poorest people in rural communities in developing countries?
8. Why are joint and extended families still important in many developing countries?
9. Why are farm and household decisions often inseparably linked in developing countries?

10. Why are institutional changes often as important as technological changes for agricultural development?
11. Why do farmers practice mixed cropping? Are agricultural diversification and mixed cropping synonymous?
12. Why is hired labor often important in traditional or semi-subsistence agriculture?
13. Why are new biological technologies often more important than new mechanical technologies for fostering agricultural development?
14. Why is agricultural diversification becoming increasingly important in many Asian countries?

RECOMMENDED READING

Hayami, Yujiro. *Anatomy of a Peasant Economy: A Rice Village in the Philippines* (Los Banos, Laguna, Philippines: International Rice Research Institute, 1978).

Hopper, W. David. "Allocation Efficiency in a Traditional Indian Agriculture." *Journal of Farm Economics,* vol. 47 (1965), pp. 611–25.

Norman, David W. "Economic Rationality of Traditional Hausa Dryland Farmers in the North of Nigeria," in *Tradition and Dynamics in Small-farm Agriculture,* ed. Robert D. Stevens (Ames: Iowa State University Press, 1977).

Schultz, Theodore W. *Transforming Traditional Agriculture* (New Haven, Conn.: Yale University Press, 1964).

Siegel, Paul B. and Jeffrey Alwang, *An Asset Based Approach to Social Risk Management.* SP Discussion Series 9926, Human Development Network, Social Protection Unit (Washington, D.C.: World Bank, October 1999).

Skjonsberg, Else. *Change in an African Village: Kefa Speaks* (West Hartford, Conn: Kumarian Press, 1989).

Wolgin, J. M. "Resource Allocation and Risk: A Case Study of Smallholder Agriculture in Kenya," *American Journal of Agricultural Economics,* vol. 54 (1975), pp. 622–30.

CHAPTER 8

Agricultural Systems and their Determinants

The agricultural pattern that has emerged in each area is in part the result of ecological factors — a particular combination of climate and soil — and in part the result of economic and cultural factors in the society that grows the crops.   — Robert S. Loomis[1]

This Chapter

1. Identifies factors that influence the agricultural systems found in a particular country at a point in time
2. Explores the differences in farming systems found in various parts of the world
3. Presents economic concepts that help explain input and output choices in farming systems.

MAJOR DETERMINANTS OF FARMING SYSTEMS

Farming systems in each region of the world show considerable variety, and are differentiated by how production is organized, by the nature of technologies employed, and by the types of crops and livestock produced. Each system consists of a small number of dominant crops (or livestock) and numerous minor crops (or livestock). We must understand agricultural systems if we are to improve them, and therefore let us examine the primary determinants of the prevailing systems before classifying and describing them.

Technical, institutional, and human factors affect the type of agricultural system that predominates in a region. These factors interact at each location and point in time to provide a unique environment for agricultural production (Fig. 8-1). When these factors remain constant for several years, the farming system that evolves represents a long-term adaptation to that environment. Different farming systems have

---

[1]Robert S. Loomis, "Agricultural Systems," *Scientific American* (September 1976), p. 69.

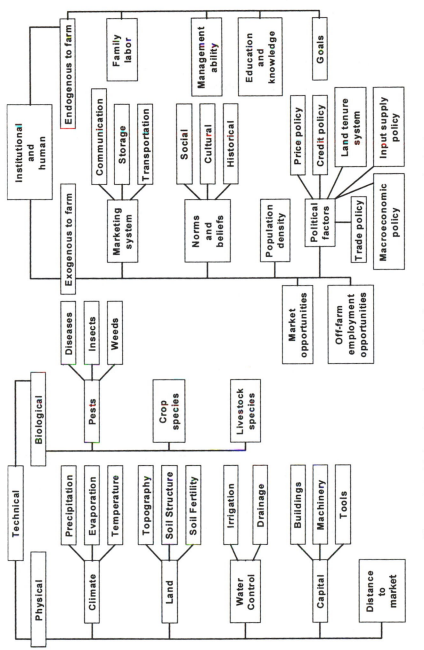

Figure 8-1. Major determinants of the farming system at a point in time.

different needs for public support such as infrastructure, legal systems, and market-related rules and norms. They also have different impacts on the natural environment. Economic development can introduce rapid changes in the underlying factors, thus placing pressure on an existing system.

Technical Determinants of Farming Systems

Technical elements, including both physical and biological factors, help determine the potential types of crop and livestock systems. Physical factors — including climate, land, water access, capital items, and distance to markets — are unique to each location; although water access and other capital items can be altered through investments and new technologies over time. Similarly, investments in road infrastructure can alter the relationship between physical distance and travel time. For example, the nomadism, discussed below, that prevails in many arid regions of the world represents an adaptation to harsh climates. However, the introduction of wells has encouraged more settled farming or ranching in parts of nomadic areas.

Biological factors including pests, crop species, and livestock species are even more susceptible to modification. In the short run, however, these factors play a major role in defining the prevailing agricultural system. The existence of the tsetse fly in areas of the African humid tropics has created farming systems that are dramatically different from those in similar climates where the fly does not exist. Animal traction is not an option in areas where the tsetse fly is common. Technologies to control the fly can help spread animal traction and alter traditional farming relations.

Institutional and Human Determinants of Farming Systems

Institutional and human elements influencing farming systems are characterized by both exogenous (externally controlled) and endogenous (internally controlled) factors. Factors largely outside the control of individual farmers include social and cultural norms and beliefs, historical factors, population density, market opportunities and marketing systems, and off-farm employment opportunities. For example, high population densities in many South Asian countries are partly responsible for the very different farming systems there as compared to the systems found in the relatively low-density areas of Sub-Saharan Africa and Southeast Asia.

Politically determined institutions such as pricing policies, credit policies, macroeconomic policies, trade policies, and land-tenure systems affect the farming system. Land ownership is highly skewed in many countries. In areas of Central America, for example, large commercial farms and plantations exist alongside small peasant subsistence and semi-subsistence farms. The farming practices used in these areas are significantly influenced by the distribution of land: plantations rely on landless and near landless workers as suppliers of labor, and the laborers mix off-farm incomes with food crops grown on their own holdings. The prevailing mix of land uses, crops produced, and technologies used on the different-size farms is clearly affected by the distribution of land holdings. In many areas of the world, people have only use-rights over the land they farm. In much of Africa, for example, families are given land to farm but they cannot rent or sell it to others, and cannot use it as collateral for credit. Such land-use institutions influence incentives for investments in land improvements, which, in turn, influence the prevailing farming system. The political system itself may dictate collectives, communes, or private property as the primary means of organizing land use in agriculture.

Endogenous or farmer-controlled determinants of agricultural systems include family labor, management ability, education, and knowledge, as well as the goals for which farmers are striving. Investments in education affect the value of time used on and off the farm, and as educational levels change, farming systems change in response. The risk associated with agricultural production, particularly in arid, rain-fed regions, has forced farmers to adapt their practices to ensure survival. These adaptations are determined, in part, by the farmers' degree of risk aversion, which is affected by income, education, and so on. Any of these exogenous or endogenous factors can change over time. New technologies and population growth are two particularly important determinants of how and in what direction agricultural systems change over time.

MAJOR TYPES OF FARMING SYSTEMS

While the specific type of farming system in use depends on a large number of factors (Fig. 8-1), many years ago Duckham and Masefield grouped farming systems into three basic types: shifting cultivation, pastoral nomadism, and settled agriculture (Fig. 8-2).[2]

Settled agriculture includes many subtypes. Let us briefly examine each of these systems.

---

[2] See Alec N. Duckham and G. B. Masefield, *Farming Systems of the World* (London: Chatto & Windus, 1970). Substantial variation is observed within these highly stylized farming system typologies.

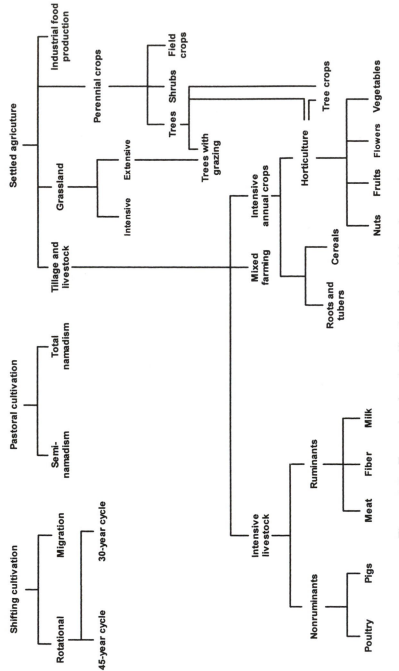

Figure 8-2. Example of a classification of world farming systems.

Shifting Cultivation

Shifting cultivation is an old form of agriculture still practiced in many parts of the world. As the name implies, it involves shifting to a new piece of land when the fertility of the original patch runs out or when weeds and other pests take over. The movement may be fast or slow, and animal manure may extend the use of one location. Migration from one piece of land to another may be random, linear, or cyclic. When cyclic, the rotation frequency can last as long as 30 to 45 years.

Shifting cultivation has also been called *slash and burn* because usually new areas are cleared by slashing the brush with a machete and burning it to clear the fields and release nutrients into the soil. Capital investment in the farm is low, with machetes, digging sticks, and hoes being the primary tools. Typical crops include corn, millet and sorghum, rice, and roots. Usually the crops are mixed. Occasionally the cleared area is used for perennial crops. Shifting cultivation is still practiced on about 15 percent of the world's exploitable soils, particularly in Africa and Latin America. It is popular where population pressures are not too severe.

Shifting cultivation is frequently associated with insecure control over the land, either because of absentee (or government) ownership or unclear tenure status. It has been linked to soil erosion and other environmental problems in several developing countries, partly because there are few incentives to invest in practices that maintain soil fertility.

Pastoral Nomadism

Pastoral nomadism involves people who travel, more or less continuously, with herds of livestock. Pastoral nomads have no established farms, but often follow well-established routes from one grazing area to another. Although probably only about 15 million pastoral nomads are found in the whole world, they move through an area almost as large as the entire cultivated area in the world. They are especially prevalent in the arid and semi-arid tropics. Some examples include the Masai of Kenya and Tanzania, the Hima of Uganda, the Fulani of West Africa, the Bedouin of the Eastern Mediterranean, and the nomads of Mongolia.

Pastoral nomadism can be total or partial. In the latter case, the nomads have homes and some cultivation for part of the year. Typically five or six families travel together with 25 to 60 goats and sheep or 10 to 25 camels. Sometimes they own cattle as well. The livestock eat natural pasture and their productivity is low.

Pastoral nomadism can lead to a variety of problems. Because grazing takes place on common land, there is a tendency for overgrazing because every individual farmer wishes to maximize his or her number

Nomads are common in the northern half of Africa (photo: Mesfin Bezuneh).

of animals. As the number of animals increases, the grazing areas deteriorate and incomes shrink. This problem is known as the "Tragedy of the Commons," and ample evidence shows that traditional legal systems and norms have evolved in response to it. Little scope for technical improvement exists in pastoral nomadic systems, and serious problems arise in years of drought. As the human population grows, additional pressures are placed on the resource base supporting the nomadic system.

Settled Agriculture

Settled agriculture represents a variety of agricultural systems including mixed farming systems, intensive annual crops, intensive and extensive livestock systems, and perennial crops. The dominant farming systems result from an enormous amount of human experimentation. The systems we see most often produce a relatively high and certain return in storable products per unit of effort. They have spread from farmer to farmer, replacing other settled systems that are far less productive. As discussed above, the environment and distance to market are major factors influencing the choice of the system as well as the individual commodities for particular locations.

Mixed farming usually involves a mixture of crops and livestock. Few farming systems in developing countries consist of just one commodity. However, what is meant by mixed farming is the integration of crops and livestock production such that multiple commodities are grown at the same time on the same land. As mentioned in Chapter 7,

144

mixed farming is common in traditional agriculture because it produces relatively high returns while helping to manage risk, makes efficient use of labor and land, and helps maintain soil fertility. Good management is required to coordinate the various farming activities.

Intensive annual crops are extremely important in the world. About 70 percent of the cultivated area of the world is planted to the major grain crops, which include wheat, rice, and corn. Other important annual crops are barley, millet, sorghum, roots, tubers, vegetables, and pulses. Pulses include such crops as beans, soybeans, peas, and peanuts.

Perennial crops are grown and harvested over several years and include crops such as cocoa, coffee, bananas, and sugar cane. Some are grown in large plantations but often on very small farms as well, even in the same country. On small-scale farms, perennial crops are often *interplanted* with annual crops such as corn and beans. Perennial crops tend to be high-valued and are frequently exported. They can also help prevent soil erosion and preserve biodiversity in ecologically fragile areas, such as the Andean slopes in South America.

Intensive livestock systems include both *ruminants* (for example, cattle, buffalo, sheep, and goats) that produce milk, meat, fiber, dung, and other products, and *non-ruminants* (for example, pigs and poultry) that are particularly important for their meat and eggs. These animals are often fed grains in addition to pasture and forage. In a few countries, intensive livestock systems involve carefully managed grasslands or pasture.

Extensive livestock systems include a variety of grazing systems on semi-arid range, high and cool mountain pastures, wet lowlands, and more. Livestock may graze on leaves as well as grasses.

In summary, a large number of crop and livestock systems exist, many of which have been relatively productive or at least well suited to their environment. As population expands and other conditions change, a particular system may no longer be adequate and is forced to change (see Box 8-1). Few systems are static for very long today, and several offer potential for improved productivity.

The Influence of the Political System

In Figure 8-1, political factors were listed as significant determinants of farming systems, including land tenure systems. The political system can dictate how property rights are allocated, including a variety of collective, commune, or other types of land tenure arrangements. When systems such as collectives and communes restrict individual farmers' responsibilities and rights to manage farm resources in response to

BOX 8-1
POPULATION DENSITY AND AGRICULTURAL SYSTEMS

The intensity of land utilization varies worldwide, and there is a close relationship between this intensity and the density of population in a particular region. Boserup hypothesized that pressure from increasing population has caused a shift in recent decades from more extensive to more intensive systems. This classification scheme traces a continuum from shifting cultivation to settled agriculture:

1. Forest fallow cultivation: one- to two-year planting of plots followed by a 20- to 25-year fallow period.
2. Bush fallow cultivation: six- to ten-year fallow period. Periods of uninterrupted cultivation may be as short as one to two years, or as long as five to six years.
3. Short fallow cultivation: fallow lasts one or two years.
4. Annual cropping: land is left uncultivated only between the harvest of one crop and the sowing of the next.
5. Multi-cropping: the most intensive system of land use; the land bears two or more successive crops every year.

Boserup hypothesized that increased population densities put pressure on food production systems to increase outputs. Successively more intensive systems require increased labor inputs for weeding and cultivation, and more varied farming implements. In forest fallow cultivation, only an axe is needed, and as the fallow period is shortened, implements such as hoes, plows, and even irrigation systems are used.

Different patterns of land use exist within similar agro-climatic zones. For example, the land used for intensive cultivation in parts of Nigeria is remarkably similar to the land used for long fallow cultivation in the same country. Thus, Boserup concluded that humans not only adapt to the climatic conditions they face, but actually change the relationship between the conditions and agricultural output by using methods that enhance soil fertility. These adaptations are mostly influenced by rates of population growth.

Source: Ester Boserup, *The Conditions of Agricultural Growth* (London: Allen & Unwin, 1965), esp. ch. 1, pp. 15-22.

market signals, the result has usually been inefficiency and waste of those resources. Political systems that allow independent family farms to operate in competitive markets have generally yielded higher productivity levels and faster growth rates over time. A particularly important example of this is the reform of communist China's collective land tenure system.

Beginning in 1979, China allowed individual farmers to respond more freely to market incentives and since then has experienced significant increases in agricultural production (see Box 8-2). Adoption of new technologies and use of purchased inputs such as fertilizer have increased substantially. These changes have occurred rapidly in China, and have caused important changes in world markets. Remember that China has more than 1.3 billion people and less than one-half hectare per farm worker. Rice production predominates, but livestock, particularly pork, is very important. Agricultural growth in China has, over time, stimulated broad-based increases in income, and this income growth will have profound implications for food demands. A challenge for the world food system is to make adjustments to meet these emerging demands.

Government policies other than rules governing land tenure also affect farming systems. Price policies that favor certain products over others or promote the use of different inputs can have a strong impact on the types of crops planted, on how long they are grown, and even on the degree to which traditional farmers interact with markets. Policies affecting the value of the land create incentives for more or less investment in land. For instance, policies that discriminate against agriculture, such as export taxes, are quickly reflected in lower values for agricultural land. These effects can change farming practices dramatically. Population and family planning programs can affect population densities, which influence the nature of the agricultural system.

In summary, the major types of farming systems in the world include shifting cultivation, pastoral nomadism, and several types of settled agriculture. These systems, particularly settled agriculture, can be affected in a major way by the political system in the country, which dictates private or public control over land use. Other government policies influence agricultural systems both directly and indirectly.

ECONOMIC DETERMINANTS OF INPUT USE AND CROP AND LIVESTOCK MIX

As noted above, policies can influence the evolution of farming systems by changing relative prices of inputs and outputs. Let us examine more carefully how economic factors affect the choice of inputs and, more broadly, the type of farming system. In Chapter 5 we introduced the concept of an *isoquant* to illustrate that the same level of output can be produced with more than one combination of two inputs. The concept of *allocative efficiency* relates to how well farmers choose the correct amounts of inputs to apply and outputs to produce, given the available technology, assuming they are trying to maximize profits. While farm

---

BOX 8-2
CHINESE AGRICULTURAL SYSTEMS

In rural areas of China prior to 1979, the agricultural production system was organized according to guidelines established in the national agricultural plan. Farming operations were organized into collective teams of 20 to 30 households; these teams were required to sell fixed quantities of output to the government at set prices. Quantities produced in excess of the quotas were also surrendered to the government. The collectives had some freedom to adjust inputs, but the acreage planted to each crop was determined by government planners.

This rigid system led to stagnation in agricultural output. Between 1957 and 1978, per capita grain production grew at a 0.3 percent annual rate, while soybean and cotton production per capita *declined*, respectively, by 3.0 and 0.6 percent annually. In 1978, rural incomes were virtually identical to levels of 20 years earlier. This poor performance of the agricultural sector had important implications in a country where 80 percent of the population resides in rural areas.

In 1978, the government decided to introduce the *Household Responsibility System*, which restored individual households as the basic unit of farm operation. Under this system, a household leases a plot of land from the collective, and, after fulfilling a state-set grain procurement quota, can retain additional output. This output can be consumed or sold to the government. The households have flexibility to determine acreage for individual crops. At the same time, the government prices of agricultural commodities were increased, and the prices paid for above-quota grain production were increased substantially above quota prices. Agricultural output began to grow rapidly following these reforms, and agricultural growth averaged 6 percent per year from 1978-2003. These reforms led to a wholesale change in the Chinese agricultural system; by 1983 over 97 percent of the collective teams in China had been converted to the new system.

Sources: Justin Y. Lin, "The Household Responsibility System Reform and the Adoption of Hybrid Rice in China," *Journal of Development Economics*, vol. 36, 2 (1991), pp. 353–73; Ehou Junhua, "Economic Reform: Price Readjustment (1978–87)," *Chinese Economic Studies*, vol. 24, no. 3 (1991), pp. 6–26.

---

and family decisions are intermingled, the family's success and even survival depends in part on how efficiently they allocate their productive assets. Efficient farmers are able to combine inputs in a way that reflects their relative prices. Efficient farmers also choose the most profitable output levels. The farming systems described in this chapter vary in terms of intensity of input use and productivity, but they all represent long-term adjustments to prevailing conditions. As a result we can conclude that they are efficient. As relative scarcity (and hence prices) of inputs and outputs change, these efficient producers will make adjustments to input mixes and amounts of output.

In Figure 8-3, the curved isoquant represents the combinations of labor and animal power that can be used to produce a specific amount of output, with a given level of all other inputs; for example, two tons of corn on one acre of land. We expect all farmers to produce somewhere along this curve. Production to the right or above the curve would use more inputs than needed, and would be technically inefficient. Production to the left, or below the curve, is technically impossible, given the other resources and technology available to the farmer. The slope of the isoquant is known as the *Marginal Rate of Technical Substitution* (MRTS) between the two inputs. In this case, the isoquant's slope is the additional animal time needed to save one hour of labor time. We expect farmers to adjust their use of animals and labor until the value of that labor time just equals the cost of adding animal time. In terms of the graph, we expect farmers to adjust until the slope of the isoquant just equals the price of labor relative to the price of animal power. That ratio is the slope of the relative price line, also called an iso-cost line because it traces a line of constant total cost. The economically efficient

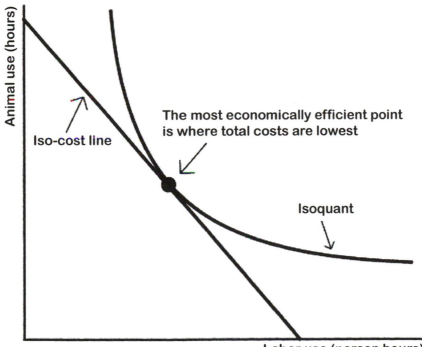

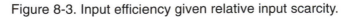

Figure 8-3. Input efficiency given relative input scarcity.

149

input combination is the point where the isoquant and the iso-cost curves are tangent or where the MRTS equals the input price ratio. If the price of labor goes down relative to the price of animal power, the iso-cost line would become flatter, tangency would occur at a point farther down the isoquant, and more labor and less animal power would be used. Thus, the drive to be efficient leads to changes in input mixes and, over time, this drive can alter the farming system. As an example, compare differences in farming systems between Africa, where labor is relatively scarce and land is relatively abundant, and South Asia, where labor is relatively abundant and land is relatively scarce.

Similar trade-offs occur between different kinds of output. In Figure 8-4, the curved line represents the *production possibilities frontier* (PPF) or the combinations of corn and beans that can be produced with available resources. As with Figure 8-3, we expect farmers to produce somewhere along this curve. Production inside the curve would generate less output than is possible, and so be technically inefficient. Production outside the curve would be technically impossible, given these resources and the technology. The slope of the PPF is known as the *Marginal Rate of Transformation* (MRT). In this case, that slope is the amount

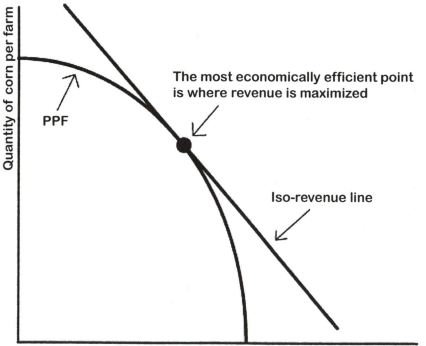

Figure 8-4. Output efficiency, given the technology and resource base.

of additional corn that can be produced with one less unit of beans. Again, the allocatively efficient combination of outputs depends on the relative prices of these two outputs. For example, if the price of beans rises relative to the price of corn, the iso-revenue line becomes steeper and it pays to shift more resources into producing beans and away from corn.

Input and output combinations observed in farming systems around the world are heavily influenced by technologies, resource bases, and relative prices. Farmers allocate resources to maximize their families' well-being, taking into account expected costs and revenues. Economic profitability is just one factor they consider in their decision-making, but usually an important one.

SUMMARY

Farming systems in the world exhibit considerable variability. Both technical and human factors determine the types of farming systems. Technical factors include both physical and biological factors. Institutional and human factors are characterized by both externally and internally controlled forces. The major farming systems of the world can be grouped into three classes: shifting cultivation, pastoral nomadism, and settled agriculture. Settled agriculture represents a variety of agricultural systems including mixed farming systems, intensive annual crops, intensive and extensive livestock systems, and perennial crops.

IMPORTANT TERMS AND CONCEPTS

Farming systems	Perennial crops
Human determinants of farming systems	Political determinants of farming systems
Intensive annual crops	Production possibility frontier
Iso-costs and iso-revenues	Settled agriculture
Mixed farming system	Technical determinants of farming systems
Pastoral nomadism	

Looking Ahead

In this chapter, we briefly examined the nature and diversity of existing agricultural systems in developing countries and the determinants of farming systems. In the next chapter, we focus on environmental or natural resource problems that can influence the ability of a farming system to improve and achieve sustainable development.

QUESTIONS FOR DISCUSSION

1. What are the major technical determinants of farming systems?
2. Describe the major human determinants of farming systems. Be sure to distinguish exogenous from endogenous factors.
3. How might the political system affect the nature of the farming system?
4. What is shifting cultivation, and why is it found more commonly in Africa and Latin America than in Asia?
5. What is pastoral nomadism, and what problems might be present in this type of system?
6. Distinguish among the major types of settled agriculture.
7. How do the optimal quantities of inputs and outputs change as iso-cost and iso-revenue lines become flatter, and why?
8. In what sense is the Boserup argument, presented in Box 8-1, consistent with the discussion of economic determinants of input use and output choice?

RECOMMENDED READING

Duckham, Alec N., and G. B. Masefield, *Farming Systems of the World* (London: Chatto & Windus, 1970).

Lin, Justin, "Agricultural Development in China," Chapter 31 in *International Agricultural Development*, ed. Carl K. Eicher and John M. Staatz (Baltimore, Md.: Johns Hopkins University Press, 1998).

Loomis, Robert S., "Agricultural Systems," *Scientific American* (September 1976), pp. 98–105.

Ruthernberg, Hans, *Farming Systems in the Tropics* (Oxford: Oxford University Press, 1980).

CHAPTER 9

Resource Use and Sustainability

[P]overty compels people to extract from the ever-shrinking remaining natural resource base, destroying it in the process. In fact, the major characteristic of the environmental problem in developing countries is that land degradation in its many forms presents a clear and immediate threat to the productivity of agricultural and forest resources and therefore to the economic growth of countries that largely depend on them.
                                            — Gunter Schramm and Jeremy J. Warford[1]

This Chapter
1. Examines the nature of environmental or natural resource problems that influence the sustainability of agricultural development in developing countries
2. Identifies the principal causes of environmental problems in developing countries
3. Discusses some potential solutions to environmental problems in developing countries.

NATURE OF ENVIRONMENTAL PROBLEMS
Sound environmental management is generally recognized as essential for sustainable agricultural and economic development. Yet the effects of environmental degradation and poor natural resource management are increasingly evident throughout the world. The wide-ranging yet often interrelated problems of soil erosion, silting of rivers and reservoirs, flooding, overgrazing, poor cropping practices, desertification, salinity, water-logging, deforestation, energy depletion, climate change, loss of biodiversity, and chemical pollution of land, water, and air, appear to be increasing problems in developing countries. The poorest

---

[1] Gunter Schramm and Jeremy J. Warford, eds, *Environmental Management and Economic Development* (Baltimore, Md.: Johns Hopkins University Press, 1989), p. 1.

countries tend to be most dependent on their natural resource base and thus have the potential of being the most vulnerable to environmental degradation. These countries find environmental problems particularly difficult to solve because rapid population growth, outmoded institutional relationships, poverty, and a lack of financial resources conspire against solutions. The poorest people within these countries usually suffer the most as a result of environmental degradation.

As agricultural and economic development occur, forces are set in motion, some reducing and others increasing the pressures on the environment. Changes in the rate of population growth, new technologies, social and institutional relationships, the increased value of human time, and shifts in the weight placed on future as opposed to current income — all influence the relationship between human activity and the natural resource base. Economic and natural resource policies and other institutional changes can either alleviate or aggravate natural resource problems. The nature of particular types of environmental problems is discussed below, followed by a description of causes and potential solutions.

Global Versus Local Problems

Many environmental problems are local in cause, effect, and potential solution. Others are regional or even global. Problems such as erosion may be local, but also have more widespread implications if soil is deposited in rivers and transported to neighboring countries where the silt raises river levels, causing flooding. Others such as deforestation may appear local but can affect global temperatures as carbon is released into the atmosphere. Environmental problems affect every nation in the world, can hinder the long-term sustainability of farming systems, and are a growing concern. The following is a brief description of some of the more serious natural resource and environmental issues.

Soil Degradation, Erosion, Silting, and Flooding

Topsoil is one of the world's most important natural assets. Farmers frequently invest heavily in trying to improve its quality and abundance, through a wide range of soil fertility amendments and soil conservation techniques. But these investments are costly, and the poorest farmers are often less willing or able to wait for the future benefits they provide. As a result, we observe that the lowest-income farmers often draw down their "soil capital," applying insufficient soil amendments to fully replenish the nutrients removed at harvest or generated through natural processes. In effect, they are "mining" soil nutrients. The result-

Farming an erodible hillside in Ecuador.

ing soil degradation is usually reversible, if and when farmers find it profitable to apply more nutrients than plants withdraw.

An irreversible kind of degradation is soil erosion, caused by the exposure of soil to wind and water runoff. The extent of the world erosion problem is difficult to assess because few nations have systematically surveyed the condition of their soil resources. Nevertheless, the amount of agricultural land being retired due to soil erosion is estimated to be at least 20 million hectares per year.[2] An erosion rate of 50 tons per hectare is common in upland watersheds in many developing countries, while soil can regenerate somewhere between only 0 and 25 tons per hectare.[3]

A loss of 50 tons per hectare represents a loss of only about 3 millimeters from the top of the soil, yet often the gullies and exposed bedrock from uneven erosion scar the landscape. The effects on productivity are potentially serious. Eroded soils typically are at least twice as rich in nutrients and organic matter as the soil left behind. Soil nutrient losses can be partially replaced by increased use of chemical fertilizers, but only up to a point, and fertilizer can be expensive. At any rate, the yields with fertilizers are lower than they would be in the absence of erosion, so that erosion reduces productivity below its

[2] See Norman Meyers, "The Environmental Basis of Sustainable Development," in *Environmental Management and Economic Development*, ed. Gunter Schramm and Jeremy J. Warford (Baltimore, Md.: Johns Hopkins University Press, 1989), p. 59.
[3] See Alfredo Sfeir-Younis, "Soil Conservation in Developing Countries," Western Africa Projects Department (Washington, D.C.: World Bank, 1986).

potential. It is estimated that erosion of good soils in the tropics may be resulting in maize-yield reductions of 10 to 30 percent.[4] In Guatemala, 40 percent of the productive capacity of the land may have been lost through erosion.[5]

Watersheds are seriously deteriorating in the hills of the Himalayas, on the steep slopes of the Andes Mountains, in the Yellow River basin in China, in the Central American highlands, in the Central Highlands of Ethiopia, and on densely populated Java. The worst erosion in terms of average soil loss per hectare is found in the crescent from Korea to Turkey, in Eastern Europe and the former Soviet Union, followed by the Central American highlands and the Sahel in Africa. Differences are due to the intensity of cultivation on highly erodible soils.

The indirect or off-site effects of erosion through silting of rivers and reservoirs are perhaps more serious than the on-site effects. When reservoirs fill with sediment, hydroelectric and irrigation storage capacity is lost, cutting short the useful lives of these expensive investments. When rivers silt up, flooding occurs during rainy seasons. For example, soil erosion in the hills of Nepal causes flooding in the plains of Nepal, India, and Bangladesh. Flooding in the Yellow River basin in China is another example. This flooding causes both direct human suffering and crop destruction.

While flooding is a serious periodic problem in certain countries, not all or even most flooding is due to silt. Low-lying countries such as Bangladesh and parts of Egypt, Indonesia, Thailand, Senegal, The Gambia, and Pakistan are particularly vulnerable to flooding due to high river levels during the rainy season and sea surges during storms. About 80 percent of Bangladesh, for example, is a coastal plain or river delta. In 1998, approximately two-thirds of this country of 130 million people was flooded. While a certain amount of normal flooding can have a positive effect on agricultural production, excessive flooding results in substantial loss of life from disease as well as drowning.

Desertification

Excluding real deserts, potentially productive drylands cover about one-third of the world's land surface. About one-sixth of the world's population lives in dryland areas that produce cereals, fibers, and animal products. In the arid regions, averaging 200 to 300 mm of annual rainfall, vegetation is sparse and nomadic herding of such animals as goats

[4] See Meyers, "Environmental Basis of Sustainable Development."
[5] See Robert Repetto, "Managing Natural Resources for Sustainability," in *Sustainability Issues in Agricultural Development*, ed. Ted J. Davie and Isabelle A. Shirmer (Washington, D.C.: World Bank, 1987).

and cattle predominates. In the semi-arid regions, with 300 to 600mm of rain, dryland farmers grow cereals such as wheat, sorghum, and millet in more settled agriculture. The semi-arid regions are smaller in area but more densely populated than are arid regions.

The term *desertification* applies to a process occurring in these arid and semi-arid regions. Desertification involves the depletion of vegetative cover, exposure of the soil surface to wind and water erosion, and reduction of the soil's organic matter, soil structure, and water-holding capacity. Intensive grazing, particularly during drought years, reduces vegetative cover; the loss of vegetation reduces organic matter in the soil and thus changes soil structure. After a rain, the earth dries out and becomes crusted, reducing the infiltration of future rains. Then, even more vegetation is lost for lack of water, the surface crust is washed or crumbles and blows away, and the soil that is left is less fertile and unable to support much plant life.

Cropping, particularly when very intensive and when combined with drought, is another major cause of desertification. If soil organic matter is depleted by intensive farming practices and not replaced, a process similar to that described above occurs. As supplies of firewood dwindle, people use dried manure for fuel rather than fertilizer. As the soil loses its fertility, crop yields fall and wind and water erosion accelerates. Eventually the land may be abandoned.

Flooded houses in Bangladesh.

157

Moderate desertification may cause a 25 percent loss of productivity, while severe desertification can reduce productivity by 50 percent or more. It is estimated that 65 million hectares of productive land in Africa have been abandoned to desert over the last 50 years. Desertification is particularly a problem in the Sahel region of Africa and in parts of the Near East, South Asia, and South America. In terms of people directly affected, approximately 50 to 100 million people are currently dependent on land threatened by desertification. This number is less than 2 percent of the world's population, making the problem globally less important than the more general problem of soil erosion, except of course to the 50 to 100 million affected people. Areas where desertification is a problem also tend to be areas with rainfall that is both low and unpredictable. The ensuing periodic droughts create short-term severe food crises in those areas.

Salinity and Waterlogging

Irrigation, one of the oldest technological advances in agriculture, has played a major role in increasing global food production. However, bringing land under irrigation is costly, and degradation of irrigated land through questionable water management practices is causing some land to lose productivity or be retired from production completely. The major culprits are waterlogging and salinity.

Seepage from unlined canals and heavy watering of fields in areas with inadequate drainage can raise the underlying water-table. Almost all water contains some salts. High water tables concentrate salts in the root zones and also starve plants for oxygen, inhibiting growth. Inadequate drainage also contributes to salinization when evaporation of water leaves a layer of salts that accumulate and reduce crop yields. A typical irrigation rate leaves behind about 2 to 5 tons of salt per hectare annually, even if the water supply has a relatively low salt concentration. If not flushed out, salt can accumulate to enormous quantities in a couple of decades.

No one knows for sure how large an area suffers from salinization, but the Food and Agriculture Organization of the United Nations (FAO) estimates that one-half the world's irrigated land is so badly salinized that yields are affected. Others put the figure closer to one-quarter of the world's irrigated land. Regardless, the figure is large. Some 20 to 25 million hectares are affected in India, 7 to 10 million hectares in China, and 3 to 6 million hectares in Pakistan. Other developing areas severely affected include Afghanistan, the Tigris and Euphrates river basins in Syria and Iraq, Turkey, Egypt, and parts of Mexico.

Deforestation has led to soil erosion in Nepal.

Deforestation and Energy Depletion

Forests play a vital role in providing food, fuel, medicines, fodder for livestock, and building materials. They provide a home for innumerable and diverse plant and animal species. They protect the soil, recycle moisture, and represent a sink for atmospheric carbon dioxide. But forests are being cleared at a rapid rate throughout the world. During the 1980s, about 15 million hectares of tropical forest were being cleared each year; the rate has fallen somewhat since then, but today far more than 10 million hectares per year are being lost.[6] The Earth's forested areas have declined by about one-half in the last century. Deforestation continues at a rapid pace in countries such as Cote d'Ivoire, Paraguay, Nigeria, Nepal, Costa Rica, Haiti, El Salvador, Nicaragua, Brazil, Indonesia, and Colombia.

Deforestation creates environmental problems on land and in the air. Forest clearing can degrade soils and increase erosion in tropical watersheds. Soils in tropical forests tend to be fragile and unsuited for cultivation; their fertility is quickly depleted by the erosion that follows tree clearing. In semi-arid areas, deforestation contributes to loss of organic matter, increases wind and water erosion, and speeds up the rate of desertification. As forests are burned to clear land, carbon dioxide and carbon monoxide are emitted into the atmosphere, contributing to

[6] See Arild Angelsen and David Kaimowitz, "Rethinking the Causes of Deforestation: Lessons from Economic Models," *The World Bank Research Observer*, vol. 14, no. 1 (February 1999), pp. 73–98.

global warming. Brazil, for example, adds some 336 million tons of carbon to the air each year through deforestation, more than six times as much as it does by burning fossil fuels.[7]

In developing countries, seven out of ten people depend on fuel wood for meeting their major energy (cooking and heating) needs. The FAO estimates that three out of four people who rely on fuel wood are cutting wood faster than it is growing back. When people cannot find fuel wood, they turn to other sources of organic matter such as dung for fuel, thereby depleting soil fertility and aggravating soil erosion and desertification.

Deforestation also threatens the world's biological diversity. Tropical forests cover only 7 percent of the world's landmass, yet they contain more than 50 percent of the plant and animal species.[8] In Madagascar, for example, there were, until recently, 9500 documented plant species and 190,000 animal species, most of them in the island's eastern forest. More than 90 percent of the forest has now been eliminated, along with an estimated 60,000 species.[9]

Climate Change

The Earth's climate is undergoing change. Surface temperatures increased by 1° F during the twentieth century, and the 1990s were the hottest decade of the century. Projections of future increases range from 2.5° F to 10.4° F by 2100.[10] This problem is both affected by and felt by all countries, but projected impacts vary from region to region. At least some of the warming is due to buildup of carbon dioxide and other changes in the atmosphere, much of it the result of fossil-fuel combustion in developed countries as well as changes in land use and burning of forests. As carbon dioxide and other gases accumulate in the atmosphere, they trap heat, creating the so-called greenhouse effect. As world temperatures rise, average sea levels also rise, thus threatening coastal lands. Violent storms, monsoons, droughts, floods, and generally increased weather variability are likely. While a warmer world is not necessarily less favorable to agriculture, regional impacts are harder to predict. And global warming could alter disease prevalence and be very hard on certain animal species because their ecosystem may shift while the property-line boundaries of their preserves do not.

[7] See Christopher Flavin, "Slowing Global Warming," in *State of the World 1990*, ed. Lester Brown *et al.* (New York: W.W. Norton, 1990), p. 20.
[8] See E. O. Wilson, "The Current State of Biological Biodiversity," ch. 1 in *Biodiversity*, ed. E. O. Wilson (Washington, D.C.: National Academy Press, 1988), p. 8.
[9] See Repetto, "Managing Natural Resources for Sustainability," p. 174.
[10] See Tom M. L. Wigley, "The Science of Climate Change," *Pew Center on Climate Change Report* (2005).

Chemical Pollution

Misuse of chemical pesticides and fertilizers has contaminated the land and water in many developing countries, damaging the health of producers and consumers, stimulating the emergence of pests resistant to pesticides, destroying the natural enemies of pests, and reducing fish populations or rendering them unsafe for human consumption. Acute pesticide poisonings are common, and little is known about potential long-term health effects. Few developing countries have established effective pesticide regulatory and enforcement systems.

Hundreds of pests have become resistant to one or more chemicals, and the number is growing. World pest populations have increased as pesticides kill natural predators of pests. Fertilizer runoff increases nitrate levels in ponds and canals, reducing oxygen levels and killing fish. Excessive pesticide levels often destroy fish in irrigated rice paddies.

Heavy use of pesticides and fertilizers tends not to hurt agricultural production in the short run. However, as resistance to pesticides builds up and predators are reduced, future production potentials are jeopardized. And society bears the cost of off-farm pollution.

CAUSES OF ENVIRONMENTAL PROBLEMS

Environmental degradation can result from physical, economic, and institutional factors. Many environmental problems are interrelated; for example, deforestation, erosion, and silting of rivers and reservoirs are all linked. Natural resource degradation usually has both direct and indirect causes. For example, desertification can directly result from overgrazing and poor cropping practices, but indirectly result from poverty and population growth. Understanding the underlying causes of environmental degradation requires searching for and analysis of complex direct and indirect physical, economic, and institutional linkages.

Physical or technical causes of natural resource degradation are often the most visible and direct, even though a series of complex linkages may be involved. Land clearing for timber, fuel wood, cattle ranching, or farming causes deforestation. Deforestation results in loss of biodiversity, loss of soil, and diminished soil fertility, since soil uncovered in tropical forests loses its fertility quickly. If the forest is burned, carbon dioxide enters the atmosphere. If the area is semi-arid, loss of forests can contribute to desertification. Desertification can also result from overgrazing, which itself is caused by too many cattle eating grass in an area subject to dry spells or droughts. Intensive cropping in semi-arid areas contributes to desertification. Many other examples of physical causes of natural resource degradation can be cited. Salinity and water-logging result from poorly constructed and poorly managed irrigation systems. Chemical pollution results from excessive fertilizer and pesticide use. Silting of rivers and tidal surges during storms cause flooding.

It is important to identify physical causes of environmental problems. However, it is even more important to identify the underlying economic and institutional causes, including social, cultural, and policy-related causes.

Economic Causes of Natural Resource Degradation

Poverty and environmental degradation go hand in hand. Poverty drives people to farm marginal lands intensively, to seek fuel wood, and to follow other agricultural practices that produce food at the potential sacrifice of future production. As discussed in Chapter 4, poverty reinforces population growth, which is a major contributor to deforestation, overgrazing, and farming on steep slopes, drylands, and flood plains.

The concern of the poor for the present, implying heavy discounting of future costs and benefits (see Box 9-1), is matched by the needs of governments in developing countries to deal with internal and external debt problems. Indeed, the existence of debt problems in many countries reflects previous decisions to spend on current consumption rather than save for the future. Governments follow policies that encourage natural resource-based exports to pay off debts and import capital goods. They lack the financial resources to address environmental problems.

Countries implementing economic development programs usually find high rates of return to many types of capital investment. The high interest rates often characteristic in these cases encourage current consumption and may place demands on natural resources. Interest rates in developing countries are also influenced by interest rates in major developed countries, due to linkages through international financial markets.

Market failures due to externalities and transactions costs are an additional source of natural resource problems. Actions of farmers that influence soil erosion or pesticide pollution create social costs that are not borne by the farmers themselves. These external costs may not be considered by farmers when making decisions (see Box 9-2). Furthermore, a lack of information or concern about environmental damage creates transactions costs that facilitate environmentally destructive behavior. In summary, a variety of economic factors are responsible for natural resource problems.

BOX 9-1
THE DISCOUNT RATE AND THE ENVIRONMENT

Positive interest and discount rates are caused by two factors: a preference for benefits now rather than later, and the productivity of capital. If current consumption is sacrificed and invested, this capital investment will provide greater potential for consumption in the future. The effect of a positive discount rate is to place more value on present than on future consumption. While there is continual discussion about the ethics of using discount rates to gauge the environmental impact of projects, nevertheless individuals, businesses, and governments all use discounting to make decisions.

Discount rates, along with the length of time in the future that a payoff or cost will be incurred, have a large influence on the current value of that payoff or cost. Most decision-makers make their decisions based on the difference between the present value of benefits and the present value of costs. If this difference is positive, then the action will be undertaken, and if negative, it will not. Smallholders who perceive current benefits, such as food production, from their actions and distant or uncertain future costs, such as productivity losses from erosion, are unlikely to invest heavily in erosion control devices because they heavily discount these future costs. Similarly, governments deciding to promote policies that boost current incomes at the expense of future costs will be more likely to decide in favor of those policies if future costs are discounted. The higher the discount rate, the lower the present value of future costs.

Interest rates do not, however, have an unambiguous effect on the environment. High interest rates clearly slow investments that will conserve natural resources for future use. They lower the present value of future environmental costs. Low interest rates, on the other hand, stimulate investments in industries, roads, irrigation systems, and so on. Often, future environmental costs are not taken into account, either by private decision makers or governments, and the low cost of capital induced by low interest rates permits investments that degrade the environment.

Present Value of $100 Given in the Future

Years in future	Discount rate		
	5%	10%	12%
10	61	38	32
20	38	15	10
30	23	6	3

Note: $PV = V_t/(1+r)^t$, where r = discount rate, t = years, V_t = value at time t of item being considered—in this case, $100.

BOX 9-2
EXTERNALITIES AND PRIVATE DECISIONS

One commonly cited cause of environmental degradation is the divergence between private and social costs of actions. This divergence is caused by external costs. An external cost exists when an activity by one agent causes loss of welfare to another agent *and* the loss is not considered by the author. The effect of externalities on private decision-making is illustrated in the following figure:

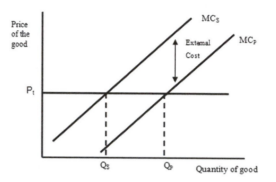

A farmer who cannot influence market prices will produce a good up until the point where the private marginal cost of its production (MC_p) equals the market price. In the figure, this point is shown where $MC_p = P_t$ and Q_p units are produced. An external cost is represented by the social marginal cost curve (MC_s), which exceeds the private cost curve. From society's point of view, the desirable production level is Q_s (where $MC_s = P_t$). Thus, the externality leads to more production of the good than is socially desirable.

Institutional Causes of Natural Resource Degradation

A major cause of environmental degradation is institutional failure, both private and public. Institutions are rules of behavior that affect private incentives. Existing social structures and local customs may not be adequate to preserve the environment as population growth and economic development continue. Or, environmentally constructive social structures and customs may be destroyed by national policies or by increased transactions costs and collective action. In some cases, inadequate institutions are the legacy of colonial interference or the result of more recent international influence.

Market institutions help determine values, and market failures are a chief cause of environmental degradation mainly because they cause natural resources and environmental services to be undervalued. Market failures may be due to inadequately defined property rights, transactions costs associated with monitoring and enforcing property rights, and weak enforcement institutions.

Inadequate property rights in forest, pastures, and ground and surface waters can undermine private or local collective incentives to manage resources on a sustainable basis. In some areas, the land or water resource was traditionally held in common. Under a common-property regime, people in the village or community had access to use the resources but did not own or rent them privately. When the local society could maintain authority over the resource, or when population pressures were such that the resource was in abundant supply, then this common property could be managed in a socially optimal manner. However, as population increases and as national policies usurp local authority, breaking down traditions and customs, incentives for resource preservation and traditional means of controlling access are often destroyed. If one person does not cut down the tree for fuel wood, another will. Or, if one person's goat does not eat the blade of grass, another person's goat will. Or, if one person does not use the water or catch the fish, another will. The result is that incentives exist for each individual to over-exploit resources because otherwise someone else will.

Common-property regimes do not have to cause resource mismanagement if local institutions create incentives to efficiently manage the resource. In many areas of Africa, common-property institutions were said to cause overgrazing on rangelands. However, attempts by the government to replace these institutions with private ownership schemes were largely counterproductive, contributing to more rapid degradation of resources and leading to increased economic inequality. Common-property institutions can certainly be a viable means of managing resources.[11]

In areas of frontier colonization, poorly defined and inadequately enforced property rights can create incentives for over-exploitation of natural resources. For example, the Peten Region of northern Guatemala is currently undergoing high rates of deforestation, particularly in its western extremities. These areas are part of the ecologically sensitive and culturally important Mayan Biosphere. In the western Peten, the Guatemalan government established the Laguna de Tigre national park in 1990. It was hoped that a national park would slow settlement and lead to conservation of the forest in its original state. However, the

---

[11] See Daniel W. Bromley, ed., *Making the Commons Work* (San Francisco, Calif.: Institute for Contemporary Studies Press, 1992).

government does not have the resources to monitor and discourage settlement on these isolated public lands, and a weak legal system prevents enforcement of laws prohibiting illegal settlement. As a result, illegal settlers have deforested the area and converted lands into cattle rangelands; population pressures are growing, water is increasingly scarce, and ecological integrity has been destroyed.

Public policies are another major institutional cause of natural resource degradation. Agricultural pricing policies, input subsidies, and land-use policies often discourage sustainable resource use. Governments in developing countries intervene in agricultural markets to keep food prices artificially low. This discrimination against agriculture causes land to be undervalued, reducing incentives for conservation. In addition, low incomes make the investment required for sustainable output difficult. On the other hand, higher agricultural prices raise the value of land, and, as a result, contribute to increased deforestation.[12] These competing impacts of agricultural prices on the environment make it important that policy impacts be explored as a part of government decision-making.

Governments frequently subsidize fertilizer and pesticides, in part to compensate for keeping farm product prices low. If fertilizer or pesticide use causes an externality, then subsidies, because they increase input use, will increase the level of the externality. Subsidies may be indirect in the form of roads or may be export subsidies that encourage deforestation. Road access is strongly associated with deforestation in all regions of the world. Subsidized irrigation water can encourage its wasteful use.

Land tenure and land-use policies may cause exploitation of agricultural and forest lands with little regard for future productivity effects. Short leases, for example, create incentives to mine the resource base for all it is worth in the short run and, as just noted, it is an error to think that local incentive problems can be entirely corrected by national policies. Bromley and Chapagain point out that, in Nepal, national policies on forests have destroyed local conservation practices and incentives.[13] A common policy in Latin America has been to require that in colonized areas land needs to be developed, which usually means cleared of trees, prior to receiving title to the land. A large part of the deforestation in the Brazilian Amazon is associated with these types of titling rules.

[12] See Angelsen and Kaimowitz, "Rethinking the Causes of Deforestation."
[13] See Daniel W. Bromley and Devendra P. Chapagain, "The Village Against the Center: Resource Depletion in South Asia," *American Journal of Agricultural Economics*, vol. 68 (December 1984), pp. 868–73.

Land-use patterns are sometimes affected by colonial heritage or other international influences. In parts of Latin America and the Caribbean, large sugar cane, coffee, and banana plantations, and even cattle ranches, are found in the fertile valleys and plains, while small peasant farms intensively producing food crops dot the eroding hillsides. The low labor intensity of production in the valleys depresses job opportunities and forces the poor to rely on fragile lands to earn incomes. These patterns are the legacy of colonialism. Colonial powers in Africa changed the cropping system to cash cropping in areas where cash cropping could not be supported by the natural resource base. Peasants have been forced on to marginal lands, reducing lands for nomads. Traditional nomadic trading patterns were also disrupted.

These and other institutional policies have contributed to natural resource problems as they exist today. Institutional change is therefore one of the potential solutions to these problems, as described below.

POTENTIAL SOLUTIONS TO NATURAL RESOURCE PROBLEMS

Solutions to environmental problems contain technical, economic, and institutional dimensions. Technical solutions are needed to provide the physical means of remedying natural resource degradation, while economic and institutional solutions provide the incentives for behavioral change.

Technical Solutions to Natural Resource Degradation

A variety of technical solutions are available to solve deforestation, erosion, desertification, flooding, salinity, chemical pollution, and other environmental problems. Where technical solutions are lacking, government-sponsored research and education can develop new natural resource-conserving practices and facilitate their adoption.

Windbreaks, contour plowing, mulching, legume fallow crops, alley cropping, deferred grazing, rotational grazing, well-distributed watering places, and revegetation or reforestation are all examples of physical practices that could help reduce soil erosion, silting, and desertification. Solar pumps, biogas generators, and more efficient cooking stoves can provide or save energy, thereby reducing fuelwood consumption, deforestation, and desertification. Embankments can provide protection from flooding for limited areas, and dams can be built on rivers to control water flows.

Irrigation canals can be better lined to reduce waterlogging and salinity and conserve water resources. Integrated pest management techniques can be developed that involve increased biological and cultural

A brush fence in Kenya being used to facilitate rotational grazing
(photo: Mestin Bezuneh).

control of pests to reduce pesticide pollution. Germplasm banks can be used and conservation reserves established to preserve endangered plant species.

These are just a few potential technical or physical solutions to environmental problems. In many cases these technical solutions are already known, but in others additional research is essential for success. In the pest management area, for example, much work still needs to be completed on biological controls for major pests in developing countries. Integrated pest management (IPM) is a family of pest management techniques that lowers dependence on toxic pesticides; in many countries, these techniques are technically feasible, but have not been widely spread to farmers due to limited extension and agricultural outreach services in many developing countries.[14]

The availability of technical solutions to natural resource problems is essential for reducing environmental degradation. In almost all cases, however, these solutions must be combined with economic and institutional changes that create incentives for behavioral change. Without these incentives, it is unlikely that the technologies will be widely adopted, since they usually imply increased costs to their users.

[14] See *Globalizing Integrated Pest Management: A Participatory Research Process*, ed. George W. Norton, E. A. Heinrichs, Gregory C. Luther, and Michael E. Irwin (Ames, Iowa: Blackwell Publishing, 2005).

Economic and Institutional Solutions to Natural Resource Degradation

International and natural agricultural research systems can generate new technologies that increase food production and incomes. As incomes grow, population pressures are reduced, and the demand for environmental protection increases. New institutions may be formed (or existing institutions may evolve) in response to this demand, and incentives for resource conservation are created.

As countries develop, the major source of growth is not the natural resource base, but new knowledge (see Chapter 5). This knowledge can, to some extent, substitute for natural resources and is less subject to the diminishing returns associated with more intensive use of natural resources. Increases in agricultural productivity resulting from the new knowledge or technologies not only raise incomes, but also the value of human time; as the value of human time increases, population growth rates decline with favorable implications for natural resource problems.

Economic development means more resources for servicing external debts and addressing environmental problems. The poorer the country, the fewer the resources that tend to flow toward solving environmental problems.

Many of the economic solutions to environmental problems described above are long term or indirect. A series of direct institutional changes may hold greater promise for more immediate improvements in the natural resource base. Changes in taxes, subsidies, regulations, and other policies can influence local incentives for conservation,

Reducing the discrimination against agriculture in pricing policies should help. Low returns to agriculture depress farmland prices and the returns to investments in land conservancy practices, as noted earlier. Low returns reduce the demand for labor and therefore labor income. If returns to agriculture were raised, subsidies on inputs such as agro-chemicals could be eliminated. However, increasing returns to agriculture are also likely to put additional pressure on forest resources, so institutional mechanisms to reduce deforestation must accompany changes in agricultural pricing policies.

Several means are available for addressing the underlying market failures associated with environmental degradations. Subsidies and taxes can be used as "carrots" or "sticks" to reduce externalities or off-site effects associated with agricultural and forestry use (see Box 9-3). An example of a conservation subsidy (i.e., a "carrot") might be a program in which the government shares the cost of building terraces, windbreaks, and fences, or of planting trees. In some cases, local workers

BOX 9-3
THE IMPACT OF AN EXTERNALITY
ON CONSERVATION ACTIVITIES

Externalities cause the private benefits of resource conservation activities to diverge from social benefits. These private benefits determine the demand for, or use of, conservation activities. In the figure below, the net private benefits (NPB), or the private marginal returns from conservation activities, are shown to be lower than the social benefits (NSB). The returns are lower because the benefits of conservation to society (such as the off-farm benefits of erosion reduction) are not considered by the private decision-maker. Thus, the private decision-maker may employ fewer conservation techniques than are socially optimal (Qp < Qs).

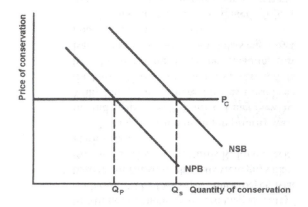

In the case of inputs that create external costs, such as pesticides, the reverse situation arises, and NPB exceeds NSB. In this case, Qp will exceed Qs, and more pesticides are applied than is socially optimal.

Price incentive policies such as taxes and subsidies may be used to move the NPB curve so that it corresponds to NSB. In this case, Qs = Qp and the external costs or benefits are said to have been internalized.

can be paid in kind with food from internationally supplied food aid. An example of a "stick" is a sales tax on chemical pesticides. Such subsidies are designed to "internalize" the externality, so that the economic actor considers the social costs associated with his or her decisions.

Institutional change that creates secure property rights will help address some problems of environmental degradation. Ownership of land titles increases the returns to long-term investments in land. On the other hand, the removal of institutions that guarantee land titles only if forests are cleared will help stop deforestation (see Box 9-4). The provision of property rights does not necessarily imply privatization. There are numerous examples of common-property regimes managed

BOX 9-4

INSTITUTIONS AND DEFORESTATION IN THE BRAZILIAN AMAZON

Brazil contains 3.5 million square kilometers of tropical forests, some 30 percent of the world's total. Most of the forests are found in the Brazilian Amazon Basin. Deforestation of this rich reserve of plant and animal species has increased in recent years, raising concerns for its effects on atmospheric carbon levels and on the maintenance of global biodiversity.

The Brazilian government made a conscious decision in the 1960s to develop the Amazon as a means of relieving population pressures, providing territorial security, and exploiting the region's wealth. Ambitious road-building programs, other infrastructure development, agricultural colonization projects, and policies providing tax and other incentives for agricultural and industrial development were begun. These projects had the effect of opening access to the Amazon, and promoting environmentally unsound development.

Tax exemptions and cheap credit spurred the creation of large-scale livestock projects, whose economic and environmental suitability to the region was questionable. The National Integration Program established a network of villages, towns, and cities and cleared lots for in-migrating settlers. The plans for these settlements were made without regard for soil fertility or agricultural potential, and the cleared forest lands were quickly eroded and otherwise degraded.

Environmentally destructive settlement practices are promoted throughout the Amazon by the Brazilian government's practice of awarding land titles only for deforested lands. A migrant in either an official settlement project or an invaded area can obtain title to the land simply by clearing the forest. Once the title is granted, the migrant can sell or transfer it to someone else, and proceed to clear additional lands. Calculations show that it is more profitable to clear land, plant subsistence crops for two years, and then sell and move than it is to remain as a permanent settler.

Clearly, the rate of deforestation in Amazonia is directly influenced by government policies and other institutional arrangements. It is just as clear that policy reform and institutional adjustments can slow down, or even reverse, this process.

Source: Dennis J. Mahar, "Deforestation in Brazil's Amazon Region: Magnitude, Rate, and Causes," ch. 7 in *Environmental Management and Economic Development*, ed. Gunter Schramm and Jeremy J. Warford (Baltimore, Md.: Johns Hopkins University Press, 1989).

Spraying pesticides
in the Philippines.

in environmentally sound fashions, and it is only when population growth or other changes put pressure on group management that the effectiveness of the management is diminished. Institutional changes that reinforce these common-property management schemes may be more effective than privatization.

Many successful examples may be found of assigning property rights and creating markets for environmental quality. In eastern Peten, Guatemala, community organizations were granted contracts for sustainable use of forest resources. These organizations, because they have the rights to the natural resources, control access to the forest and "police" extractive activities, such as timber harvest. As a result, the eastern Peten is still heavily forested, especially in comparison to the west, where inadequate property rights and high enforcement costs have contributed to heavy deforestation (see above). In Zimbabwe, local villagers were given rights to harvest elephants, and sell these rights to foreign hunters. The money from these sales is kept and used for development purposes in the villages. The villagers now see the elephants as a valuable resource and protect them from poachers. As a result, elephant populations are growing rapidly in areas where 15 years ago the elephant was practically extinct.

Certification is a process whereby international markets recognize and reward products that are sustainably produced. For example, wood in the eastern Peten is harvested in an environmentally sustainable manner and is certified as "green" by Smartwood, an international organization. The wood is favorably received in international markets and receives a price premium. Other products — such as coffee, cocoa, and bananas — can also be certified as being produced in an environmentally and socially sustainable manner. Certification is a process that creates markets for environmental quality.

Regulation is an alternative institutional mechanism for influencing environmental behavior. Although difficult to enforce, regulation can play a role when combined with other economic incentives. For example, burning of crop stubble, farming of particularly erosive lands, or logging in certain areas can be prohibited in conjunction with a program that also provides other government economic benefits to farmers or forest owners. Families can be restricted from settling in flood-prone areas, perhaps with the provision of funds for resettlement. Experience shows that without incentives for changing behavior, regulations tend to be ineffective, since enforcement is costly and there are private incentives to cheat.

Physical restrictions on grazing, land reform programs that distribute land to small farmers, revised leasing arrangements, and many other government-sponsored institutional changes can improve natural resource sustainability if certain principles are followed. First, there is a need for careful assessment of the economic benefits and costs, including externalities, resulting from the policies. Second, local input is needed in the decision-making process. Third, compensation is often required for any losers. That society as a whole will be better-off following these institutional changes is not enough. Losers need to be compensated or they may oppose any change.

These three principles hold for institutional changes at various levels — local, regional, national, and international — and they are not always easy to apply. If developed countries want developing countries to reduce carbon-dioxide emissions associated with forest burning, developed countries must be willing to foot part of the bill. The Kyoto Protocol for climate change, adopted in 1997, reflected this need for mutual sacrifice to limit greenhouse gas emissions. It was the product of several years of intense negotiations and reflected developing-country energy needs for economic development. The agreement, although not ratified by the U.S. government, entered into force in early

2005 and sparked creation of markets for trading emission allowances. New markets for formerly unvalued environmental goods (such as carbon sequestration) represent opportunities for producers in developing countries.

Similarly, if national governments want deforestation reduced, they cannot just pass a national decree. They must involve local decision-makers in designing an institutional solution that provides individual incentives for appropriate behavior. In addition, someone may need to estimate the costs and benefits associated with alternative institutional mechanisms.

In many cases, the presence of transactions costs and collective action has created institutional environments that are destructive to the natural resource base. Imperfect information, corrupt government officials, and the absence of new institutional arrangements to replace previous social and cultural norms that constrained behavior harmful to the groups are serious problems. Improvements in information flows and creation of markets to reflect environmental values are essential if such corrupt behavior and reductions in other transactions costs are to be reduced. Education also becomes vitally important. Thus, focusing on communications infrastructure and human-capital development are two keys to environmental improvement.

SUMMARY

Sound environmental management is essential for sustained agricultural and economic development. Yet environmental degradation is evident throughout the developing world. Soil erosion, silting of rivers and reservoirs, flooding, overgrazing, poor cropping practices, desertification, salinity and waterlogging, deforestation, energy depletion, loss of biodiversity, and chemical pollution have become major problems. Poverty, high rates of return to capital, debt problems, rapid population growth, and misguided public policies conspire against solutions. Environmental problems are interrelated, and understanding their causes requires sorting out complex physical, economic, and institutional linkages. Technical solutions are needed for each of these problems, but economic and institutional changes must provide the incentives for behavioral change. As incomes grow, population pressures are reduced, and the demand for environmental protection increases. Economic development means more resources in the long run for addressing environmental problems. Changes in taxes, subsidies, regulations, and other policies can influence local incentives for conservation. Balancing benefits with costs, obtaining local input in the decision-making process, and compensating losers are activities needed for effective solutions to local and global environmental problems. Because transactions costs must be reduced for natural resource conservation to occur, information flows must be improved and human capital must be developed.

IMPORTANT TERMS AND CONCEPTS

Biodiversity	Flooding
Chemical pollution	Greenhouse effect
Climate change	Institutional change
Common property	Natural resource management
Deforestation	Overgrazing
Desertification	Regulations
Discounting of costs and benefits	Salinity and waterlogging
Global warming	Soil erosion
Environmental degradation	Subsidies and taxes
Externalities	Sustainable resource use

Looking Ahead

In this chapter, we examined the nature and causes of environmental problems in developing countries. Potential technical, economic, and institutional solutions were considered so that agricultural development can be sustainable. In the next major section of the book we consider what it takes to improve agriculture more generally from both a technical and an institutional perspective to contribute to sustainable development. However, first, in Chapter 10, we consider how human resources, including family structure and gender issues, influence standards of living in developing countries.

QUESTIONS FOR DISCUSSION

1. What are the major natural resource problems facing developing countries?
2. Are the poorest countries the most vulnerable to environmental degradation? Why, or why not?
3. How are flooding and soil erosion related?
4. What is desertification?
5. How are waterlogging and salinity problems interrelated?
6. How are deforestation and energy problems interrelated?
7. How are deforestation and global climate change interrelated?
8. What are the major technical or physical causes of natural resource degradation?
9. What are the major economic causes of environmental degradation?

10. What are the major institutional causes of environmental degradation?
11. What common market failures lead to environmental degradation in developing countries?
12. What are some of the technological solutions to natural resource problems?
13. What are some of the economic and institutional solutions to natural resource problems?
14. What are three key principles that must hold if institutional changes are to successfully solve environmental problems?
15. Why are reductions in transactions costs important for sustainable natural resource use?

RECOMMENDED READING

Copeland, B. R., and M.S. Taylor, "Trade, Growth, and the Environment," *Journal of Economic Literature*, vol. 42 (March 2004), pp. 7–71.

Markandaya, Anil, Patrice Harou, Lorenzo Giovanni Bellu, and Vito Cistulli, *Environmental Economics for Sustainable Growth* (Northhampton, Mass.: Edward Elgar, 2002).

Pearce, David W., and R. Kerry Turner, *Economics of Natural Resources and the Environment* (Baltimore, Md.: Johns Hopkins University Press 1990), pp. 61–9, 342–60.

CHAPTER 10

Human Resources, Family Structure, and Gender Roles

Women account for 70 to 80 percent of household food production in Sub-Saharan Africa, 65 percent in Asia, and 45 percent in Latin America and the Caribbean. They achieve this [production] despite unequal access to land, to inputs such as improved seeds and fertilizer, and to information.— Lynn R. Brown *et al.*[1]

This Chapter

1. Discusses the role of human resources in agricultural and economic development
2. Examines differences in family structure and gender roles in farm households in developing countries
3. Considers determinants of gender roles in farm households.

Poor agricultural households in developing countries generally have few assets. Some own small parcels of land, and all households have human assets. The productivity of these human assets helps determine prospects for accumulation of other assets and increased income over time. Productivity of labor assets can be improved through investments in education, health care, and acquisition of skills. Decisions about investments in education, how household labor is deployed, and about the size and structure of families are made by families. These decisions depend on policy-based and other incentive structures, cultural norms, and gender roles; such decisions have major impacts on productivity, asset accumulation, and household well-being. In some societies, for example, girls are less likely to attend school than are boys; in others, women are less likely to receive health care and have shorter life expectancies than do men. We examine the role and determinants of

[1] Lynn R. Brown, Hilary Feldstein, Lawrence Haddad, Christina Peña, and Agnes Quisumbing, ch. 32, p. 205 in *The Unfinished Agenda: Perspectives on Overcoming Hunger, Poverty, and Environmental Degradation*, ed. Per Pinstrup-Andersen and Rajul Pandya-Lorch (Washington, D.C.: International Food Policy Research Institute, 2001).

investments in education, how human resources affect household well-being, and the roles of men, women, and children in making household decisions and participating in household activities.

ROLE OF EDUCATION

The overall productivity of the economy depends on the quantity and quality of inputs into production. Better education, health care, and acquisition of skills are clear means of improving labor productivity. Evidence continually shows that better-educated individuals earn higher incomes and these higher incomes reflect more productivity.[2] Education can be an important contributor to improved agricultural productivity; underutilization and low productivity of human resources in agriculture is a serious problem in many developing countries. Better-educated farmers are more able to adopt new technologies, are better able to understand price and market information, and have more access to credit and other forms of capital. Countries that fail to improve the skills and knowledge of farmers and their families find it difficult to develop anything else.

Objectives and Benefits of Education

Rural education is an investment in people that has as its objectives: (1) improving agricultural productivity and efficiency, and (2) preparing children for non-farm occupations if they have to leave farming. Education may help motivate farmers toward change, teach farmers improved decision-making and farm management methods, provide farmers with technical and practical information, and lead to better marketing of higher-valued farm outputs. Agricultural extension is complementary to other sources of information because it speeds up the transfer of knowledge about new technologies and other research results (see Chapter 12 for more details on extension systems).

A country with a literate people in rural areas will have better information flows than one without, due simply to better written communications. Communications help reduce the transactions costs that hold back development; they improve the timing of productive activities and lower risk. Education helps farmers acquire, understand, and sort out technical, institutional, and market information.

The result is that investments in education yield returns not just for the farmer, but for society as a whole. As education levels increase in a village, all villagers tend to gain from more productive neighbors,

---

[2] Paul Glewwe, "Schools and Skills in Developing Countries: Education Policies and Socioeconomic Outcomes," *Journal of Economic Literature*, vol. 40, no. 2 (June 2002), pp. 436–82.

better information flows, and more experimentation and innovation — education is a public good. Because rural education results in a more productive and efficient agriculture and in a more productive labor force for non-farm employment, and because of its public good characteristics, most countries — both developed and developing — finance education, particularly at the primary and secondary levels. As countries develop, the social benefit from education becomes so great that the scope of rural education grows. T.W. Schultz has argued that education helps people to deal with economic disequilibria. Thus, as agriculture in a country shifts from a traditional to a more dynamic, science-based mode, the value of education increases.

Education is important not just for the farmer and for children who will continue farming, but for those who leave agriculture. Education for non-farm jobs is particularly important for agricultural development if children of farmers acquire jobs as agricultural extension agents, managers of cooperatives and other business firms supplying inputs to farmers or marketing their products, agricultural scientists, or government officials who administer agricultural programs. Educated children who do not choose agricultural occupations often send remittances back home; these remittances are an important source of investment capital for farmers.

Education of girls can be particularly important for development prospects. As women become more educated, they tend to live longer and healthier lives, the value of their time increases, the health and nutrition of family members improves, and total fertility declines.[3] They have fewer, healthier, and better-educated children. They also earn more in farming and off the farm.[4] Payoffs to women's education are found in the short run through improved productivity, and long-run payoffs include the reduction in intergenerational poverty. Although progress has been made in improving access of girls to schooling, gaps remain, particularly in the poorest countries, where girls are only 80 percent as likely to attend school as are boys.

Major Types of Education

Three basic types of education exist: (1) primary and secondary education, (2) higher education, and (3) adult education. Most countries have a goal of almost universal primary education and eventually secondary education as well. Primary education provides the basic literary

---

[3] See T. Paul Schultz, "Women's Role in the Agricultural Household," ch. 8 in *Handbook of Agricultural Economics*, ed. Bruce L. Gardner and Gordon C. Rausser (New York: Elsevier, 2001).

[4] See World Bank, *Engendering Development* (New York: Oxford University Press, 2001).

and computational skills. Secondary education provides training for students going on to higher education, and technical education for those who seek immediate employment.

The need for higher education related to agriculture depends in part on the growth of employment opportunities in agricultural research, extension, agribusiness, and government. Undergraduate agricultural programs have expanded in many African, Asian, and Latin American countries in recent years. Some of these colleges, such as the Pan-American Agricultural School in Zamorano, Honduras, require a mix of academic and practical training and draw students from several countries.

Postgraduate programs have also expanded in several larger developing countries such as India, the Philippines, Brazil, and Mexico. The quality of these programs is variable, but the programs have a better track record of having their students return home after completing their degrees than do graduate programs in developed countries. Foreign academic training in developed countries also has the disadvantage that the training and research may be less relevant to the home country of the student.

In adult education, often called *extension* education in agriculture, farmers are the primary clientele and the programs are mostly oriented toward production problems facing farmers. Extension accelerates the dissemination of research results to farmers and, in some cases, helps transmit farmers' problems back to researchers. Extension workers provide training for farmers in a variety of subjects and must have technical competence, economic competence, farming competence, and communication skills. Thus extension workers require extensive training and retraining to maintain their credibility with farmers.

Education Issues in Developing Countries

Because education is critical for a country's development prospects, several interrelated issues must be addressed by education policymakers. These issues include: finance questions, such as measures to recover costs in K-12 education; use of resources to retain students through higher grade levels versus expanding basic coverage to all; decisions about educational curricula, such as providing technical versus more general education; and gender and economic barriers to participation in basic education. Cost recovery measures such as school fees were introduced in many developing countries as part of structural adjustment programs in the 1980s. They are based on the idea that since some of the benefits of education are private and are captured by the individual, the beneficiary (the student or his or her family) should bear some of the costs. They also broaden the financial base of support

Female education is as important as male education,
yet it is often neglected.

for the educational system and provide resources to cash-poor local educational districts. However, increasing evidence shows that such fees represent major barriers to participation in education, and countries that have abolished fees have seen remarkable growth in school participation. The World Bank, which was a strong proponent of cost recovery in basic education, now has a blanket policy opposing such fees. Elimination of fees will help reduce gender and economic barriers to participation in education.

Developing countries face choices about the design of their educational curricula in rural areas. While most schools provide basic literacy and mathematics, choices need to be made about technical content. The experience has been mixed relative to agricultural education at the K-12 level in developing countries. While some argue that such schools need to provide useful skills and thus should focus on training in agriculture, evidence shows that design of an agricultural curriculum is difficult and costly. Often training methods do not correspond to conditions faced by poor farmers, and time spent in such training reduces time available for other subjects. When rural schools focus too closely on rural-specific skills, graduates face disadvantages when seeking higher education or finding work in urban areas.

FAMILY STRUCTURE AND GENDER ROLES

Family structures vary significantly around the world, and that variation implies differences in specific roles played by individual family members in household affairs, in agricultural production and market-ing, and in income generation in and out of agriculture. For example, in many West African countries, families live in compound households that include more than one generation, and individual family members are assigned specific parcels of land to farm. In much of Latin America, the basic household is a nuclear family with parents and children, and family members have specific responsibilities within the household and in farming. In many parts of Asia, nuclear families predominate, and in some cases family members work side by side in fields, but in others males and females undertake different tasks. Regardless of the region, women have key roles to play in farming systems. Women are involved not only in household chores and child-rearing but are also a major source of labor for food production and account for a large proportion of economic activity.

Gender Roles

The term "gender" refers to non-biological differences between women and men, and roles in farming and household decisions in developing countries differ by gender. With the notable exception of strongly Is-lamic societies, women play two major roles in the rural areas of most developing countries. First, they have household responsibilities for child-rearing, food preparation, collecting water and firewood, and other chores. Second, they are paid or unpaid workers in agriculture or off the farm. They produce, process, preserve, and prepare food. They work in the fields, they tend livestock, they thresh grain, and they carry pro-duce to market. In many areas, women manage the affairs of the house-hold and the farm. They sell their labor to other farms and sometimes migrate to plantations. Involvement in farm production may be sea-sonal, particularly in Asia where, in many countries, women assume major responsibilities for weeding and harvesting, both on their own farms and as paid labor on other farms. Women also work in small industries and in the informal sector, producing goods and services for sales locally or beyond.

Women are important to agriculture in most areas of the world, but they tend to play the largest role in farming in Africa. In many coun-tries nearly all the tasks connected with food production are left to women. Men may tend livestock or produce cash crops, but food crops are generally the purview of women. In Malawi, for example, over two-thirds of those working full-time in farming are women.[5] In some areas

[5] Janice Jiggins, *Gender-related Impacts and the Work of the International Agricultural Re-search Centers*, Consultative Group for International Agricultural Research (CGIAR) Study Paper No. 17 (Washington, D.C.: World Bank, 1986).

Women threshing wheat in Nepal.

of Africa where men migrate to work elsewhere, the entire administration of the household is left to women (Box 10-1). Similar cases exist in the Central American highlands where men migrate seasonally to participate in coffee harvests and to coastal plantations. Households headed by women make up 20 to 25 percent of rural households in developing countries, excluding China and Islamic societies.[6] In Latin America, women typically care for animals, particularly chickens and pigs, while tending garden vegetables and other food crops. In sugar- and fruit-producing areas, especially in the Caribbean, women work as cash laborers on plantations, and provide a substantial proportion of household income. In Asia, many examples of female farming systems are known. In Nepal, it is estimated that women on subsistence farms produce 50 percent of household income; men and children produce 44 and 6 percent, respectively.[7]

Even though they tend to work much longer days than men, the true extent of involvement of women in agriculturally related activities is often underestimated and misunderstood by policy-makers. When surveys are taken, men frequently respond as heads of households, and both men and women tend to describe the woman's principal occupation as housewife. In many areas women do not view themselves as

[6] Ibid.

[7] Meena Acharya and Lynn Bennett, *Women and the Subsistence Sector. Economic Participation and Household Decision Making in Nepal*, World Bank Staff Working Paper No. 526 (Washington, D.C.: World Bank, 1982).

BOX 10-1
GENDER DIVISION OF LABOR IN BOTSWANA

A study of traditional farms in Central Botswana uncovered illuminating differences in the division of labor by gender. Because men have opportunities to work in mines, a large proportion of rural households are headed by females (40 percent in this study). In agricultural areas, land is held communally by the village, and both men and women can obtain rights to cultivate the land. Mostly sorghum, but also maize, cowpeas, and melon varieties are grown on 4- to 5-hectare plots. Livestock, particularly cattle, is very important.

In all aspects of economic activity there is a stark differentiation between male and female roles. In crop production, men traditionally plow and maintain the fields, women sow the seeds, weed, harvest, and thresh. Men and boys almost exclusively tend and milk livestock (mostly cattle and goats), while women manage the chickens, used mostly for home consumption. Women brew and sell sorghum beer, and beer sales can produce substantial amounts of household income.

Women provide virtually all of the household maintenance. Time spent gathering firewood, fetching water, cooking, and in other household chores accounts for 68 percent of the women's total time. Men allocate only 10 percent of their total time to household chores. Even so, women provide 38 percent more time for agricultural fieldwork than do men. Women provide 48 percent of the total hours worked by members of the household, men account for 22 percent, and the children the rest.

Source: Doyle C. Baker with Hilary Sims Feldstein, "Botswana: Farming Systems Research in a Drought-prone Environment, Central Region Farming Systems Research Project," ch. 3, in *Working Together Gender Analysis in Agriculture*, Vol. I: *Case Studies*, ed. Hilary Sims Feldstein and Susan V. Poats (Westford, Conn.: Kumarian Press, 1989), pp. 43–7.

"farmers" even when they work long hours on the farm and have large influences over farming-related decisions[8] (see Box 10-2). They are then counted as economically inactive. This "invisibility" of female employment has led to policies and programs that ignore women and sometimes adversely affect them.

One impact of the "invisibility" of women has been to lower their status. Within the household, this lower status may mean less power to make decisions, less food, fewer heath-related investments, and a

---

[8] Sarah Hamilton, Keith Moore, Colette Harris, Mark Erbaugh, Irene Tanzo, Carolyn Sachs, and Linda Asturias de Barros, "Gender and IPM," ch. 14 in *Globalizing Integrated Pest Management: A Participatory Research Process*, ed. George W. Norton, E.A. Heinrichs, Gregory C. Luther, and Michael E. Irwin (Ames, Iowa: Blackwell Publishing, 2005).

BOX 10-2
GENDER AND INTEGRATED PEST MANAGEMENT

A recent study by Hamilton and others examines how gender roles in different regions of the world affect the use of pest management practices in agriculture. Studies show that improper use of pesticides can lower household incomes and have negative health consequences for household members. Women have a special interest in pesticide use, as they frequently shoulder responsibility for the health of the family, particularly children. Evidence shows that women have to overcome unique barriers if they or their families are to adopt integrated pest management (IPM) practices, which are usually associated with less use of pesticides. Lack of recognition means that women are often excluded from information about IPM practices; they have less access to extension services, are less likely to participate in training, and are less frequently members of producer organizations, which transmit information to their members. Women also have less access to labor, due either to excessive time demands on their own or limited access to hired labor markets. IPM practices tend to be labor intensive. Women have less access to land and, because of uncertainty associated with IPM, most adopters of IPM have larger holdings; they adopt IPM on part of their lands and use conventional techniques on others.

Despite these constraints, the experience from West Africa, the Philippines, and Central and South America found involvement of women to be a key determinant of whether households use IPM or not. Women's participation in field-level trials, in identifying constraints so that research could address them, and in training programs helped spread IPM adoption in all the countries studied. Women are especially receptive to IPM messages because they play a major role in managing household finances and easily recognize the health consequences of mishandled pesticides.

Source: Sarah Hamilton, Keith Moore, Colette Harris, Mark Erbaugh, Irene Tanzo, Carolyn Sachs, and Linda Asturias de Barros, "Gender and IPM," ch. 14 in *Globalizing Integrated Pest Management: A Participatory Research Process*, ed. George W. Norton, E.A. Heinrichs, Gregory C. Luther, and Michael E. Irwin (Ames, Iowa: Blackwell Publishing, 2005).

---

heavier work and disease burden. In times of household crisis, women and female children may bear a heavier burden; in southern Ethiopia, for example, research shows that women suffer more from shocks to income and health.[9] In society, lower status of females has been shown to lead to weaker control over household resources, less access to information and public services such as education and health, discrimination

[9] Stephan Dercon and Pramila Krishnan, "In Sickness and in Health: Risk Sharing within Households in Rural Ethiopia," *Journal of Political Economy*, vol. 108, no. 4 (August 2000), pp. 688–727.

in employment, and unequal rights to land and other important assets.[10] Women are less likely to be members of producer and marketing organizations, are less likely to have title to land (and thus, access to many forms of credit) and have less capital available to invest. These factors affect the women's own nutritional and health status and that of their children.

The role and rights of women vary by region and with size of farm, as discussed below.

DETERMINANTS OF GENDER ROLES IN AGRICULTURE

Social, cultural, and religious factors; population pressures; farming techniques; off-farm job activities; colonial history; income levels; and many other factors determine the role of women in farming systems. Sometimes in areas with apparently similar physical conditions, women assume very different roles. As off-farm job opportunities, population pressures, income levels, and farming techniques change, so too does the role of women.

Shifting cultivation with hand labor tends to lend itself more to female labor than does settled cultivation with a plow. For countries with low population densities, adequate food could be raised without using male labor in farming. Men used to spend their time felling trees, hunting, and in warfare. In most areas, agriculture has changed from shifting cultivation to settled agriculture and cash crops, resulting in a greater role for men, but often the role of women in farm work still dominates.

The shift to the plow and draft animals has made a difference in the amount of male labor used in some areas, and long-standing differences in farming techniques undoubtedly account for many of the regional gender differences in farm labor. In regions of intensive cultivation on small, irrigated farms, for example, in several Asian countries, men, women, and children must work hard to generate enough production on a small piece of land to support themselves. Work is mostly done by hand. In contrast, on some larger farms, more tasks may be mechanized and women may devote a higher percentage of their time to housework. In some areas, mechanization has displaced female labor and has tended to lower their status, since housework is often under-appreciated compared to farm work.

---

[10] Lisa C. Smith, Usha Ramakrishnan, Aida Ndiaye, Lawrence Haddad, and Reynaldo Martorell, *The Importance of Women's Status for Child Nutrition in Developing Countries*. International Food Policy Research Report No. 131 (Washington, D.C.: IFPRI, 2003).

Increased integration of peasant farmers into the labor market has increased the importance of women's role in agriculture, because it is often the males who find outside wage work. In some countries, males may work away from the household for several weeks or months at a time. In Lesotho, for example, the result has been that 70 percent of the households are headed by women.[11]

Policy Implications

Why is it important to address gender inequities in society? First, as a normative concept, gender equality is important in its own right. Women ought to have equal legal and social status because social justice is an important indication of development. Second, many recent studies have shown that gender inequities slow down the process of economic development. Lower status of women is associated with less schooling, lost earnings, inefficient allocation of labor, and poor health of women and their children.[12] Over time, gender inequities lead to lower nutritional and health status of children, less educational attainment, and slower growth. In agriculture, gender is important as one of the several socioeconomic characteristics that influence the adoption of new technologies. If women are important in agriculture, their opinions must be sought when designing new technologies. The impact of these technologies on the relationship between men and women should be considered during this design. If women are making production decisions, they must receive education and guidance from extension services. They must also have access to credit and to inputs.

Women often have inadequate access to credit for a number of reasons. First, in many societies, women lack the legal status necessary to enter into contracts. Second, only very infrequently do women hold title to land, often necessary as collateral for loans. Third, there seems to be a bias against women in the administration of credit programs.

It is likely that most new agricultural technologies are relatively gender neutral, and we see some efforts on the part of certain public extension systems to reach women farmers.[13] However, lack of female access to credit and purchased inputs in many countries makes many new technologies gender-biased. Furthermore, women often grow food

[11] Jiggins, *Gender-related Impacts and the Work of the IARCs.*

[12] World Bank, *Engendering Development.*

[13] In The Gambia, research on rice was expected to increase women's income, since women were the primary producers. Instead, following the introduction of new technologies, men took over this production. See Joachim von Braun, Detlev Puetz, and Patrick Webb, *Irrigation Technology and Commercialization of Rice in The Gambia: Effects on Income and Nutrition,* International Food Policy Research Institute, Research Report No. 75 (Washington, D.C., 1989).

crops that are minor in terms of value of production but are important in the diets of families on small farms. Agricultural research often neglects these crops, and this neglect may have adverse effects on nutrition. In addition, because extension services are still highly male in most countries, communication with female farmers can be inhibited. Even in Africa where women comprise the majority of farmers, males have greater contact with extension services.

The impacts of credit, technology, and other agricultural policies on women have been exacerbated by discriminatory land reform and settlement policies. In Latin America, where land reform and settlement schemes have often been designed to benefit "heads of households," women have been, by convention, largely excluded. In Ethiopia and Tanzania, rights to lands have been bestowed on men. In Asia, specifically the settlement schemes in Indonesia, Papua New Guinea, and Sri Lanka, land was given only to male heads of households. Inadequate access to land, worsened by government policies, when combined with problems of access to credit can hinder women's ability to participate in agricultural development. Given the large role that women play in developing country farming systems, efforts that ignore or discriminate against women have distorting effects and diminish chances of success. Studies have found that farm fields controlled by women often have lower yields due to lack of access to fertilizer and other resources.

Economic development itself can have positive impacts on gender equality. The process of development expands job opportunities, and the presence of more capital raises productivity. These changes raise the value of time — women's time as well as men's. Development also is typically accompanied by more investments in infrastructure, such as water lines, roads, and electricity. These changes can lower work burdens of women, leaving more time for other duties. Higher incomes leave more resources for investments in assets such as human capital. As incomes grow, gender disparities in education and health status tend to shrink. Public investments in schools and health facilities lower the cost of investing in human capital and help shrink gender inequalities. In fact, gender disparities in education are most acute in the lowest-income countries and almost non-existent in high-income countries.[14]

Despite strong empirical links between economic growth and gender equality, equality is not an automatic by-product of growth and the path of development can have important implications for gender relations. Governments that encourage equal participation and foster rights of women often find that growth and greater gender equality march

[14] World Bank, *Engendering Development.*

188

Colombian women receiving instructions on how to vaccinate a chicken.

hand in hand. Gender equality has beneficial growth effects, and growth enhances women's rights. Governments can be proactive by reforming institutions to establish equal rights and opportunities for women and men, they can strengthen policy and institutional incentives for more equal access to resources and participation, and they can take active measures to confront disparities.[15]

Role of Children

Children represent the future human resource base of a country. Economic growth and development over time depend on how resources are invested in children. As noted in Chapter 4, children represent current sources of pleasure for parents, and they are a source of investment for future income gains and security in old age. Children are a major source of farm labor in every region of the world, and their tasks expand with each year of their age. They typically begin by following a parent or sibling into the field and rapidly become involved in hoeing, weeding, harvesting, and other tasks. They feed and otherwise care for animals. They, particularly boys, may work as low-paid farm laborers on other farms. Young girls often care for younger brothers and sisters to free their mother for other work. Farm children throughout the world take on major farm responsibilities at a very young age.

---

[15] Ibid., particularly ch. 6.

Teenage child and her mother sorting the potato harvest
on a farm in Ecuador.

At times, conflicts occur between the use of children in farm duties and providing income to the family and longer-term investments.
For example, in times of household crisis, such as drought or crop failures, children may be pulled out of school to lower expenses (such as
school fees) or increase incomes. Such means of informal risk management can have long-term adverse consequences. Gender inequalities in
investments in children also have long-term consequences, but depend
on social norms and other factors. For example, in many societies, in
times of crisis, decreased spending on girl's education and even health
care and food is a common means of coping with household financial
stress. Such actions lower the status of girls and their quality of life, but
are the product of long-standing cultural norms.

SUMMARY

The overall productivity of the economy depends on the quantity and
quality of labor. Better-educated individuals earn higher incomes and
these higher incomes reflect greater productivity. The underutilization
and low productivity of human resources in agriculture is a serious
problem in many developing countries. Better-educated farmers are
more able to adopt new technologies, are better able to understand price
and market information, and have more access to credit and other forms
of capital. Education also prepares children for non-farm occupations.

Women and children play important roles in agriculture, and these roles vary by region, by stage of development and other factors. Social, cultural, religious, technological, off-farm employment, historical and other factors determine the role of women in farming systems. Women's roles in agriculture have implications for credit and input policies, for the generation and extension of new technologies, and for land reform policies. Gender inequities can have adverse implications for long-term development inside and outside of agriculture. Compelling evidence shows that governments should take proactive steps to lower gender inequalities.

IMPORTANT TERMS AND CONCEPTS
Constraints faced by women farmers
Determinants of the role of women in agriculture
Human capital
Impacts of education on development
Implications of the role of women in agriculture
Multiple roles of women
Regional differences in the roles of women

Looking Ahead
In this chapter, we briefly examined the role of human resources, family structure, and women and children in the process of agricultural and economic development. In Part 4 we consider means for improving those systems to increase agriculture's contribution to human welfare. We begin in Chapter 11 by providing an overview of agricultural development theories and strategies before exploring in detail the individual components of those theories and strategies.

QUESTIONS FOR DISCUSSION
1. How do investments in human capital affect productivity inside and out of agriculture?
2. What is the purpose of education for the farmer and his or her family?
3. What are the major types of education?
4. What roles do women and children play in agriculture?
5. In which region of the world is the role of women in agriculture the greatest?
6. What factors determine the roles of women in agriculture?
7. What are some important implications of the roles of women in agriculture?

8. Why might census statistics and other data undercount female participation in farming?
9. Why do women from near-landless and smallholder households participate more in agriculture relative to those from larger farms with more land ownership?
10. How might gender inequality slow down the process of development?
11. What steps might governments take to address problems of gender inequality?

RECOMMENDED READING

Baker, Doyle C. with Hilary Sims Feldstein, "Botswana Farming Systems Research in a Drought-prone Environment, Central Regional Farming Systems Research Project," in *Working Together: Gender Analysis in Agriculture,* Vol. I: *Case Studies,* ed. Hilary Sims Feldstein and Susan V. Poats (West Hartford, Conn.: Kumarian Press, 1989), ch. 3.

Deere, Carmen D. "The Division of Labor by Sex in Agriculture: A Peruvian Case Study," *Economic Development and Cultural Change,* vol. 30 (1982), pp. 795-811.

Folbre, Nancy, "Engendering Economics: New Perspectives on Women, Work, and Demographic Change," in *Annual World Bank Conference on Development* (Washington, D.C.: World Bank, 1995).

PART 4

Getting Agriculture Moving

International Rice Research Institute in the Philippines.

CHAPTER 11

Theories and Strategies for Agricultural Development

The process of agricultural growth itself has remained outside the concern of most development economists.
—Yujiro Hayami and Vernon W. Ruttan[1]

This Chapter

1. Describes how the sources of agricultural growth tend to change as development occurs, and considers how theories of agricultural development have changed over time
2. Presents the theory of induced innovation as applied to agriculture, and its implications for the types of technologies generated and for institutional change
3. Discusses the possibility that transactions costs and collective action may alter the direction of technical change, with implications for asset distribution.

THEORIES OF AGRICULTURAL DEVELOPMENT

We have discussed the importance of agricultural development for solving the world food-income-population problem. We have considered the nature and diversity of existing agricultural systems in developing nations. We now need to consider means for improving these systems to increase agriculture's contribution to human welfare. In this chapter, we provide an overview of agricultural development theories and strategies. In subsequent chapters we examine in more detail the individual components of the basic strategies outlined here. Our overriding concern is to identify strategies that facilitate growth with equity. We explore why agricultural development has occurred in some countries and why it has not (or has proceeded very slowly) in others.

---

[1] Yujiro Hayami and Vernon W. Ruttan, *Agricultural Development: An International Perspective* (Baltimore, Md.: Johns Hopkins University Press, 1985), p. 41.

Many theories have been suggested to explain how the basic sources of growth (labor, natural resources, capital, increases in scale or specialization, improved efficiency, education, and technological progress) can be stimulated and combined to generate broad-based agricultural growth.[2] It is clear from historical experience that the relative importance of alternative sources of growth changes during the development process and has changed over time for the world as a whole. It is also clear that institutional arrangements such as marketing systems, price and credit policies, a well-functioning legal system, and transparently enforced property rights play an important role in stimulating or hindering development. Let us examine agricultural development theories and evidence to see what lessons they provide for operational strategies.

Expand the Extensive and Intensive Margins

One means of generating increased agricultural production is to expand the use of land and labor resources. The development of agriculture in North America, South America, Australia, and other areas of the world during colonization was based on using new lands. In some cases indigenous labor was also exploited. The opening up of forests and jungles by local populations in parts of Africa, Latin America, and Asia provides additional examples of expanded resource use. Economists call this increased use of land and labor *expanding the extensive margin*.

In many of these historical cases, surplus lands and labor were used to produce commodities for both local consumption and export. Reductions in transportation costs facilitated exports. In Thailand, for example, rice production increased sharply in the latter half of the nineteenth century, and much of the increased production went to export markets. In many colonies, exports of primary production were extracted for use in more developed countries, and often a large share of the benefits of these exports was not realized by the local countries but was transferred to the developed countries.

Expansion of unutilized land resources provides few opportunities for substantial growth in developing countries today. In areas of Latin America and Africa where additional land does exist, disease, insect, and soil problems prevent its use in agriculture. Abundant labor is available in many countries, and continued growth of the labor force

---

[2] Hayami and Ruttan (*Agricultural Development*) have characterized previous agricultural development theories into six basic approaches: (1) resource exploitation, (2) resource conservation, (3) location, (4) diffusion, (5) high payoff input, and (6) induced innovation. The first part of this chapter draws heavily on their ideas.

Agriculture in Asia is intensive even in hilly regions.

will generate increases in total agricultural output. However, most growth in per capita agricultural output will have to come from more *intensive* use of existing resources.

Many methods can be used to achieve more intensive resource use. Early efforts in England, Germany, and other European countries included more intensive crop rotations, green manuring, forage live-stock systems, drainage, and irrigation. In many developing countries, these same factors increased land productivity. Terracing is an effective means of conserving soil productivity in hilly areas of Asia. In the moun-tainous regions of Central America, grass strips have been used to cre-ate terrace-like structures that conserve soil and enhance productivity. Crop rotations are frequently used to enhance soil productivity and control pests. Hayami and Ruttan estimate that agricultural develop-ment based on similar types of "conservation" has been responsible for sustaining growth rates in agricultural production in the range of 1 per-cent per year in many countries, including developing countries, for long periods of time.[3]

While scientists are gaining additional knowledge of the technical and institutional considerations that can lower the cost of conservation efforts, population pressures are creating a need for better ways of sus-taining the natural resource base. Hence, conservation is likely to play an increasingly important role in maintaining if not expanding agricul-tural production in the future.

[3] Hayami and Ruttan, *Agricultural Development*, p. 52.

Another means of intensifying agricultural production is to produce more crops per unit of time through altering cropping patterns or using shorter-season varieties so that two and three crops can be produced per acre per year where one or two were produced before. Such production changes usually require scientific input to develop the required seeds, tools, or other inputs to make the double or triple cropping possible. Access to irrigation or surface water sources can facilitate this intensification.

Yet another means of intensification is through a process of diversification and production of higher-valued commodities. This means of intensification is likely to become more important as development proceeds and incomes grow, creating increased demand for higher-valued vegetables and meats. Intensity of production can be changed as well by improving transportation systems to bring higher-valued commodities to urban centers. It has long been recognized that the pattern and intensity of agricultural production vary in relation to the proximity of urban-industrial centers and to the quantity and quality of transportation.[4] Closeness to cities and transport matters because of differences in transportation and marketing costs, in effects on labor and capital markets, in the ease of obtaining new and more productive inputs, and in ease of information flows.

One implication of this "location" theory of agricultural development is that countries should encourage decentralized industrial development, particularly in the middle and late stages of development. During these stages, strong linkages between agriculture and markets for inputs (fertilizers and pesticides) and outputs can help stimulate the local economy. Developing nations should improve transportation infrastructure in rural areas.

Diffuse Existing Knowledge

Agricultural development can be stimulated by diffusing knowledge among farmers more rapidly within or across national borders. Existing technologies and economic knowledge can be transferred from the more progressive to the lagging farmers, thereby increasing productivity. This idea has provided part of the rationale for agricultural extension systems, particularly in farm management. Unfortunately, in some cases diffusion theory has led to unrealistic expectations of the size of potential productivity gains under the existing level of technology.

---

[4] Today, economists still draw on theories proposed by Heinrick Von Thunen (1783–1850), who studied the optimal intensity of farm enterprises in relation to their distance from urban areas.

Diffusion theory has also led to attempts to directly transfer knowledge and technologies from more-developed to less-developed countries. More success has been achieved with transferring knowledge than with transferring agricultural technologies. Adoption of transferred technologies has been limited except where efforts have been made to adapt the technologies to the new setting.

Develop High-payoff Inputs

More recent agricultural development theory builds on these earlier approaches but adds the important dimension that the process can be accelerated through provision of new and improved inputs and technologies (particularly improved seeds, fertilizers, pesticides, and irrigation systems). This approach, articulated by Schultz in *Transforming Traditional Agriculture,* is based on the idea discussed in Chapter 7 that farmers in traditional agriculture are rational and efficient given their current resources and technologies.[5] What these farmers need are new high-payoff inputs and technologies to increase their productivity.[6]

The need for high-payoff inputs has been widely accepted because of the success achieved by modern wheat, corn, and rice varieties beginning in the 1950s and 1960s. These varieties are highly responsive to fertilizer, pesticides, and water management and have resulted in substantial growth in agricultural output in many developing countries. Some have argued that the relative absence of these inputs has been one factor holding back agricultural development in Africa compared to other developing regions. The distributional or equity effects and environmental impacts of these inputs, however, have been the subject of much debate and are discussed in more detail in Chapter 12.

Hayami and Ruttan argue that the high-payoff input theory is incomplete because it fails to incorporate the mechanism that induces these new inputs and technologies to be produced in a country. The theory also fails to explain how economic conditions stimulate the development of public agricultural experiment stations and educational systems. It does not attempt to identify the process by which farmers organize collectively to develop public infrastructure such as irrigation and drainage systems. In the next section we explore the induced innovation theory proposed by Hayami and Ruttan to address these issues.

---

[5] Theodore W. Schultz, *Transforming Traditional Agriculture* (New Haven, Conn.: Yale University Press, 1964).

[6] Hayami and Ruttan have labeled Schultz's approach the "high-payoff input" model.

THEORY OF INDUCED INNOVATION

Induced innovation theory helps explain the mechanism by which a society chooses an optimal path of technical and institutional change in agriculture.[7] The theory says that technical change in agriculture represents a response to changes in resource endowments and to growth in product demand. Changes in institutions are induced by changes in relative resource endowments and by technical change.[8]

Induced Technical Innovation

Technical change in agriculture can follow different paths. Technologies can be developed that facilitate the substitution of relatively abundant and low-cost factors of production for relatively scarce and high-cost factors. A rise in the price of one factor relative to others will induce technical change that reduces the use of that factor relative to others. For example, if the price of land goes up relative to labor and fertilizer, indicating that land is becoming relatively scarce, technologies such as improved seeds will be developed that can be combined with labor and fertilizer to increase production per unit of land.

This process of induced technical change is illustrated graphically in Figure 11-1. The range of possible technologies in time period 0 can be represented by what Hayami and Ruttan call the *innovation possibilities curve*, I_0^*. The specific technology employed in that time period is represented by the isoquant I_0. Production occurs at point A with N_0 units of land and L_0 units of labor, the least-cost combination of those resources given the price ratio P_0. Now, if over time labor becomes more abundant relative to land so that the price of labor is reduced relative to the price of land (the new price ratio is represented by P_1), incentives are created to adopt a more labor-intensive technology. If there were no technical change, production might occur at point B on isoquant I_1. However, the theory of induced innovation says that incentives are created not only to select a new technology from the current technology set (that is, move to point B on I_1), but also to develop new technologies to save scarce resources and use abundant resources more intensively.

---

[7] Induced innovation theory was developed originally by John R. Hicks, *Theory of Wages* (London: Macmillan, 1932). Hayami and Ruttan during the 1960s were the first to apply the theory to agricultural development. Their underlying assumption is that technological and institutional changes are vital to agricultural development.

[8] Hayami and Ruttan (*Agricultural Development*, p. 94) define institutions as "the rules of society or of an organization that facilitate coordination among people by helping them form expectations which can reasonably hold in dealing with others. They reflect the conventions that have evolved in different societies regarding the behavior of individuals and groups relative to their own behavior and the behavior of others."

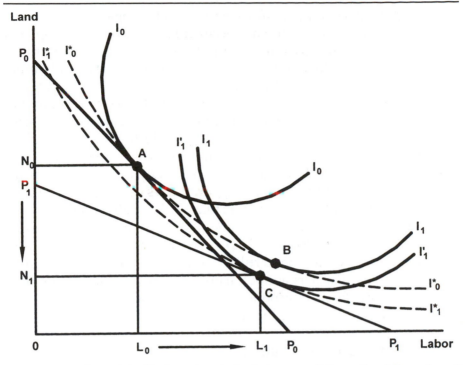

Figure 11-1. A model of induced technical change. If the ratio of the price of land to labor changes from P_0 to P_1, incentives are created not only to substitute labor for land and to move from technology I_0 at point A to technology I_1 at point B, but also to develop a new technology I'_0 at point C. Innovation possibility curves I^*_0 and I^*_1 represent the range of potential technologies that can be applied in period 0 and period 1 (Source: Yujiro Hayami and Vernon W. Ruttan, *Agricultural Development: An International Perspective*, Baltimore, Md.: Johns Hopkins University Press, 1985).

The new technology set is represented by the new innovation possibility curve I_1^*. As the innovation possibility curve moves toward the origin, the same quantity can be produced at lower cost. Following the generation of this new technology set, farmers can adopt the new least-cost technology 1 and employ N_1 of land and L_1 of labor at point C.

Hayami and Rattan compare the agricultural development histories of Japan and the United States to illustrate the validity of the theory. Japan experienced increasingly higher priced land compared to labor and stressed the development of biological technologies such as improved seeds and fertilizers. These technologies tend to save land and use labor more intensively. The United States, on the other hand, has approximately twice as much land per worker as does Japan. As the U.S. frontier was moved west, land became relatively abundant compared to labor, and the development of mechanical technologies that saved labor was stressed. The result was successful agricultural development in both countries, but agricultural output per worker is 10 times greater in the United States than in Japan while output per hectare is 10 times greater in Japan than in the United States.[9]

Changes in output price relative to an input price can also induce technical change, as illustrated in Figure 11-2. The curve u represents the range of current and possible production technologies in a given time period. Hayami and Ruttan call this the *meta production function*. Specific production technologies are represented by v_0 and v_1. At the initial fertilizer output price ratio (P_0), producers use technology v_0 and produce at point A. If the price of fertilizer falls relative to the price of output (P_1), then incentives are created to move to point B on the existing technology. If the price ratio P_1 is expected to continue, farmers press scientists to develop a more fertilizer-responsive variety, v_1, if it does not already exist. Farmers adopt the new variety and move to point C. In the long run, the meta-production function itself may shift as more basic scientific advances are made.

Induced Institutional Change

Incentives are created for technical change, but where do these new technologies come from? How do farmers acquire them? What determines whether technologies are developed that are suitable for all farmers or only for *some* of the farmers? All of these questions are addressed by the theory of induced *institutional* change.

Farmers demand new technologies not only from private input suppliers but from the public sector as well. Hayami and Ruttan argue that public research scientists and administrators are guided by price signals and by pressures from farmers. The more highly decentralized the research system, the more effectively these pressures work. Research systems that welcome and facilitate inputs from farmer groups and that engage in participatory planning and research are also more responsive. The development of the research systems themselves can be the result of pressures from farmers who are responding to market forces.

Many other types of institutions (rules of societies or organizations) affect technical change and agricultural development. The rights to land, marketing systems, government pricing and credit policies, and laws governing contracts are just a few. The theory of induced

---

[9] Hayami and Ruttan, *Agricultural Development*. Many developing countries, particularly in Asia, are finding the Japanese path of technical change more appropriate than the U.S. path, given their relative resource endowments and the nature of changes in those endowments.

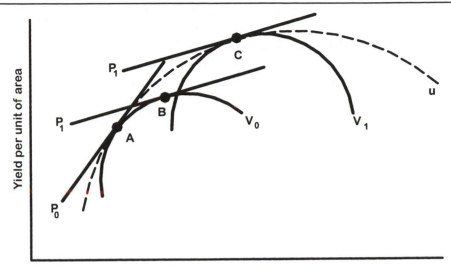

Fertilizer input per unit of area

Figure 11-2. Shift in fertilizer response curve as price ratio changes. If the output/fertilizer price ratio changes from P_0 to P_1, incentives are created not only to apply more fertilizer and increase output from A to B using the traditional variety v_0, but to develop and adopt a new variety v_1 and to move to point C. Curve u represents the "envelope" of a series of available and potential crop varieties (Source: Yujiro Hayami and Vernon W. Ruttan, *Agricultural Development: An International Perspective* (Baltimore, Md.: Johns Hopkins University Press, 1985).

institutional innovation recognizes that institutions can become obsolete and in need of adjustment over time. It says that new technologies and changes in relative resource endowments or price changes provide incentives for a society to demand new institutional arrangements (see Box 11-1 for an example).

Examples of institutional changes induced by technological change may be found in the shift from share tenure to more fixed-payment leases, which has occurred in several countries as new varieties and irrigation systems have increased yields while reducing risks.[10] An example of an institutional change due to a change in relative resource endowments is the switch from communally owned land to more private forms of property rights as population pressures increase land scarcity.

In some countries we observe what appear to be socially desirable institutional changes, technical changes, and relatively rapid and broad-

---

[10] Share tenure is an arrangement whereby a farmer who is renting land pays the rent with a fixed percentage of the farmer's output.

BOX 11-1
INDUCED INSTITUTIONAL INNOVATION IN JAVA

In Java, customary rules have governed both land rights and labor exchange for many centuries. With traditional technologies, these rules have helped allocate resources so that subsistence levels of foods have been available to all village members. These communal institutions have been put under stress by modern technologies that increase the productivity of labor and the returns to landowners. These changes induce changes in the institutions governing resource allocation.

An example of an institutional innovation is the disappearance of the *bawon* rice-harvesting system. This traditional system allowed everyone, whether they were from a particular village or not, to participate in the harvest and share the output. As population grew with traditional technologies, this purely open *bawon* system gradually evolved into various forms, some of which limited harvest rights to village residents, while others limited harvest rights to a set number of participants, or to people who were invited by the farmers.

The widespread diffusion of fertilizer-responsive rice varieties created sharply higher returns to harvest labor, and induced a remarkable change in harvest-contract institutions. One such innovation was the introduction of the *tebasan* system, in which standing crops are sold to middlemen who hire contract labor for harvesting and thus reduce the harvester's share while increasing returns to the landowners. Another institution is the *ceblokan* system, which limits harvesting rights to those workers who perform extra services such as transplanting and weeding without pay. A study shows that in a village where *ceblokan* was first adopted in 1964 by seven farmers, by 1978, 96 out of 100 farmers had adopted the system.

These innovations in harvest-labor institutional arrangements were largely spurred by increased incomes and higher wages accompanying technological innovation. Increased incomes and wages created incentives for farmers to change their labor-contracting system. These changes are now widespread in Java.

Source: Masao Kikuchi and Yujiro Hayami, "Changes in Rice Harvesting Contracts and Wages in Java," ch. 6 in *Contractual Arrangements, Employment and Wages in Rural Labor Markets in Asia*, ed. Hans P. Binswanger and Mark R. Rosenzweig (New Haven, Conn.: Yale University Press, 1984).

based agricultural development. However, in others we observe what seems to be perverse institutional change, agricultural stagnation, or agricultural growth with the benefits received by only a small segment of the population. Of course, many countries fall between these extremes or may move from one group to the other over time. Why do we see these differences in institutional changes that influence agricultural performance, and how do they relate to the theory of induced innovation? The answer lies partly with transactions costs and with the incentives for and effects of collective action by groups of people with common interests.

IMPLICATIONS OF TRANSACTIONS COSTS AND COLLECTIVE ACTION

The induced innovation theory presented above implicitly assumes well-functioning markets for all products and factors. Prices are assumed to convey all the relevant information to decision-makers, and resources are allocated efficiently and independently of the distribution of assets (such as land) in society. Price-responsive producers are assumed to possess knowledge about alternative technologies, and be able to lobby agricultural scientists to develop improved technologies to save scarce resources. Assuming no economies of scale in production, there is one optimal path for technological change.

Transactions costs

Unfortunately, transactions costs affect both factor and product markets, creating the possibility of differing optimal paths of technical change and of institutional change, depending on farm size or other factors. Transactions cost refer to the costs of adjustment, of information, and of negotiating, monitoring, and enforcing contracts.[11] These costs arise because assets are fixed in certain uses in the short run, because there is a lack of perfect information, because there are differences in the ability to use information, and because people are willing to benefit at the expense of others.[12]

---

[11] A succinct discussion of transactions costs is found in Douglas C. North, "Institutions, Transactions Costs, and Economic Growth," *Economic Inquiry*, vol. 25 (1987).

[12] William J. Baumol, "Williamson's 'The Economic Institutions of Capitalism'," *Rand Journal of Econometrics*, vol. 17 (1986), p. 280, points out that if there were no fixed or sunk costs in land, capital, or people, resources could easily be transferred to optimal uses. If information were perfect or if people could always figure out how to design contracts to cover any contingency, fixed costs would not matter. If people did not try to profit at others' expense, contracts could be drawn loosely and adjustments made as conditions change.

The presence of transactions costs may mean, for example, that the cost of credit decreases as farm size increases, that labor costs per hectare increase as farm size increases (because of supervision costs), and the cost of land transactions declines as farm size increases. Therefore, as farm size grows, labor use per hectare may decline while machinery use per hectare and the demand for capital-intensive technologies may increase. Owners of large farms may also be quicker to adopt new technologies, because they have fewer credit constraints affecting input purchases.

The presence of transactions costs means that the distribution of assets matters for the direction of technical and institutional change.[13] Because the demand for particular types of technical and institutional changes will vary by farm size, the potential is created for conflicting demands on the public sector. Politicians and other public servants respond to the demands of competing groups by considering their own personal gains and losses. Consequently, a change that would benefit society as a whole may not occur if a politician receives greater private gain from an interest group that does not want the change than from a group that does.

Collective Action

When producers of a commodity are few, economically powerful, and regionally concentrated, they may find it easier to act collectively to influence public decisions in their favor than if these conditions do not hold. Even if the conditions do not hold, if a commodity is very important in the diets of people in urban areas or if it earns substantial foreign exchange, the public sector may still act to help its producers. However, if producers are neither organized into a powerful collective lobby nor producing an important commodity for urban consumption or export, they will seldom receive public help such as new technologies. This fact may explain why peasant farmers with small land holdings are often neglected when agricultural research priorities are set.

Implications for Induced Innovation

The implications of transactions costs and collective action for the induced innovation model presented earlier are illustrated in Figure 11-3. Changes in the underlying resource base for the country as a whole may imply that the least-cost path of technical change would occur in

---

[13] See Alain deJanvry, Marcel Fafchamps, and Elisabeth Sadoulet, "Transactions costs, Public Choice, and Induced Technological Innovations," in *Induced Innovation Theory and International Agricultural Development: A Reassessment*, ed. Bruce M. Koppel (Baltimore, Md., and London: Johns Hopkins University Press, 1995).

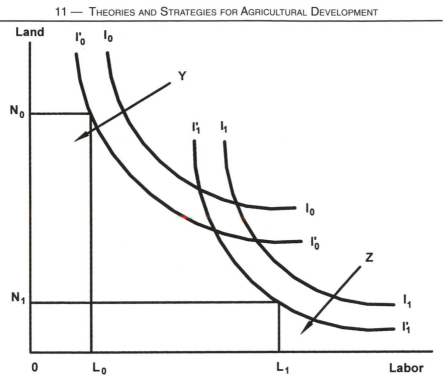

Figure 11-3. Induced technical innovation in the presence of transactions costs. The direction of technical change as dictated by changes in relative factor prices might call for cost-reducing path Z. However, transactions costs and collective action may create pressures to follow path Y, reducing the rate of overall economic growth.

the direction of arrow Z (i.e., a path that would use relatively abundant labor and save relatively scarce land). Following path Z might be facilitated by the development of new labor-intensive, biologically-based technologies. However, if a few large-scale producers, due to the presence of transactions costs and collective action, were able to influence public officials so that technology I'_0 were to be developed rather than I'_1, then technical change might occur in the direction of arrow Y (perhaps through the development and adoption of capital-intensive, mechanically based technologies) rather than arrow Z. Benefits to the large farmers would be maximized but overall economic efficiency gains might be reduced.

The concern over the existence of transactions costs and collective action is not just a concern over the distribution of the benefits of agricultural development. Rather, it is a concern that the rate of economic growth itself will be diminished as well. If, in the previous example, the farmers demanding path Y were few in numbers, and their total value of production compared to the farmers demanding path Z also was small, then the decision to develop technology along path Y would mean a growth rate below the country's potential.

Policy Implications

The above discussion illustrates that technological progress is important for agricultural development, but so too are institutional arrangements and information. Although the theory of induced innovation provides an optimistic look at how market forces can work, almost like an invisible hand to stimulate technological and institutional change, the presence of transactions costs and collective action sounds a cautionary note that there is an invisible foot out there eager to stomp on that hand. The reality that agricultural and overall economic development has progressed steadily in some countries while stagnating in others, demonstrates that development is neither automatic nor hopeless. An operational agricultural development strategy is needed that recognizes (1) the role that relative prices can play in guiding technical and institutional change, and (2) that imperfect information and other transactions costs can sidetrack development unless domestic and international institutions are proactively developed to constrain inappropriate collective action. Inappropriate here is defined as actions that impose gross inefficiencies on the sector or that fail to meet the equity goals of a society. In the sections below, several of these institutions are briefly mentioned; they are discussed more thoroughly in subsequent chapters.

Domestic Institutions

Land, credit, pricing, marketing, and research policies are all critical to development and adoption of appropriate technologies and for agricultural development in general. Sources of agricultural growth change over time, and few countries today are able to achieve substantial production increases by expanding their land bases. In addition, land currently in production is being degraded in many countries due to population and other pressures on a fragile natural resource base. Ownership of land and other assets is highly unequal in many countries and fragmented in others. Hence one institutional component of an operational agricultural development strategy is to re-examine the arrangements governing land ownership and use and to make any needed adjustments.

Improved transportation, marketing, and communications systems also become critical as development proceeds. Lower transportation, marketing, and communications costs can reduce transactions costs and

Improved transportation to reduce transactions costs becomes critical as development proceeds.

improve information flows, and thereby facilitate broad-based agricultural growth. Isolated regions tend to be poor regions.

Provision of high-payoff inputs and credit to finance their purchase are additional components of a successful agricultural development strategy. Farmers are rational and relatively efficient, given their current resources. Consequently new inputs embodying improved technologies are needed to improve the productivity of farmers in developing countries. Research and technology-transfer policies can facilitate the development and adoption of these technologies. In addition, pricing policies should be designed so as not to discourage the use nor encourage the abuse of improved inputs.

Educational levels of farmers must also be increased to improve their ability to recognize the benefits of and to use the technologies. Education improves the capacity of people to assimilate and use information and thus can help reduce transactions costs.

Macroeconomic and International Institutions

Agricultural development is affected by macroeconomic and trade policies that arise outside the agricultural sector. The levels and types of taxes, spending, and government borrowing can dramatically influence farm prices and input costs. Exchange rates, or the value of the country's currency relative to currencies in other countries, can have major effects on domestic agricultural prices and trade.

In some countries, foreign debt repayments significantly constrain growth and reduce domestic consumption. Internationally influenced interest rates and prices vary substantially over short periods of time, adding an additional measure of unpredictability to debt levels and national incomes. International labor markets for agricultural scientists mean that high salaries draw some of the brightest and most educated scientists to more developed countries and international agencies. Foreign aid is a source of capital and technical assistance for some countries, but is often unreliable and usually comes with strings attached. Developing countries must carefully design macroeconomic and trade policies that do not discriminate against their agricultural sector if they expect it to grow.

Enlightened Self-interest

Any operational agricultural or economic development strategy must (1) recognize individual incentives, (2) consider the lack of perfect information, and (3) include institutional arrangements to offset externalities and other market imperfections. Individuals must feel it is in their self-interest before necessary institutional changes will occur.

Information is valuable, imperfect, and costly to acquire, and can exhibit economies of scale in acquisition. These attributes of information provide the incentives and the means for some people to use the advantage they have from asset ownership, military power, or willingness to engage in unscrupulous behavior to acquire information before others.

In fact, even if all assets were initially distributed equally, unless information were available equally to all or unless enforceable rules were instituted to constrain dishonest behavior, the willingness of some to gain "unfair" advantage would eventually lead to unequal distributions of assets. In primitive societies, information is basically available to all and inappropriate activities are constrained by social and cultural norms. However, as societies become more complex concurrently with economic development, information becomes more imperfect, and new institutions are needed to replace the rules that no longer constrain behavior.[14]

People must feel it is in their interest to design and enforce particular institutional changes; and they need to know the implications of

---

[14] These ideas are similar to those expressed by North ("Institutions, Transactions Costs, and Economic Growth"), pp. 420–5. North notes that impersonal exchange with third-party enforcement is essential for economic growth. Third-party enforcement implies that legal institutions exist.

those changes. Institutional change involves costs because some people benefit from current arrangements and will fight any change.

The following six suggestions might help lower the cost of institutional change through enlightened self-interest:

- First, in those countries where asset ownership has become so unequal that inefficiencies in property rights are retarding agricultural development, asset redistributions (particularly land) are needed, usually with compensation arrangements (so that the changes will in fact occur).
- Second, improvements in education, communications, and transportation can improve information flows and the ability of a large number of people in the country to act on information.
- Third, decentralized industrial growth should lower labor adjustment costs (and facilitate employment), reduce externalities associated with urban crowding, improve market performances in rural areas, and help stimulate agricultural growth.
- Fourth, social science research can help lower the cost of designing and examining the implications of alternative institutional changes affecting agriculture.
- Fifth, a government structure is needed that includes enforceable laws to protect citizens from each other and from the government itself. Government policies and regulations can also be used to reduce market failure. Well-functioning and transparent legal systems with independent judiciaries can help facilitate transition toward enhanced institutions.
- Sixth, improved and enforceable international laws and other institutions are needed to reduce incentives for international abuses of power.

SUMMARY

Several theories of agricultural development have been proposed over time. Expansion or conservation of resources, diffusion, use of high-payoff inputs, and induced innovation are some of the major ones. Technical and institutional changes are key components of any operational agricultural development strategy. These changes can be induced by relative price changes resulting from change in resource endowments and product demand. Because of transactions costs, collective action, and the realities of human behavior, agricultural sectors may not follow an economically efficient development path. The distribution of assets has important implications in the presence of transactions costs and collective action. If land is unequally distributed, then, because of transactions costs, the demands (for technologies, inputs, policies) of one group of producers are likely to be very different from those of others. Collective action can then pull the development process from its optimal path. Institutional changes to improve information flows and constrain exploitive behavior can become critical to agricultural development.

IMPORTANT TERMS AND CONCEPTS

Agricultural research and extension	Innovation possibilities curve
Asset distribution	International factors
Asset fixity and adjustment costs	Invisible hand
Communications	Location theory
Compensation schemes	Macroeconomic factors
Diffusion theory	Market failure
Enlightened self-interest	Meta-production functions
Externalities	Perfect information
High-payoff inputs	Resource conservation
Induced institutional innovation	Resource exploitation
Induced technical innovation	Transactions costs

Looking Ahead

In this chapter we considered theories of agricultural development and suggested a broad framework for operational agricultural development strategies. In the following five chapters we consider sector-specific means of generating particular technical and institutional changes to stimulate agricultural growth. In later chapters we consider macroeconomic and international factors. We begin in Chapter 12 by focusing on agricultural research and extension.

QUESTIONS FOR DISCUSSION

1. Contrast the resource exploitation, resource conservation, and diffusion theories of agricultural development.
2. Why is the resource exploitation theory of agricultural development less useful today than it has been historically?
3. Why has the importance of resource conservation increased in recent years?
4. What are the limitations of the diffusion theory of agricultural development?
5. Why has the high-payoff input theory become widely accepted?
6. What criticisms do Hayami and Ruttan make of the high-payoff input theory?

7. Describe the theory of induced technological innovation. Be sure to identify both the importance of relative input price changes and changes in the relative prices of inputs to outputs.
8. Describe the induced institutional innovation theory.
9. Contrast transactions costs and collective actions.
10. What are the implications of transactions costs and collective action for institutional innovation?
11. What do we mean by the term *enlightened self-interest*?
12. How might information be made more accessible to farmers?
13. What are the implications of a grossly unequal asset ownership pattern for economic growth?
14. Why are improved international institutions needed for agricultural development?
15. Why does Japanese agriculture have much higher output per hectare than U.S. agriculture, but much lower output per worker?

RECOMMENDED READING

Binswanger, Hans P., Vernon Ruttan, *et al.*, *Induced Innovations: Technology, Institution, and Development* (Baltimore, Md.: Johns Hopkins University Press, 1978).

Boserup, Ester, *Population and Technological Change* (Chicago, Ill.: University of Chicago Press, 1981).

Hayami, Yujiro, and Vernon W. Ruttan, *Agricultural Development: An International Perspective* (Baltimore, Md.: Johns Hopkins University Press, 1985), ch. 3 and 4.

Koppel, Bruce M., ed., *Induced Innovation Theory and International Agricultural Development: A Reassessment* (Baltimore, Md., and London: Johns Hopkins University Press, 1995).

North, Douglas, "Institutions, Transactions Costs, and Economic Growth," *Economic Inquiry,* vol. 25 (1987), pp. 415–18.

Ruttan, Vernon W., *Technology, Growth and Development: An Induced Innovation Perspective* (Oxford: Oxford University Press, 2001).

Williamson, Oliver, *The Economic Institutions of Capitalism* (New York: Free Press, 1985).

CHAPTER 12

Research, Extension, and Education

The man who farms as his forefathers did cannot produce much
food no matter how rich the land or how hard he works.
— Theodore W. Shultz[1]

This Chapter

1. Discusses the role of public and private agricultural research in
   generating improved technologies and institutions and the effects
   of those technologies on income growth and distribution and on
   food security
2. Describes the major types of agricultural research, and factors influencing the transfer of research results from one country to another
3. Examines the role of technology and information transfer mechanisms such as agricultural extension.

THE ROLE OF AGRICULTURAL RESEARCH

A major determinant of growth in agricultural production is the effectiveness of agricultural research. Through research, the productivity of
existing resources is increased, new higher-productivity inputs and ways
of producing food are developed, and new or improved institutional
arrangements are designed. Examples of research outputs include
higher-yielding plant varieties, better methods for controlling insects
and diseases, increased knowledge about methods for manipulating
plant or animal genes, and designs for improved agricultural policies.
Research creates the potential for increased agricultural production,
moderated food prices, increased foreign exchange, reduced pressure
on the natural resource base, and many other positive results. Let us
consider in more detail the nature of these effects and the possibilities
for negative as well as positive outcomes.

---

[1] Theodore W. Schultz, *Transforming Traditional Agriculture* (Chicago, Ill.: University of
Chicago Press, 1964), ch. 1, p. 3.

Over time, agricultural research has been associated with improvements in incomes and reductions in poverty. It is estimated that without the productivity improvements generated through agricultural research, an additional 350 million hectares of land, about the size of India, would have been needed to feed the world's population growth since 1960. Productivity gains have thus saved highly erosive fragile soils, reduced deforestation and helped preserve biodiversity.[2] Specific research successes include a new African rice variety that is more productive and better suited to harsh environmental conditions, cassava varieties that are resistant to cassava mosaic virus and raise yields by 10 tons per hectare, and enhanced strains of tilapia fish that grow 60 percent faster than traditional strains.[3] Despite consistent evidence of high rates of return to agricultural research, pressures to reduce funding for it are frequent.

Impacts on Agricultural Productivity

Productivity increases generated through agricultural research imply a shifting upward of agricultural production functions. The simple example of increasing the output per unit of an input, say, fertilizer, is illustrated in Figure 12-1. If a more responsive seed variety is made available through research, output produced per kilo of fertilizer may increase. The research which produced that higher-quality seed may be either public or private or both. Public research is conducted in national research institutions, public universities, or government-sponsored research in private entities. Private research is financed by private companies.

Research and subsequent technical change in agriculture raises returns to producers. The value of agricultural production added per worker is shown in Table 12-1 for India, China, Indonesia, Nigeria, and Brazil (five of the more populous countries of the world) from 1980 to 2003. Despite rapid population growth, which might be expected to push production on to more marginal agricultural lands, agricultural productivity per worker rose substantially in each of these countries, more than doubling in China and almost tripling in Brazil. This same pattern is found in most other developing countries, although output per capita for the total population has declined in several Sub-Saharan Africa countries where rapid population growth has outpaced slow productivity improvements.

---

[2] See CGIAR Science Council, *Science for Agricultural Development: Changing Contexts and New Opportunities* (Rome, Italy: Science Council Secretariat, 2005).
[3] See CGIAR Science Council, *Science for Agricultural Development*, for details.

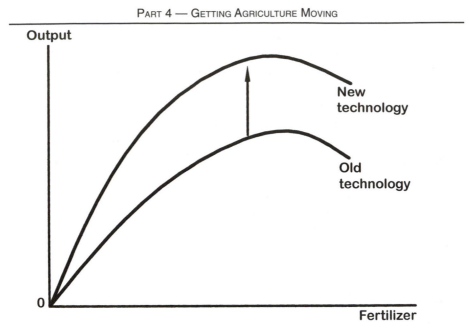

Figure 12-1. The effect of research on input productivity. New technologies generated through research can shift the production response function upward.

The examples shown in Figure 12-1 and Table 12-1 are oversimplified in the sense that most new technologies require different mixes of inputs; not all other inputs are held constant. Measurement of total productivity gains due to research requires netting out the cost of any additional inputs employed with the improved technologies. The resulting total net-cost reduction per unit of output produced can then be used to summarize the total productivity effect. This total productivity effect is illustrated in Figure 12-2. New or improved technology shifts the original commodity supply curve (S_1) downward to S_2 because the supply curve is a marginal cost curve and the new technology has

Table 12-1. Agricultural Value-added Per Worker (2000 dollars)

Country	1980	1990	2000	2003
India	273	348	389	406
China	163	245	346	349
Indonesia	425	480	522	574
Nigeria	491	578	774	871
Brazil	1,113	1,628	2,585	3,227

Source: World Bank, *World Development Indicators Online Database*.

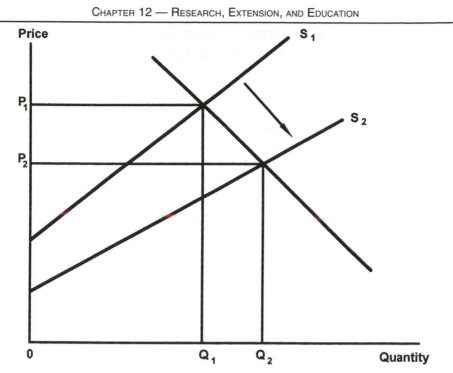

Figure 12-2. Effect of research on supply. Agricultural research reduces the cost per unit of output, thereby causing the supply curve to shift down to the right.

reduced the cost of production. The new lower cost of production per unit of output means that more output is produced at a lower price. This lower price is good for consumers of the product, but producers might be hurt.

Many studies have been conducted to estimate the economic returns to society from public research investments aimed at achieving these productivity increases. A recent study found more than 1,700 distinct estimates of the returns to various research programs around the world.[4] A summary of their results is presented in Table 12-2. Individual programs and projects vary widely in their estimated returns, but on the whole agricultural research has been a highly profitable investment for the societies that undertake it. Both mean and median annual rates of return are well above government cost of capital, or earnings on alternative investments. National leaders have a responsibility to invest scarce public resources in activities that yield high returns.

[4] See Julian M. Alston, Connie Chan-Kang, Michele C. Marra, Philip G. Pardey, and T. J. Wyatt, "A Meta-analysis of Rates of Return to Agricultural R&D," *IFPRI Research Report 113* (Washington, D.C.: IFPRI, 2000).

Table 12-2. Rates of Return to Agricultural Research by Commodity Orientation

Commodity orientation	Number of estimates (count)	Mean rate of return (percent)	Median rate of return (percent)
Multicommodity[a]	436	80	47
All agriculture	342	76	44
Crops and livestock	80	106	59
Unspecified[b]	14	42	36
Field crops[c]	916	74	44
Maize	170	134	47
Wheat	155	50	40
Rice	81	75	51
Livestock[d]	233	121	53
Tree crops[e]	108	88	33
Resources[f]	78	38	17
Forestry	60	42	14
All studies	1,772	81	44

a Includes research identified as "all agriculture" or "crops and livestock," as well as "unspecified."

b Includes estimates that did not explicitly identify the commodity focus of the research.

c Includes all crops, barley, beans, cassava, sugar cane, groundnuts, maize, millet, other crops, pigeon pea, chickpea, potato, rice, sesame, sorghum, and wheat.

d Includes beef, swine, poultry, sheep, goats, all livestock, dairy, other livestock, and pasture.

e Includes "other tree" and "fruit and nuts."

f Includes fishery and forestry.

Source: Julian M. Alston, Connie Chan-Kang, Michele C. Marra, Philip G. Pardey, and T.J. Wyatt, *A Meta-analysis of Rates of Return to Agricultural R&D*, IFPRI Research Report No. 113 (Washington, D.C.: IFPRI, 2000), Table 15, p. 58.

Increased agricultural productivity not only creates the potential for higher real incomes to producers through lower costs and to consumers through lower food prices, but can also help a country's agriculture become more competitive in world markets. Efficiency gained through higher agricultural productivity can be turned into foreign exchange earnings or savings as a result of additional exports or reduced imports.

The fact that agricultural research has yielded high returns in many countries in the past does not imply that these returns are guaranteed for all research systems or types of research. Each country must carefully consider the appropriate type of research organization and portfolio of activities, given its resource base and special needs (see Box 12-1 for an example of a research portfolio). This issue is discussed in more detail below.

Distributional and Nutritional Effects

Agricultural producers at different income levels, with different farm sizes, in different locations, and with different land tenure arrangements can gain or lose as a result of new technologies and institutional changes generated through research. These gains and losses depend on market conditions, among other factors. Consumers are major beneficiaries of agricultural research due to falling product prices, but the benefits they receive vary as well by income level and are influenced by the nature of the research portfolio. Returns to land versus labor are also influenced by research. Nutritional implications follow from these differential producer, consumer, and factor-income effects.

Farm Size and Tenure: The issue of whether improved agricultural technologies benefit large farms more than they do small farms has been the subject of substantial debate. Farm size is not a major impediment to adoption of new biological technologies such as improved seeds, which are the major focus of developing country agricultural research. However, larger farms do tend to be among the first adopters of many new technologies, probably because it pays large farms more to invest in obtaining information about the technologies. Owners of large farms may have more formal education that helps them process the information, and a greater ability to absorb risk. Large farms often have better access to credit needed to purchase modern inputs. Most small farms in the same region as large farms do eventually adopt the technologies, but the first adopters typically receive greater income gains from them. Late adopters are often faced with lower producer prices because supplies shift outward as early adopters increase output. Of course even if all producers in a given region adopted a scale-neutral technology at the same time, absolute income differences would widen because the increased returns per hectare are spread over more hectares on larger farms.

As noted in Chapter 11, not all technologies and institutional changes are scale-neutral. For example, certain types of mechanical technologies can be used profitably on large but not small farms. With differences in transactions costs, large farmers may press research systems for research results suitable for them even if the country's resource base

BOX 12-1
MAJOR TYPES OF RESEARCH IN THE NATIONAL AGRICULTURAL RESEARCH INSTITUTION IN ECUADOR

The listing below of the major types of agricultural research activities in Ecuador provides an example of a typical applied research portfolio for a small developing country. Given its limited research budget, the country must decide which commodities to concentrate on and how much to emphasize each type of research.

1. **Plant breeding:** development of new lines and varieties that yield more and are resistant to insects and diseases; maintenance of a germplasm collection
2. **Cultural practices:** determination of optimal planting densities, improved harvesting methods
3. **Crop protection:** improved methods for control of insects, diseases, weeds, nematodes, including biological, cultural, and chemical methods
4. **Soils and fertilizers:** development of improved soil conservation methods, chemical analysis of soils including macro- and micro-element analysis, toxicity studies, economic analysis of soil conservation and fertilization practices
5. **Water management:** studies of water needs, improved irrigation methods, salinity control
6. **Mechanization:** design of improved agricultural implements
7. **Socioeconomics:** diagnosis of constraints to technology adoption, monitoring and evaluation of research, analysis of farm management practices and opportunities
8. **Technology validation:** on-farm transferring, testing, and validation of new technologies
9. **Seed production:** basic and registered seed production, technologies for seed production, improved vegetative propagation
10. **Post-harvest technologies:** improved methods for storage, drying, cleaning, packaging, and transporting agricultural products
11. **Agro-forestry:** improved systems of agro-forestry and of pasturing forests
12. **Animal improvement:** animal breeding, introduction and selection of animals from outside the country, adaptation of animals to different climates
13. **Animal health:** prevention and cure of diseases and external and internal parasites
14. **Animal nutrition:** improved forages, analysis of concentrates and other supplementary feeding programs, evaluation of nutritional deficiencies, nutritive value of feeds.

Source: Julio Palomino, Planning Director, National Agricultural Research Institution, Ecuador.

on average would dictate a different type of technology. Also, because many technologies are scale-neutral and some are biased toward large farms, it may be difficult to generate technologies biased toward small farms. All this implies that reducing transactions costs through improved information is important, but it also implies that research may not be the best policy tool for achieving distributional objectives.

Tenant farmers represent an important producer group in many countries. It is difficult to generalize about the effects of research on the incomes of tenants versus landlords. One might expect that improved biological technologies would make labor more productive and thus help tenants, but the distribution of income gains is influenced by other factors as well. If each landlord has several tenants, so that the average size of landlord holdings is greater than the average size of tenants' farms, then the average landlord would gain relative to the average tenant if each received equal shares of income gains per hectare.

Contractual arrangements influence the distribution of research benefits, and the arrangements may change as well as a result of new technologies.[5] If the tenant pays the landlord a fixed *share* of the output, the division of any income gains after adopting the new technology depends on the relative sharing of both output and production costs. But if the tenant pays a fixed *amount* to the landlord, the tenant can keep the income gains until the landlord raises the rent. Often, increases in land productivity are bid into land rents, and landowners are able to capture these rents by changing tenancy agreements.

Regional Disparities: Regional differences in resource endowments and basic infrastructure can influence the distribution of research gains among producers. In fact, interregional disparities in the net benefits from research tend to be larger than intraregional disparities. Data from India indicate that the new rice and wheat varieties which increased production so dramatically in that country in the late 1960s benefited primarily the more productive wheat and rice states. Productivity increased dramatically in the country's northern region. At the same time, during 1967 to 1976, the central and eastern regions actually had decreasing rice yields. These inter-regional yield differentials diminished over time, but the technologies clearly benefited certain regions more than others.[6] The introduction of modern crop varieties has exacerbated

---

[5] See Julian Alston, George W. Norton, and Philip G. Pardey, *Science Under Scarcity: Principles and Practice for Agricultural Research Evaluation and Priority Setting* (Ithaca, N.Y.: Cornell University Press, 1995), ch. 3.

[6] J. S. Sarma and Vasant P. Gandhi, *Production and Consumption of Foodgrains in India: Implications of Accelerated Economic Growth and Poverty Alleviation*, International Food Policy Research Institute Research Report No. 81 (Washington, D.C., 1990), pp. 17-34.

interregional disparities in many countries because those technologies have often required irrigation and greater use of farm chemicals. Producers in dryland areas and regions with poor infrastructure for transporting fertilizer have been disadvantaged. Broadening the scope of agricultural research and decentralizing the research structure should help reduce regional disparities, although rates of return on research aimed at more productive regions are consistently higher than those for marginal areas.

Producers and Consumers: The impacts of technological change on the distribution of income between producers and consumers depend to a large extent on the degree to which quantity demanded responds to price changes. If producers face an elastic demand for their output, increased supplies will place little downward pressure on prices, so producers rather than consumers capture most of the benefits of the innovation. Export crops, for example, tend to have relatively elastic demands, and thus new technologies for the production of these commodities tend to favor producers. Many commodities that are basic staples in the diet have relatively inelastic demands, as discussed in Chapter 3. The benefits of research on these commodities flow largely to consumers through lower prices.

The poor spend a higher proportion of their income on food, and so benefit more than others from any decline in food prices due to research-induced increases in food supplies. This benefit is received by both the urban and the rural poor. The rural poor are often landless laborers, who purchase food, or small owner-operators or tenants, who retain a large part of their output for home consumption. Scobie and Posada found in Colombia, for example, that while the lower 50 percent of Colombian households received about 15 percent of total national household income, they captured nearly 70 percent of the net benefits of the rice research program.[7] These benefits to consumers flow across regions, especially where adequate transportation exists, and dampen the interregional disparities to producers mentioned above.

Land, Labor, and Capital: New technologies allow the same output to be produced with fewer resources, thus freeing up those resources to be used elsewhere in the economy. The dual-economy model described in Chapter 6 illustrated the potential for labor released from agriculture to become a fundamental source of industrial growth. However, the effect of technical change on the demand for resources is influenced

---

[7] Grant M. Scobie and Rafael Posada T., "The Impact of Technical Change on Income Distribution: The Case of Rice in Colombia," *American Journal of Agricultural Economics*, vol. 60, no.1 (February 1978), pp. 85-92.

by the inherent nature of the technology and by the nature of product demand.

Some new technologies result in proportionate savings of all inputs, while others save labor and use land or vice versa. For example, a new machine to cultivate the land may save labor and require a farmer to use more land to justify the cost. A higher-yielding rice variety may require more labor but produce more per unit of land. If a technology is neutral with respect to its effect on land and labor use, and if the demand for the product is elastic, the demand for both land and labor may grow proportionately following adoption of the technology. The reason is that, with elastic demand, total revenue increases with a shift out in the supply curve, providing increased returns to all resources. On the other hand, if product demand is inelastic, a neutral technical change can reduce the demand for all inputs proportionately.

Most new technologies are biased toward the use of one resource or another. Many of the higher-yielding varieties that comprised the green revolution (see Box 12-2) require significantly more labor input per unit of land. As a result, strong poverty-reducing impacts of the green revolution were transmitted through labor markets. In countries where markets are highly competitive and input prices reflect true input scarcity, the induced-innovation model presented in Chapter 11 predicts that new technologies will be developed to save the relatively scarce resources. However, if input prices are distorted, externalities exist, or transactions costs are high, technical change will not necessarily be biased in a direction that saves the scarcest resources; this "inappropriate" bias will thus reduce the rate of overall agricultural growth below its potential.

Because so many factors influence the effect of new technologies on resource use, it is difficult to generalize about the effect of research on employment, on the long-run returns to land, and so on. One implication is that agricultural research is a relatively blunt instrument for implementing a policy of distributing income to particular resources.

Nutritional Implications: Agricultural research can influence human nutrition through several mechanisms. First, if new technologies are aimed at poor farmers, a high proportion of the resulting income streams will be spent on improving the diet. If the technologies are aimed at commodities produced and consumed at home, the effect will be direct. If the technologies affect export crops produced by small farms, the extra income may be spent on buying food from others. Even if the new technologies are suitable only for large farms producing export crops, the influence on nutrition of the poor may be positive if the

---

BOX 12-2
THE GREEN REVOLUTION

The term *green revolution* was coined in 1968 by William S. Gaud, former Administrator of the U.S. Agency for International Development, to describe the dramatic wheat harvests that had been achieved in 1966 to 1968 in India and Pakistan. The term gained further publicity in 1970 when Norman Borlaug was awarded the Nobel Peace Prize for his research that produced the high-yielding, semi-dwarf Mexican wheats that had performed so well in Asia and Latin America. At the same time that the semi-dwarf wheats were making their dramatic entry, IRRI released new semi-dwarf rice with the same dramatic effect.

The big innovation of the green revolution was developing varieties of wheat and rice that would not fall down (lodge) when nitrogen fertilizers were applied. These new lines of plants also tended to be earlier maturing, to produce many shoots (tillers), and to be less sensitive to day length.

Source: Donald L. Plucknett, "Saving Lives through Agricultural Research," Consultative Group on International Agricultural Research, Issues in Agriculture, No. 1 (Washington, D.C., May 1991), pp. 9-10.

---

demand for labor increases. However, this employment effect is not at all certain and depends on the factor biases discussed above.

An important nutritional effect of research comes from the increased availability of food at lower prices. As supply shifts out against a downward-sloping demand curve, all consumers benefit from lower food prices that improve their real wages.

Research can be used to reduce fluctuations in food supply, prices, and income and thereby alter nutrition. Some of the severest malnutrition occurs in rural areas during years of low incomes due to lower than normal production. Research on drought-tolerant varieties can help reduce production fluctuations and help lower malnutrition.

It is difficult to draw conclusions about the nutritional implications of a particular portfolio of research activities because the sources of nutritional impacts identified above can act counter to one another. For example, a labor-saving technology used to produce export crops might lower wages and not induce changes in food supply, thus making landless laborers worse off. Some concern has been voiced about the nutrition effects of research devoted to export-crop production. If numerous producers switch from food crops to export crops, then there is potential for domestic food prices to rise, and such a rise would hurt the urban and landless poor. However, there is little empirical evidence of this switch, and nutritional levels are perhaps most influenced by research that generates the largest income gains, particularly if those gains are realized by low-income producers. Therefore, focusing research disproportionately on commodities with high nutritional content may result in less income than if the research were focused on other commodities. For example, improving the productivity of a vegetable export crop in Guatemala may improve the family's nutrition more than improving the productivity of its maize crop, because the former will lead to a greater increase in farm income and therefore the family's ability to buy food.

Environmental Effects of Research

Concerns over environmental degradation in developing countries were discussed in Chapter 9. Deforestation, soil erosion, desertification, pesticide pollution, and so on., have become serious problems in many countries, and research can play a significant role in their solution.

First, new technologies for mitigating soil erosion, providing alternative energy sources, and substituting for chemical pesticides can be generated through research. Second, research can be used to design improved government policies that provide increased incentives to adopt management practices and help sustain the integrity of the natural resource base. Third, the higher incomes generated through research-induced productivity increases will put downward pressure on population growth in the long run. Fourth, higher income streams will also reduce the pressures to abuse the environment in the short run just to obtain food and fuel. Finally, income growth will create more demand for environmental quality. Thus agricultural research is critically important for encouraging environmentally sound and sustainable agricultural growth.

Research organizations have been criticized in the past for devoting too many resources to research related to modern inputs such as fertilizer, pesticides, and irrigation. Excessive and improper use of these inputs can cause environmental damage. An additional criticism has been that too little research is aimed at resource-conserving technologies, such as integrated pest management and methods for reducing soil erosion. There is some truth in these claims, although research on sustainable farming practices has accelerated (see Box 12-3). Also, market failures tend to cause an undervaluation of environmental services, as discussed in Chapter 9. Because of this undervaluation, producers and consumers often do not demand resource-conserving technologies. In the long run, one of the best ways to combat forces leading to environmental degradation is to raise incomes and reduce poverty. Research can be an effective means of raising incomes, though in the

---

BOX 12-3
RESEARCH AND THE ENVIRONMENT:
THE CASE OF THE CASSAVA MEALYBUG

The cassava mealybug was accidentally introduced from Latin America into Africa in the early 1970s and soon began causing severe damage to cassava crops. Because some 200 million Africans depend on cassava as a staple food, this damage became a deep concern.

Researchers at the International Institute of Tropical Agriculture (IITA) in Africa, in collaboration with those at the International Center for Tropical Agriculture (CIAT) in Latin America, found a means of biological control. Importation and distribution of the parasitic wasp *Epidinocasis lopez*, a natural enemy of the mealybug from Latin America, has led to dramatic reductions in African mealybug populations with biological methods. No extensive pesticides are required, and the small-scale African farmers are freed from a damaging pest by nature itself.

Source: John Walsh, *Preserving the Options: Food Productivity and Sustainability*, Consultative Group for International Agricultural Research, Issues in Agriculture No. 2 (Washington, D.C., 1991), pp. 7–8.

---

short run, more agricultural research should, perhaps, be aimed at conserving environmental resources.

Other Research Issues

Institutional Change: Much agricultural research results in new or improved technologies that are embodied in inputs or methods of production. However, agricultural research can be directed toward the design of new or improved policies or institutional changes. In other words, agricultural research can help lower the cost of adjusting institutions to the changing physical, natural resource, economic, and biological environments. A static or distorted institutional environment can be as great a hindrance to agricultural development as can a static technology base.

Credit policies, marketing and pricing policies, land tenure rules, and natural resource policies are examples of institutional arrangements that can be improved through research. Institutional changes that improve the flow of market information and reduce externalities are particularly important.

Public versus Private Sector Research: Just because agricultural research is important to development does not imply that the public sector must carry it out. Typically, the public sector is heavily involved in agricultural research in both developed and developing countries, but the private sector is heavily involved in many countries, and increasingly so.

Why does the private sector not provide all the needed research? There are three basic reasons. First, individual farms are too small to do all their own research, although they often cooperate with public research institutions and certainly do a great deal of experimenting. Second, and most important, for many types of research it is difficult for one firm to exclude other firms from capturing the benefits from the research; in other words, a firm may incur substantial costs in conducting research but, once the research is completed, other firms can make use of the results without incurring much cost. Thus, the firm has little incentive to do the research in the first place. Third, many types of research are highly risky, so that many firms are hesitant to take the risk for fear of incurring a substantial loss.

Certain types of research, particularly applied research related to mechanical and chemical innovations, are less risky and potentially patentable and thus attract sizable private research activity. Some types of biological and soils research, on the other hand, have historically been more difficult to patent and have thus been primarily conducted in the public sector. However, the patentability of biological research has increased in recent years and has played a major role in the development of new genetically modified crops and animals. As a country develops, the research role of the private sector typically increases in developing and marketing improved seeds as well as in mechanical and chemical innovations. However, there is often a time lag between the development of public sector research and the establishment of substantial private sector research activity. One action that a country can take to promote private research is to establish enforceable property rights (e.g., patents, licenses) over research results, not just for mechanical and chemical technologies but for biological technologies as well.

Intellectual Property Rights: Intellectual property rights (IPRs) refer to legal protections, granted for a defined period of time, to scientific, technological, and artistic inventions. Copyrights, trademarks, patents, plant breeders' rights, and trade secret laws are some of the ways that intellectual property rights are granted. Legal systems differ by country and hence the types, extent, and duration of rights granted vary as well. Patents and plant breeders' rights are the most important forms of intellectual property protection for agricultural research results and technologies. Over time, copyrights are becoming more important as well because the databases that contain information about plant genes can often be copyrighted.

Patents are the strongest type of intellectual property, as the patent holder can exclude all others from making, using, selling, or offering to sell the invention in the country while the patent is in force (unless others purchase a license to use it). To be patentable, an invention must be new, useful, not obvious, and be disclosed so that others can pay a license to use and replicate it. Plant Breeders' Rights (PBRs) grant protection to crop varieties that are new, distinct, uniform, and stable. Patents and PBRs give a monopoly on commercializing the invention or variety for a defined period of time, which allows the inventor or breeder to recover their costs. This protection therefore gives them incentives to invent or breed that they otherwise would not have.

Many developing countries are still in the process of developing and implementing an intellectual property protection system for plants and animals. Details of IPR systems vary from country to country, but those who lag behind run the danger that private firms and individuals will be reluctant to develop or sell products with new technologies embedded in them in their countries. Developing countries have grown fearful that as more and more technologies (including genes) are covered by intellectual property rights, their people and firms will be discouraged from using the technologies and resulting products because of the high costs of licensing the technologies or paying for the higher-cost products. This issue has been a topic of discussion and action in multilateral trade negotiations since the early 1990s, and is discussed more in Chapter 17.

NATURE, ORGANIZATION, AND TRANSFER OF RESEARCH

Some research is very "applied" and yields immediate practical results. Other research is more "basic" or fundamental and may not yield results for many years. Research systems themselves are organized in a variety of different ways. Let us consider the major categories of agricultural research and organizational arrangements.

Categories of Agricultural Research

Agricultural research can be categorized into basic research, applied research, adaptive research, and testing. *Basic* research develops knowledge with little or no specific use in mind. Studies of evolution, genetics, biochemical processes, and so on, may discover fundamental principles of substantial significance to more applied researchers, but the specific end use of the research results are often difficult to identify prior to the research. Most basic research is carried out in developed countries or in the largest of the developing countries.

Applied agricultural research is aimed at solving particular biological, chemical, physical, or social science problems affecting one or more countries or areas in a state or region. Development of new plant varieties, methods for controlling specific insects and diseases in plants or animals, and animal nutrition research are examples of applied research. Applied research may take place at international research centers or in national research systems.

Adaptive research takes the results of applied research and modifies or adapts them to local conditions within a country or region. A plant variety developed for a broad area may need to be modified for a specific microclimate. Fertilizer recommendations, methods for controlling soil erosion, and many other technologies require adaptation to the local setting. Most of this research takes place on local experiment stations or on farms.

Testing research is conducted on local experiment stations or on farms to assess whether research results from other locations are suitable for solving local problems. Improved pesticides, management practices, or plant varieties are examples of research results that may be tested. All countries conduct some testing research, but for very small countries with limited resources, testing may represent a large portion of total research. Much testing is conducted by farmers themselves.

These categories of research are linked and dependent on each other. A research center may be involved in several categories.

Biotechnology

Much applied and adaptive agricultural research involves what has been called *biotechnology* research. *Traditional biotechnology* research includes well-established techniques in plant breeding, biological control of pests, conventional animal vaccine development, and many other types of research. *Modern biotechnology research* includes use of recombinant DNA, monoclonal antibodies, and novel bio-processing techniques, among others.

Modern biotechnology provides new tools and strategies for increasing agricultural production. Tools for improving agricultural output range from novel approaches to cell and tissue culture to the genetic manipulation of biological material. Modern biotechnology is based on several new technologies. One of them, recombinant DNA, often called *genetic engineering*, enables the essential genetic material in cells, DNA, to be manipulated. It offers the possibility of transferring genetic material from one species to another, thereby transferring a useful genetic trait.

Another biotechnology technique, monoclonal antibodies, is used to detect individual proteins produced by cells, thereby providing a method for rapid and specific diagnosis of animal and plant diseases. A third, novel bio-processing technique involves new cell and tissue

Ecuadorian scientists recording disease data in an applied pest management experiment on plantain.

culture technologies that enable rapid propagation of living cells. These techniques provide improved methods for large-scale production of useful compounds by the microbial or enzymatic degradation of various substrates.

The types of products that modern biotechnology can potentially produce include new plant varieties, new animal breeds, plant and animal growth hormones, bio-pesticides, bio-fertilizers, diagnostic reagents for plant and animal diseases, and enzymes and food additives. They may improve the tolerance of plants and animals to particular pests and stresses such as drought, and increase the efficiency with which plants and livestock utilize nutrients, or increase a plant's nutritional quality (for example, rice with increased vitamin A or iron). They may reduce the need for agrichemicals.

Modern biotechnologies are on the cutting edge of science, and if developing countries are to be successful in developing their own modern biotechnologies, or adapting technologies produced elsewhere, they will need scientists trained in microbiology and biochemistry. These countries will need to integrate modern biotechnology into traditional biotechnology research programs. They will need to put in place or refine bio-safety rules, and resolve rules on intellectual property rights, as the rights to many aspects of the technologies are patented and owned by private companies.

Benefits and Costs of Modern Biotechnology: Concerns have been expressed by some about the health and environmental safety of modern biotechnologies, especially for the technologies that involve transferring genes across species. All technologies involve some risks, and each country needs to develop and implement a regulatory system that allows it to test and monitor the safety of its new agricultural technologies. The risks associated with modern biotechnologies are thought to be relatively low, and many of the technologies can have positive effects on health and the environment by reducing pesticide use. However, there can be no certainty that adverse health and environmental effects will not occur. Consumers, especially in Europe, have expressed strong reservations about consuming foods produced with the use of biotechnology. One concern has been the possibility that genes, say, from a herbicide-tolerant crop, might transfer through pollen to another species, creating perhaps a super-weed that would be difficult to control with a herbicide. Gene transfer across species does occur frequently, although odds of creating a super-weed are slim. Another concern is whether people may have an allergy to a transgenic crop. A third concern is whether people should be attempting to alter nature, although many types of agricultural research in addition to genetically modified organisms could be subject to this same concern.

Economic concerns have also been raised with respect to whether a few companies might end up controlling many of the intellectual property rights associated with the genes and with the transformation processes, thereby gaining some monopoly power. If they did gain such power, they might charge farmers a high price for their seeds. Countries, however, can regulate companies and have their public research systems enter into joint ventures with the private sector to ensure freedom of access to seeds at reasonable prices.

Biotechnologies are just one of many potential means of increasing agricultural productivity in developing countries, and perhaps the greatest danger is that developing countries might forgo the food and income growth that the technologies may afford them.[8] As of 2005, approximately 34 million hectares of transgenic crops were being grown in developing countries, evidence that use of biotechnologies is wide-

---

[8] See Per Pinstrup-Andersen and Ebbie Schioler, *Seeds of Contention* (Baltimore, Md.: Johns Hopkins University Press, 2000) for a more detailed discussion of issues surrounding biotechnologies and developing countries.

spread. Several studies have documented potentially large economic benefits that would accrue from increased use of biotechnologies in developing countries.[9] These benefits would be realized by consumers through lower food prices and through increased nutritional quality. Producers might also gain through lower costs of production. If developing countries do not pursue agricultural biotechnologies, they may be placed at a competitive disadvantage compared to countries that do pursue them.

Organization of Agricultural Research

Public agricultural research systems in developing countries have a variety of organizational structures. Often there is a central station with several substations located in different geo-climatic zones. Research may be conducted at universities, but the proportion of agricultural research conducted at colleges and universities tends to be much less than in developed countries such as the United States.

The structure of the research system is influenced by historical forces including, among others, colonial history and major foreign assistance projects. Much agricultural research in developing countries is organized along commodity program lines: for example, a maize program, a rice program, a wheat program, or a sheep and goats program. Other cross-cutting research areas such as soil fertility, socioeconomics, and even plant or livestock protection, may have separate programs.

Some agricultural research systems have a mandate for extension or other programs designed to reach out to farmers. Even if extension is not included in the mandate of the national research institution, that institution still needs a mechanism to obtain information on the current problems facing farmers and for testing new technologies under actual farm conditions. This mechanism may involve on-farm research.

Each research system must determine the appropriate mix of on-farm and experiment station research. Experiment station research is needed so that experiments can be run under controlled conditions that enable particular components of new technologies to be developed and tested without the confounding of numerous and possibly extraneous factors. However, the real-world robustness, profitability, and cultural acceptability of new technologies cannot be assessed without testing under actual farm conditions. Frequent contact between scientists and farmers increases the likelihood that constraints and problems facing farmers will be included in the development and evaluation of new

---

[9] See, for example, Guy Hareau, George Norton, Bradford Mills, and Everett Peterson, "Potential Benefits of Transgenic Rice in Asia: a General Equilibrium Analysis," *Quarterly Journal of International Agriculture*, vol. 44 (2005), pp. 229–46.

technologies. Because extensive on-farm interaction is expensive and scientific resources are scarce in developing countries, each research system assesses at the margin the appropriate mix of on-farm or on-station research.

International Agricultural Research Centers: The 1960s saw the emergence of a set of international agricultural research centers (IARCs) that by 2005 had grown to a network of 15 institutions located primarily in Africa, Asia, Latin America, and the Middle East, as shown in Figure 12-3. The funding and operation of these "Future Harvest" centers is coordinated through the Consultative Group for International Agricultural Research (CGIAR), headquartered at the World Bank in Washington, D.C. Although the first center, The International Rice Research Institute (IRRI), was founded in 1960, the international center model drew on the historical experiences of the colonial agricultural research institutes that were effective in increasing the production of export crops such as rubber, sugar, and tea. The model also drew on the experiences in the 1940s and 1950s of the Rockefeller Foundation's wheat and maize programs in Mexico and the Ford and Rockefeller foundations' rice program in the Philippines. The results of the research and training programs of the centers are aimed not just at the country where the center is located, but at the neighboring region or even the world.

The first IARCs, IRRI and CIMMYT (International Center for Maize and Wheat Improvement), produced new varieties of rice and wheat that substantially increased yields, especially for rice in Asia. The first of several rice varieties (IR-8), released by IRRI and cooperating national programs, responded to high rates of fertilizer and water application by producing more grain and less straw. Subsequent research has focused as well on improving grain quality, incorporating disease and insect resistance, and developing varieties for drier upland areas. The substantial yield boost experienced in parts of Asia in the late 1960s resulting from these new technologies was termed the *green revolution* (Box 12-2).

The success of the green revolution in increasing yields and incomes in many areas led to the expansion of the international agricultural research center concept to the other commodities and regions identified in Figure 12-3. Maize, millet, tropical legumes, cassava, livestock, potatoes, and many other commodities have received emphasis. The research results from these newer centers have not been as spectacular as the early gains in rice and wheat, but these centers too have made significant contributions. For example, disease-resistant beans, cassava, and millet varieties are now being grown in several countries. These centers also provide a public link between research being under-

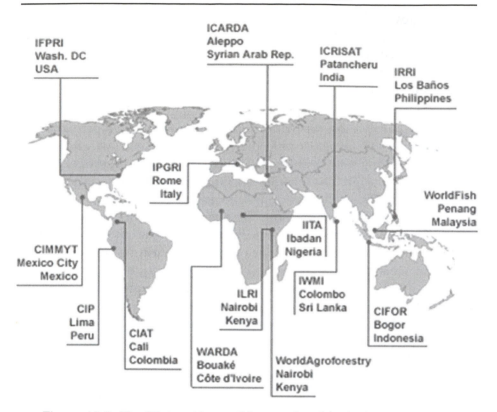

IFPRI
Wash. DC
USA

ICARDA
Aleppo
Syrian Arab Rep.

ICRISAT
Patancheru
India

IRRI
Los Baños
Philippines

IPGRI
Rome
Italy

WorldFish
Penang
Malaysia

CIMMYT
Mexico City
Mexico

IITA
Ibadan
Nigeria

CIP
Lima
Peru

CIAT
Cali
Colombia

IWMI
Colombo
Sri Lanka

ILRI
Nairobi
Kenya

CIFOR
Bogor
Indonesia

WARDA
Bouaké
Côte d'Ivoire

WorldAgroforestry
Nairobi
Kenya

Figure 12-3. The "Future Harvest" International Agricultural Research
Centers (see facing page).

taken in the private sector on modern biotechnology and national agricultural research systems in developing countries, to help ensure that these national systems are not left behind.

To some extent, the dramatic breakthroughs in yields in the early years of the green revolution created unrealistic expectations that these gains would be repeated with regularity. Agricultural research is, in fact, a continuous process that generally produces small gains from year to year.

The overall program and core funding for all 15 centers is managed by an organization called the Consultative Group for International Agricultural Research (CGIAR) whose members include the World Bank, the Food and Agricultural Organization of the United Nations (FAO), the United Nations Development Program (UNDP), and several national governments, regional banks, and foundations. These institutions provide the funds for the centers. The CGIAR, founded in 1971, is centered at the World Bank. The total budget for the 15 Future Harvest

Center — see Figure 12-3	Research Coverage
CIAT — Centro Internacional de Agricultura Tropical	Phaseolus beans, cassava, rice, tropical pastures
CIFOR – Center for International Forestry Research	Forest systems and forestry
CIMMYT – Centro Internacional de Mejoramiento de Maiz y Trigo	Wheat, barley, maize, high-altitude sorghum
CIP – Centro Internacional de la Papa	sweet potato, other root crops
ICARDA – International Center for Agricultural Research in Dryland Areas	Crop and mixed farming systems research, with emphasis on sheep, wheat, barley, broad beans
ICRAF – International Centre for Research in Agroforestry	Agroforestry
ICRISAT – International Crops Research Institute for the Semi-Arid Tropics	Sorghum, pearl millet, pigeon pea, chickpea, groundnuts
IFPRI – International Food Policy Research Institute	Food policy
IITA – International Institute for Tropical Agriculture	Farming systems: cereals, grain legumes, roots and tubers
ILRI – International Livestock Research Institute	Livestock diseases and production systems
IPGRI – International Plant Genetic Resources Institute	Conservation of plant genetic material; bananas and plantain
IRRI – International Rice Research Institute	Rice
IWWI – International Water Management Institute	Irrigation
WARDA – West Africa Rice Development Association	Rice and rice-based cropping systems in West Africa
World Fish Center	Fisheries and other living aquatic resources

Centers is now roughly $425 million when all funding sources are considered. In addition to these centers, there are a few related international research centers that play a similar role, such as the World Vegetable Center in Taiwan and the International Center for Insect Physiology and Ecology in Kenya.

Transfer of Research Results

The discussions of research categories, of national and regional experiment stations and on-farm research, and of international agricultural research centers all imply that research results may be transferred from one location to another. These transfers can occur internally in a country or across national boundaries. Let us examine the possibility and advisability of transferring new technologies or institutions.

Prior to the 1960s, little attention was focused on the importance of indigenous agricultural research in developing countries. It was thought that the possibilities for transferring technologies from developed countries were substantial and that, therefore, extension programs were needed to assist in this transfer. The relative lack of success with direct transfer of machinery, plant varieties, and other materials from developed to developing countries led to the realization that improved developing-country research capacity was essential. The desire to improve location-specific research was one of the driving forces behind the development of the IARCs mentioned above. However, many research results are regularly transferred from one country to another. What types of research results are transferable and what determines their transferability?

Materials such as improved seeds, plants, and animals; scientific methods, formulas, and designs; genes; and basic research output are all potentially transferable to some extent.[10] Each country must decide whether to simply screen these items and attempt to directly transfer them, to screen them and then modify and adapt them to their own environment, or to undertake a research program that is comprehensive enough to produce its own technologies.[11]

The choice among these transfer and research options will depend first on the relative costs of direct transfer of technology and of adaptive and comprehensive research. Transfer of research results involves

[10] See Yujiro Hayami and Vernon W. Ruttan, *Agricultural Development: An International Perspective* (Baltimore, Md.: Johns Hopkins University Press, 1985), pp. 260–2.

[11] See Robert E. Evenson and Hans P. Binswanger, "Technology Transfer and Research Resource Allocation," in *Induced Innovation: Technology, Institutions, and Development*, ed.Hans P. Binswanger and Vernon W. Ruttan (Baltimore, Md.: Johns Hopkins University Press, 1978), ch. 6. This section draws heavily on the ideas in Evenson and Binswanger.

On-farm potato variety trial of the International Potato Center (CIP).

costs of information and screening or testing. There may also be license costs or fees for patented items. Most of these transfer costs increase with the physical size and environmental diversity of the country. A country's own research costs are somewhat independent of size; for that reason, it may be more cost-effective for larger countries to conduct their own research than for smaller countries.

Second, the complementarity between screening transferred technologies and conducting in-country research can come into play. It takes some scientific capacity just to bring in and screen research results from outside the country. Therefore, it may be cost-effective to have the local scientists do some of their own adaptive research.

Third, if the natural resource base in one developing country is similar to that in another country where the new technology is produced, then the chances of transfer will increase. New wheat varieties, for example, are often transferred from Argentina to Uruguay because those countries have similar wheat-growing regions. These similarities tend to reduce the cost of transfer and to increase the likelihood that the transferred technology will be physically and economically viable.

Fourth, some technologies are more environmentally sensitive than others are. For example, new plant and animal materials may be more environmentally sensitive than more basic research results, formulas, designs, and so on. The International Agricultural Research Centers attempt to produce plant and animal materials that have broad environmental suitability. In many cases, it is necessary for the receiving country to then adapt these materials more specifically to its microclimates. Relatively basic advances in modern biotechnology have the potential for widespread applicability if developing countries create the scientific and institutional capacities needed for them to effectively utilize the research results.

Fifth, the availability of research results to transfer in is also important. For example, if a country has low labor costs and high capital and land costs, yet the technologies available to transfer in are large machines suitable for a resource environment with high labor costs and abundant land, then the country will not find the outside technology suitable.

In summary, a developing country must assess several factors in deciding whether to transfer in research results from another country or from an international center. Agricultural research is a long-term investment. Research takes time, adoption of new technologies takes time, and research results eventually depreciate as insects and diseases evolve, the economic environment changes, and so on. Developing countries often attempt to bring in research results from other countries during the early stages of development in order to shorten this process and meet critical needs. Perhaps a 1 percent productivity growth rate can be accomplished through a relatively simple transfer process, though such productivity will depend on the conditions previously mentioned.[12] However, the requirements of modern rates of growth in food demand, often in the 3 to 6 percentage range, require the coexistence of at least some indigenous agricultural research capacity, and this capacity may be a combination of public and private.

Agricultural development today requires a research system with internal and external linkages that bring in appropriate technologies; screen, adapt, and produce new technologies and institutions; and perform both on-station and on-farm testing. The major components of such a research system are illustrated diagrammatically in Figure 12-4. National and local experiment stations must interact with on-farm research and extension. This national research system must also maintain ties with the international research centers. Research in the larger national systems feeds into both the international centers and the smaller national research systems. If any of these linkages is weak or missing, agricultural productivity growth will be slowed.

---

[12] Hayami and Ruttan, *Agricultural Development*, p. 260.

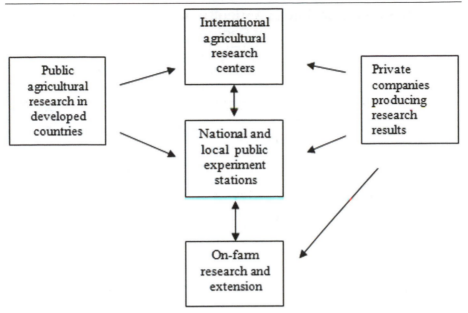

Figure 12-4. Components of a well-linked agricultural research system for developing countries.

Agricultural Research Spending

Spending on agricultural research occurs in both public and private sectors. In the developing countries, the public sector undertakes more than 90 percent of agricultural research, with the bulk of that research occurring in the Asia and Pacific region (Table 12-3). In 2000, more than $23 billion was spent in developed and developing countries on agricultural research, with roughly one-third ($7.5 billion) of the those expenditures occurring in the developing countries in the Asia and Pacific region. In contrast, only a little over 6 percent of the total was spent in Africa, and the percentage spent there has been declining since 1980. Total spending on agricultural research in Africa has increased only about 1 percent per year over the past several years despite increased malnutrition in the region.[13]

Agricultural research has become increasingly concentrated in a handful of countries worldwide.[14] The United States, Japan, France, and Germany accounted for more than two-thirds of the agricultural research in developed countries in 2000, while China, India, Brazil, Thailand, and South Africa accounted for more than half of the developing country total. These numbers place increased importance on the technology transfer issues raised above.

[13] CGIAR Science Council, *Science for Agricultural Development.*
[14] Ibid.

239

Table 12-3. Total Public Agricultural Research Spending by Region

	Agricultural R&D spending (millions 2000 dollars)		Shares in global total (percentages)	
	1981	2000	1981	2000
Asia and Pacific (28 countries)	3,047	7,523	20.0	32.7
Latin America and Caribbean (27 countries)	1,897	2,454	12.5	10.8
Sub-Saharan Africa (44 countries)	1,196	1,461	7.9	6.3
West Asia and North Africa (18 countries)	764	1,382	5.0	6.0
Subtotal developing countries (117 countries)	6,904	12,819	45.4	55.8
Subtotal high-income countries (22 countries)	8,293	10,191	54.6	44.2
Total (139 countries)	15,197	23,010	100.0	100.0

Source: CGIAR Science Council, *Science for Agricultural Development; Changing Contexts and New Opportunities* (Rome, Italy: Science Council Secretariat, 2005).

ROLE OF EXTENSION EDUCATION

Countries unable to develop the skills and knowledge of their farmers and their families find it difficult to develop anything else. The utilization of new technologies and institutions is critically dependent on a workforce that is aware of them and understands how to use them. Agricultural extension education can help motivate farmers toward change, teach farmers improved decision-making methods, and provide farmers with technical and practical information. Extension is complementary to other sources of information because it speeds up the transfer of knowledge about new agricultural technologies and other research results. It helps farmers deal with technological and economic change. Thus, as agriculture in a country moves from a traditional to a more dynamic, science-based mode, the value of extension education increases.

In extension education, farmers are the primary clientele and the programs are mostly oriented toward production problems they face. Extension accelerates the dissemination of research results to farmers and, in some cases, helps transmit farmers' problems back to researchers. Extension workers provide training for farmers on a variety of subjects and must have technical competence, economic competence, farming competence, and communication skills. Thus extension workers require extensive training and retraining to maintain their credibility with farmers.

Organization of Extension

Many types of organizational structures for extension exist in developing countries. For example, a highly structured approach encouraged by the World Bank and applied in numerous countries over several years is called the *training and visit (T&V) system*. The T&V system includes a single line of command, a set schedule of visits to farmers' groups, regular and continuous training of extension officers and workers by subject-matter specialists, and no non-extension responsibilities.[15] The T&V system facilitates discipline, accountability, and research linkages and has experienced apparent success in some countries, at least for a period of time.

Another, more common, extension structure is the village agent model that assigns extension "agents" to live in villages and provide one-on-one and group training of farmers on a variety of agricultural topics. An example that illustrates this structure is the extension service of the Colombian National Coffee Federation. That service has extension agents who operate on a fixed schedule of visits to farmers' groups, and who spend one day a week in an office receiving farmers and scheduling individual farm visits that coincide with days when they are visiting a particular location. Village extension agents are supported by regional subject matter specialists who also spend most of their time visiting farmers' groups and individual farms. Village extension agents are drawn from coffee farms and receive three years of training after high school. Extension programs are planned six months ahead and are developed jointly between the farmers' group and the agent.

Unfortunately, both the T&V and the village agent structures for extension have an uneven history of success. The T&V system suffers from its high cost and the disadvantage of not having local agents who can respond to farmer concerns as they arise. The village system suffers from the difficulty of funding a sufficient number of agents and adequately supporting them with supervision and training. Too often extension agents are poorly trained and little-motivated. They may not visit enough farms, and they become diverted into non-extension activities. The extension service itself may become politicized, corrupt, and unconnected to research. A well-functioning system needs clear lines of authority, adequate training, and financial rewards for personnel, or the

---

[15] See Daniel Benor, James A. Harrison, and Michael Baxter, *Agricultural Extension: The Training and Visit System* (Washington, D.C.: World Bank, 1989) for more details.

Extension field day in Peru.

system becomes relatively ineffective. Research and extension linkages are also essential and are facilitated if research and extension are housed in the same institution. Unfortunately, often they are not.

During the 1990s, many public extension services, especially in Latin America, were eliminated, or saw their funding cut drastically. These cuts were caused by a combination of budgetary pressures and perceptions that the extension services were not particularly effective. As a result of these and other inadequacies in publicly supported extension systems, non-governmental organizations (NGOs) are increasingly involved in agricultural extension. NGOs exist on outside financial support from governments and from private individuals and groups in other countries or on private local support. In addition, private firms that sell products such as improved seeds and chemicals are heavily involved in technology transfer associated with their specific products.

Extension Methods

A variety of methods are employed to transfer research knowledge and technologies to producers. Individual farm visits, regularly scheduled group meetings, technology demonstrations that may involve a field day when hundreds of farmers are invited to observe the latest research results, and transfer of information through mass media are just some of the methods. Some technologies are transferred more effectively through intensive methods such as regular meetings, others are amenable to transfer through less intensive (and usually less expensive) methods such as fields and mass media. Each country must decide what is most cost-effective for its public extension system, and each NGO will decide which approach allows it to best achieve its objectives. There is no one method that works best in every situation. Extension costs and effectiveness will depend on the type of technology, typography of the country, access to mass media, cultural and social factors, and many other variables.

SUMMARY

Agricultural research generates new or improved technologies and institutions that increase agricultural productivity, moderate food prices, generate foreign exchange, and reduce pressures on the natural resource base. Most studies have found that economic returns on public agricultural research investments are high. Agricultural research can have distributional effects by farm size and tenure, by region, by income level, by factor of production, and so forth. Consumers, particularly low-income consumers, are major beneficiaries of agricultural research, as the poor may spend 80 percent of any income increases on food, and food prices tend to fall as productivity increases. Agricultural research can influence nutrition by raising farm incomes, lowering food prices, and reducing the variability in food production. Agricultural research can generate technologies, institutional changes, and higher incomes that lead to reduced pressures on the environment. The public sector has a role to play in agricultural research because the private sector has inadequate incentives to conduct a sufficient amount of socially beneficial research, in part because often private firms conducting research cannot capture enough of the benefits. Intellectual property rights can help in creating incentives for private research investment.

Agricultural research can be classified into basic, applied, adaptive, and testing research. These categories are linked and dependent on each other. Research is conducted on national and local experiment stations and, to be effective, must contain an on-farm component. Since 1960, a system of International Agricultural Research Centers (IARCs) has provided new technologies and institutional changes suitable to several developing countries. These institutions helped to produce a green revolution that greatly increased the production of maize, rice, and wheat. Research can be transferred across national borders, but the ease of transfer depends on the type of research, the relative cost of transfer and indigenous research, the natural resource base, and other factors. Use of modern biotechnology has grown around the world in recent years but is yet to be widely adopted in developing countries.

Many types of extension systems exist, some more structured than others. Training for extension workers, incentives, clear lines of authority, and strong linkages to research are each critical for an effective extension service.

IMPORTANT TERMS AND CONCEPTS

Adaptive research
Agricultural education
Agricultural extension
Agricultural productivity
Agricultural research
Applied research
Basic research
Biotechnology
Experiment stations

Green Revolution
Intellectual property rights
International agricultural
   research centers
Scale-neutral technology
Technology transfer
Testing research
Training and Visit system

Looking Ahead

This chapter considered technical factors that can influence development of the agricultural sector. The following several chapters address a set of institutional issues that are equally important if agriculture is to progress in developing countries. We begin, in the next chapter, discussing land and labor policies.

QUESTIONS FOR DISCUSSION

1. What is the purpose of agricultural research in developing countries?
2. How does research influence agricultural productivity and food prices?
3. Under what conditions might research on a nonfood export crop have as much or greater positive effect on nutrition than research on a food crop?
4. Why might agricultural research tend to benefit large farms more than small farms?
5. Why might the type of lease influence the proportion of the benefits of technical change that go to the tenant compared to the landlord?
6. Why might agricultural research increase the regional disparity in income in a developing country?
7. Why are consumers, especially poor consumers, often the major beneficiaries of agricultural research?
8. What factors influence the returns to particular factors of production following research?

9. How might agricultural research help improve the environment?

10. How might research result in institutional change?

11. Why should the public sector get involved in research? Why not leave it to the private sector?

12. Distinguish among basic, applied, adaptive, and testing research.

13. What is modern biotechnology?

14. What are the International Agricultural Research Centers and how does their work tie into the agricultural research systems in developing countries?

15. What is the "green revolution," when did it occur, and where?

16. What role does extension play in agricultural development?

17. How might research, education, and extension be complementary activities?

18. How can intellectual property rights influence production of agricultural technologies?

RECOMMENDED READING

Alston, Julian M., George W. Norton, and Philip G. Pardey, *Science Under Scarcity: Principles* and *Practice for Agricultural Research Evaluation and Priority Setting* (Ithaca, N.Y.: Cornell University Press, 1995).

Benor, Daniel, James A. Harrison, and Michael Baxter, *Agricultural Extension: The Training and Visit System* (Washington. D.C.: World Bank, 1989).

CGIAR Science Council, *Science for Agricultural Development; Changing Contexts and New Opportunities* (Rome, Italy: Science Council Secretariat, 2005).

Masters, William A., "Paying for Prosperity: How and Why to Invest in Agricultural Research and Development in Africa," *Journal of International Affairs*, vol. 58, no. 2 (2005), pp. 35-64.

Pinstrup-Andersen, Per and Ebbe Schioler, *Seeds of Contention* (Baltimore, Md.: Johns Hopkins University Press, 2000).

Ruttan, Vernon W., *Agricultural Research Policy* (Minneapolis: University of Minnesota Press, 1982).

Scobie, Grant M. and Rafael Posada T., "The Impact of Technical Change on Income Distribution: The Case of Rice in Colombia," *American Journal of Agricultural Economics*, vol. 60 (February 1978), pp. 85–92.

CHAPTER 13

Land and Labor Markets

> The distribution of rights in land relates to the distribution of power, income, social status, and incentives. A land reform that changes this distribution is by definition a change that shakes the roots and not the branches of a society. — Philip M. Raup[1]

This Chapter
1. Discusses the meaning of land tenure and land reform
2. Explains why land reform and flexible land tenure systems are often necessary for agricultural development, yet difficult to achieve, and what the requisites are for a well-functioning land market
3. Describes the nature of agricultural land and labor markets in developing countries.

MEANING OF LAND TENURE AND LAND REFORM

Land and labor are fundamental production inputs, and while land is often distributed unequally in a country, labor is not. Therefore we see large farms with land concentrated in the hands of a few, and small farms with excess labor. Typically a market develops in which labor is hired to work on large farms, or land is leased to small landholders for rent or for a share of the output. An alternative is to subdivide large holdings through land reform or market-based redistribution efforts. In this chapter, we consider determinants and consequences of alternative land and labor market structures, including the redistribution of land. Well-functioning land and labor markets are crucial to agricultural development because land is a major input into agricultural production and poorly functioning markets will lead to misallocation of this valuable resource. Evidence shows that the rural poor, particularly those with little or no land, will benefit more from an additional unit of land than do the rich: efficient land allocation can increase productivity and reduce poverty.

---

[1] Philip M. Raup, *Land Reform Issues in Development*, Staff Paper No. P75-27, Department of Agricultural and Applied Economics (St. Paul: University of Minnesota, 1975), p. 1.

Land rights determine social and political status as well as the economic power of a large proportion of the population in developing countries. *Land tenure* is a term used to refer to those rights or patterns of control over land. Land rights include rights to use and to exclude use, rights to output from the land, and rights to transfer the land or its output to others.

As population density increases, farming techniques change, and markets for agricultural products grow, pressures often develop to change existing land tenure arrangements. In societies where land has been held in common, permanent and enforceable individual rights to land may evolve. In countries where ownership patterns are highly skewed with a few people owning much and many people owning little or no land, pressures are often exerted on the government to undertake a land reform or eliminate constraints to more equal land distribution by establishing well-functioning land markets. These pressures may arise from peasants who desire increased economic well-being or from those in power who hope that minimal concessions to the peasants will diffuse political unrest. A land reform is a dramatic attempt to change the land tenure system through public policies. Land reform may change not only rights and patterns of control over the land resource, but also the mode of production (whether semi-feudalistic, capitalistic, or socialist) and the agrarian class structure. Consequently, few subjects related to agricultural development are as controversial. In recent years, more attention has been focused on market-based measures to provide more access to land, and fewer land reforms have occurred.

Land Ownership and Tenure Systems

A wide array of land ownership and tenure systems exists in the world. These systems reflect differences in climate, historical influences, stage of development, culture, political systems, transactions costs, and other factors. The systems vary in size and organization of land holdings, they affect incentives to produce and invest, and they influence the distribution of benefits from agricultural growth. Examples of average size of landholdings in Africa, Asia, and Latin America are presented in Table 13-1. The larger holdings in Latin America compared to Asia and Africa are particularly evident. However, information on *average* land holdings masks important difference within countries. While average holdings, for example, in Brazil and Mexico are quite high, many landless and near landless families can be found in rural areas of both countries. A description of the impact of colonial rule on landholdings is provided in Box 13-1.

TABLE 13-1. AVERAGE SIZE OF LANDHOLDINGS, SELECTED COUNTRIES, AROUND 1970

Region and country	Average size (in hectares)
AFRICA	
Cameroon	5.2
Ghana	3.2
Kenya	4.1
Malawi	1.5
Sierra Leone	1.8
Zambia	3.1
ASIA	
India	2.3
Indonesia	1.1
Iraq	9.7
South Korea, Republic of	0.9
Pakistan	5.3
Philippines	3.6
LATIN AMERICA	
Brazil	59.7
Colombia	26.3
Costa Rica	38.1
El Salvador	4.1
Mexico	137.1

Source: Clive Bell, "Reforming Property Rights in Land and Tenancy," *World Bank Research Observer*, vol. 5 (July 1990), pp. 143–66.

Family farms, corporate farms, state farms, and group farms are major types of farm ownership, but organization of farm enterprises within these types can vary substantially. In many cases the owner of the farm is also the operator. In other cases, those who operate or work on the farm may earn a fixed wage or pay rent in cash or in a share of the farm output.

Small subsistence or semi-subsistence family farms are common in developing countries. Families often provide most of the labor, and cultivation is labor-intensive. Much of the output is consumed on the farm where it is produced. However, not all small family farms are subsistence or semi-subsistence farms; many are commercial farms producing substantial surpluses for sale. Those farmers who do consume most of what they produce are usually very poor. In some cases, family members work on other farms or in non-farm employment where available. The latter can lead to the small farms becoming part-time opera-

BOX 13-1
COLONIALISM AND LAND OWNERSHIP

Many of the land ownership patterns found in developing countries are the vestiges of colonial rule. The *latifundia* or extensive large-scale farms that currently exist in Central and South America alongside *minifundia* or very small farms are a direct descendant of colonialism. The Spanish and Portuguese colonizers allocated large tracts of land to elites who formed tropical plantations or large haciendas. Both types of landholding were made possible through the direct enslavement of indigenous populations, the importation of slaves from Africa, or the *encomienda* system that gave indirect control over local populations to certain elites.

Some of the richest agricultural lands in Africa have land ownership patterns that were established during the periods of European colonization. Because European countries coveted exotic tropical products, such as cocoa, coffee, tea, and tropical fruits, agricultural production in Africa was reorganized under colonial rule. Large landholdings were allocated to European settlers, such as the tea and coffee plantations in eastern Africa; rarely were the land's original inhabitants compensated. In many instances, the land's original inhabitants were resettled to areas with lower agricultural potential, poor rainfall, and inadequate infrastructure. Areas that had been self-sufficient in food production became exporters of goods to Europe. An adequate labor supply was maintained, sometimes through enslavement and sometimes through economic coercion.

In Asia, colonial rule led to similar forms of plantation agriculture. Japanese colonies in Korea and Taiwan produced for export to Japan; the Dutch colonized Indonesia; Spanish plantations existed in the Philippines; the British colonized much of the Indian subcontinent and other regions.

Following the end of colonial rule, many of these land-ownership patterns persisted, because of the political powers of the landed elite. Some of the anti-feudal land reforms undertaken in countries prior to 1970 were designed to remove the less desirable aspects of these landholding patterns. Implicit forms of enslavement of labor, such as through the maintenance of indebtedness, were prohibited. In many countries, the result of these reforms has been to reduce labor use, increase mechanization, and leave the distribution of land largely unchanged. The legacy of colonial landholding patterns has been pervasive rural poverty in many regions of the world.

tions, especially as development proceeds over time and job opportunities in the non-farm sector increase.

Large-scale commercial family farms sell most of what they produce. While in developed countries these farms are highly mechanized and often involve only a small amount of non-family labor, in developing countries the operations are usually more labor-intensive and use a high proportion of hired labor. The owner frequently does not live on the farm, but pays a manager to oversee day-to-day operations.

Corporate farms often produce a limited number of commodities in large-scale units. These farms may have their own marketing (including processing) systems. This type of farm is more prevalent in developed than in developing countries, but there are numerous examples of large corporate farms in developing countries. The fruit plantations found in Central American countries, banana plantations of the Philippines, and cocoa plantations of West Africa are a few examples.

State farms are usually large, owned and operated by the government, and run by hired labor. The managers are responsible to a unit of a government-planning agency that may set targets for production and direct the timing and method of key farming operations. Examples have existed recently in parts of China, the former Soviet Union, and Ethiopia. This type of farm usually suffers from inadequate incentives and ill-advised management decisions.

Group farms are communes, kibbutzim, collectives, or other types of farms that are operated by a group of people who work and manage the farm jointly. The joint income is then divided up among the members of the group. These group operations may also involve non-agricultural activities. Often, collectivized farms are characterized by over-investment in labor-saving capital-intensive technologies, since individuals do not receive the full returns from their labor. Special arrangements may be devised (for example, a points system) to provide incentives for individual members to work harder. Small individual private plots may also be allowed. The kibbutzim of Israel and the (former) Ujamaa villages of Tanzania are examples of group farm systems. *Cooperative farms* exist in some parts of the world, but cooperative purchase of inputs and marketing of outputs is a more common organization. The Mennonites in the Paraguayan Chaco are an example of privately held land whose owners organize cooperatives for input purchases and product marketing and sale.

Not only do many types of land ownership and organization exist, but also types of tenancy or leasing arrangements. Farm families may lease all or a part of their land for a specified amount of cash or for a share of the production from the land. A farmer may be allowed to farm a piece of land in exchange for his or her labor on another part of the owner's land. In some (or parts of some) countries, the village, tribe, or national government may own the land and grant use rights to part of the land to individual families. This system is common in Africa where social groups allocate land to individuals who maintain control over it and its output as long as they cultivate it.

Tenancy arrangements affect the risk and transactions costs borne by tenants and landlords and influence incentives to work or apply inputs. A share lease, for example, spreads the production risk between the landlord and tenant, while a cash lease concentrates the risk on the tenant. A cash lease implies lower transactions costs for the landlord than a share lease, since the amount received under the share lease has to be measured, and production has to be monitored. A tenant may have less incentive to apply additional fertilizer or even labor under a share lease than under a cash lease. To circumvent this disincentive, the landlord may share the cost of the fertilizer or place conditions on the amount of labor applied. In such cases, the landlord bears the transactions costs associated with monitoring input use and measuring the value of output. Thus, the use of a particular land- or labor-contracting mechanism may be a response to the presence of risk or transactions costs. These factors can explain why the land market may not dictate a single type of tenancy system even within a single country (see Box 13-2).

BOX 13-2
TENANCY, RISK, AND TRANSACTIONS COSTS:
THE CASE OF SHARECROPPING

Sharecropping is a widely practiced form of tenancy whose existence can be attributed to risk sharing and transactions costs. Because agricultural production is risky, and both tenants and landowners desire to share risks, sharecropping represents a compromise between fixed-rent contracts, where the renter bears all the risk, and wage employment, where the landowner bears all the risk. Sharecropping also represents a response to the costly supervision of workers. Since under a wage system the worker receives wages based on hours worked rather than effort expanded, there is a tendency to shirk. Supervision is necessary, yet costly. Sharecropping returns some of these incentives to the tenant and also allows risk sharing.

Cows on a kibbutz in Israel.

Types of Land Reform

Because many types of land ownership systems exist in the world, there are many types of land reform. Prior to the 1970s, many if not most land reforms involved a movement away from feudalistic and semi-feudalistic land-tenure arrangements toward capitalist or socialist ownership modes.[2] Feudalism was characterized by large-scale estates controlled by the traditional landed elite with labor bonded to the estates through peonage or extra-economic forms of coercion.

Anti-feudal land reforms have eliminated feudalism in most of the world. In place of feudalism we see the existence of small subsistence or semi-subsistence family farms, small and large commercial farms, corporate farms, state farms, and socialist or group farms. In many countries, however, this post-feudal order has resulted in some large capitalist forms or estates controlled by an elite well-to-do class and a coexisting small-farm sector. In a few countries (for example, South Korea and Taiwan), the post-reform agricultural sector consists primarily of small family farms, and in a few, such as Cuba, socialist farms predominate. The form of the post-reform agrarian structure depends largely on the motivation and political ideology behind the reforms.

[2] See Alain de Janvry, "The Role of Land Reform in Economic Development: Policies and Politics," *American Journal of Agricultural Economics*, vol. 63 (May 1981), pp. 384–92.

Land reform today generally does not refer to the types of anti-feudal reforms instituted in many countries prior to the 1970s. In those reforms, prohibition of bonded labor and reductions in labor exploitation were achieved, but in many cases a significant redistribution of land was not. As de Janvry notes, countries such as Colombia, Ecuador, and India had successful anti-feudal land reforms but very little redistribution of land.[3] Land reform as a policy issue today usually relates to seeking a shift in the dominant rural class from an elite landowning class to medium-size and smaller peasant landowners. It often involves market-based efforts to increase access to land.

In the few remaining countries where socialist or group farms predominate (as opposed to capitalist farms of whatever type), future land reforms may involve a transition to increased capitalism as these countries struggle with the incentive problems that have plagued many types of group farms (see Box 13-3). Land reforms in socialist countries are difficult to achieve without a fundamental restructuring of the political system. The change in China from the socialized agricultural system to the market-based Household Responsibility System in 1979 was accompanied by political and economic upheaval.

Transactions Costs and the Agrarian Structure

No form of land tenure is universally efficient. Differences in natural resource endowments, in the availability of new technologies, and in institutional arrangements all influence risk, transactions costs, and the farmer's opportunity to exploit his or her managerial ability. If there were well-defined private property rights and a reasonably equitable distribution of land, perfect information, and zero transactions costs (especially the cost of enforcing contracts), markets would work perfectly and it would not matter as much what type of agrarian structure prevailed. Bargaining would occur among landowners, renters, and laborers; neither the returns to labor nor the overall economic efficiency of the agricultural system would depend on the type of agrarian structure.

In the real world, however, risk varies from country to country. Markets are not perfect. The cost of acquiring information and of negotiating, monitoring, and enforcing contracts can be high. People are willing to exploit others, labor hired on a time-rate basis may shirk (increasing the cost of supervision), the price of land may decline as farm size grows (due to fixed costs associated with land transactions), and larger landowners may have better access to markets and information.

---

[3] Ibid.

BOX 13-3
LAND REFORM IN EASTERN EUROPE

Following the collapse of Soviet-style models of central economic planning and control and the movement toward democracy in the late 1980s, governments in Eastern Europe were faced with the problem of how to reform their agricultural sectors. The organization of the agricultural sectors in these countries was rather similar: approximately one-third of the farms were state farms, and two-thirds were collectives (cooperatives). Most farm employees managed a household plot of about one-half hectare, while the state farms and collectives were large, about 2,000 to 3,000 hectares. Two paths of reform are illustrative of general trends in the region: Romania and Bulgaria.

In both countries, the rights of landowners prior to collectivization were recognized by parliamentary decree in February 1991. These decrees also established procedures for reclaiming these property rights. In Romania, land redistribution proceeded quickly. Local land commissions were established to hear household claims for up to 10 hectares. Some proof of the claim was needed. Whenever possible, claimants were given back the land actually owned, and when not, an alternative of equal size and quality was returned. Once in possession, the owner could sell it immediately, or purchase more land. Thus, a market for titled land, with very few institutional restrictions, was established. There was no attempt to create farms of optimal size.

The Bulgarian redistribution proceeded much more slowly. Administrative delay hindered progress, and local commissions were very slow in forming. The laws implementing the distribution were rigid, and the construction of "appropriate size holdings" through administration was attempted. The local commissions adjudicated claims, but a planning team reassigned plots. The law prohibited the purchase and sale of land for three years, hindering development of a land market.

In both cases, most of the new landowners remained integrated into the collective management system. In Romania, the formation of a land market opened a period of holdings consolidation and resale, but actual exit from the collective was delayed until the infrastructure for individual management was developed. The slowness of the redistribution in Bulgaria guaranteed the existence of collective systems for many years.

Source: Karen Brooks, J. Luis Guasch, Avishay Braverman, and Csaba Csaki, "Agriculture and the Transition to Market," *Journal of Economic Perspectives*, vol. 5, no. 4 (Fall 1991), pp. 149–62.

Many of these risk and transactions cost factors confer an economic advantage on large farms. Larger landowners can gain additional political advantage through collective action and can reinforce their advantage through the tax laws and pressures on the types of new technologies produced by the public research system. The result can be additional gains for the elite, but reduced economic efficiency for the agricultural sector and the country as a whole. Land reform is needed in these cases.

ACHIEVING SUCCESSFUL LAND REDISTRIBUTION

A country can desire land redistribution for a variety of economic, social, and political reasons, yet land redistribution, whether it involves change in land tenancy or ownership, is always difficult to achieve. And it is frequently difficult for a society to even agree on whether a land reform, once implemented, has in fact succeeded. Let us examine briefly why land reform may be desired, why it may be difficult to achieve, how a country can measure whether a land reform has succeeded, and what factors improve the chances of achieving successful land reform.

Need for More Equitable Distribution of Land

The broad economic and development goals of most societies include desires for improved income growth (efficiency), equity (income distribution), and security (political and economic stability). A more equitable distribution of land, whether achieved through a land reform, better functioning land markets, or market-based tools such as land banks, can contribute to all three of these goals.

A skewed distribution of landholdings can hamper economic efficiency for several reasons. Large landholdings may not be farmed intensively, even in countries that are very densely populated; in fact, an *inverse* relationship between farm productivity and farm size has been found in many developing country settings.[4] Some landowners hold land for speculative reasons. Others are absentee landlords who provide little supervision of those working on the farm. If the farm is owned by the government, planning and management may be centrally and poorly controlled, and individual incentives may be stifled since farms are forced to respond to output and input quotas. Large farms may substitute machinery for labor, exacerbating an unemployment problem. Large farmers facing labor supervision or other management problems often demand capital-intensive innovations from the agricultural

[4] See Dwayne Benjamin, "Can Unobserved Land Quality Explain the Inverse Productivity Relationship?" *Journal of Development Economics*, vol. 46 (1995), pp. 51-84.

research system in the country. Thus, new technologies are generated that do not reflect the true scarcity values of land, labor, and capital in the country.

Countries with large landholdings often have a coexisting sector of farms that are too small to provide an adequate living. These very small holdings of one or two hectares or less may have labor employed to the point at which its marginal product is very low. Thus, reducing the size of large farms and increasing the size of very small farms may be the only way to raise the marginal product of labor in agriculture, and thereby raise income per worker.

As discussed earlier, other land tenure problems, including the need for tenancy reform, may have to be solved to improve entrepreneurial incentives and to reduce risks facing farmers. For example, as population density increases, property rights for land pastured in common may need to be redefined to avoid overgrazing. Share or cash rents may need to be changed as new technologies become available. Lease lengths may need to be more securely established to encourage capital investment.

Apart from growth or efficiency concerns, land redistribution is often needed for equity reasons. The number of landless laborers is growing rapidly in many countries, along with associated poverty and malnutrition. The principal resource these people control is their labor, whose value is depressed by underuse of labor on large farms. Providing land resources to these people can be an effective means of raising incomes. As large farms are broken up, even those poor who do not receive land can benefit due to increased economy-wide demand for labor. Large farms convey political power to a small group. This group may distort economic policies in a direction that hinders overall economic growth and creates severe hardship for the poorest segments of society. Thus, to achieve development as defined in Chapter 1, addressing the land distribution problem may be necessary.

In addition to growth and equity concerns, land redistribution can enhance political and economic security or stability. In fact, expropriations of land and partial land reforms experienced in many countries in the past have probably occurred primarily for purposes of political stabilization. This stabilization can have positive and negative impacts on economic growth and equity. To the extent that land redistribution dampens political unrest and reduces the chances of revolution, it reduces the chances of a country's experiencing the extremes of death and suffering that can accompany a revolution. However, to the extent that a partial land reform achieves political stability without redistributing enough land or economic power to generate widespread growth and fundamentally reduce economic hardship and hunger, it may only perpetuate a status quo of chronic suffering.

Economic and political stability facilitate economic development. With little land, peasants are forced to live very close to the margin of survival. Drought, floods, or other natural disasters can quickly push the landless or near landless below the margin.

Why Land Reform is Difficult to Achieve

Because of the political and economic power that accompanies land-ownership in many countries, it is difficult to conduct a meaningful redistribution of land. Historically, land reforms have most often been made possible only after significant social upheaval caused by revolution, the overthrow of colonial powers, or war. In the former Soviet Union and China, social revolutions destroyed the power of the landed elite prior to the institution of collectivizing land reforms. An army of occupation enforced the socialist reforms in Eastern Europe following World War II. The extreme economic, political, and social turmoil of the 1970s in China, and of the 1980s and 1990s in the Soviet Union and Eastern Europe, once again created the conditions for land reforms in those countries. In capitalist countries such as Japan, the Republic of Korea, and Taiwan, defeats in war or occupation were followed by re-distributive reforms.[5]

In countries with capitalist forms of social and economic relationships, land reforms are difficult to achieve because those holding the land rights also have strong political power. Urban consumers often align with landowners. Because large farms tend to have large marketed surpluses, consumers fear that steep food price increases may follow a dramatic reform. Small changes may be supported as a means of political stabilization, but large-scale restructuring of property rights is difficult. Occasionally, governments support redistributive land reforms in response to strong revolutionary pressures, such as in Mexico (1940 to 1977) or the Philippines (1972 to 1975). Or, land reforms result following military overthrows of the government, such as in Peru (1969 to 1975). Land reforms within capitalist agriculture are usually slow to occur because compensation is required if they are to be accepted by those losing land. Unless the government's budget has a large fiscal surplus, which is rare in developing countries, or substantial foreign aid, gainers cannot compensate the losers sufficiently for the land reform to be politically viable.

---

[5] See Clive Bell, "Reforming Property Rights in Land and Tenancy," *World Bank Research Observer*, vol. 5 (July 1990), pp. 143–66.

In countries with a socialist economic system and group farming, land reforms may involve increased privatization of land or the rights to plan and to market output. These reforms are difficult to achieve because they may imply movement toward a more market-driven economy and freer political system. Because the potential economic efficiency gains may be larger from this type of reform than from a redistributive reform within capitalist agriculture, the chances for a successful land reform may even be greater when the initial agrarian structure is socialist. The Chinese reform begun in 1979 had the wide support of the rural peasantry and urban consumers.

Many countries have had land reforms in the sense that changes in the land tenure system have occurred. The mere fact that a change has taken place does not necessarily imply that a *successful* land reform has transpired. Many national leaders are interested in land reform because the reform may lead to increased political stability. However, unless there is evidence that incentives have been created for farmers to undertake hard work and increase their capital investment and unless poverty has been reduced and social status improved for the rural poor, a successful land reform has not occurred.

A successful land reform should alter the incentive structure in rural areas. Whether this structure has been altered is perhaps best measured by evidence of increased and continuous capital accumulation by small farmers in the form of livestock, farm buildings, equipment, and other improvements in land resources. Because these investments may be small in any one year, it usually takes a generation, perhaps 25 to 30 years, to truly evaluate the success of land reform.

Agricultural productivity should also increase in the long run. However, in the first five years following a land reform, productivity may stagnate for a couple of reasons. First, the mix of commodities produced may shift toward food crops and away from a heavy reliance on cash crops. This shift and other disruptions to the normal input and output marketing channels, credit flows, changes in technologies needed from the agricultural research system and so forth, can hinder productivity growth in the short run.

Marketable surpluses may decrease because the poorer segments of the rural population, who benefit from the land reform, have a high income elasticity of demand for food. As their incomes increase through more access to land, they consume more, and the aggregate marketed surplus may decline. Thus, short-term increases in agricultural productivity or marketable surpluses are not good measures of the success of land reform.

A land reform can also affect capital formation in the public sector. Countries with land tenure systems in need of reform often have poor rural schools and other public infrastructure. Large-scale landowners typically hesitate to tax themselves to support schools, roads, and so on. Countries in need of land reform usually find it easier to collect public revenues by taxing export crops or by placing tariffs on imports. Governments implementing land reform have an opportunity to re-structure the tax system. The new owners of small plots have increased ability to pay taxes and may do so willingly if they see that the tax system is honest and the proceeds will be used for schools, roads, and other local infrastructure.

Peasant associations and other farm groups also are likely to be formed after a successful land reform. These associations can play an important role in promoting the development and adoption of new tech-nologies for agriculture, improving marketing channels, and so forth. The formation of these associations is, therefore, another test of a suc-cessful land reform.

Evaluation of the success of a land reform is a dynamic process that should be undertaken for many years following its inception. An additional indicator of success is the changing pattern of landholdings. If the reform has any prospect for success, incentives should be in place to make agriculture in the reform sector profitable. If it is not, then farm failures will occur, and, under capitalist reforms, consolidation into larger units will follow. Thus, quickly growing farm size can be an indi-cation of failure of a reform.

Requisites for a Successful Land Reform

Government commitment, government power, and administrative organization are essential for any successful land reform.[6] No land re-distribution occurs without strong government resolve and power. If transactions costs can be reduced, particularly those related to the cost of information, people may find it easier to express their views, which may help strengthen government resolve. However, in countries with corrupt leaders and little sincere desire for effective land reform, no change is likely unless peasants take land reform into their own hands. Even in countries with sincere leaders who have a strong commitment to land reform, the government must possess sufficient power to pre-vent those opposed from sabotaging it or overthrowing the govern-ment (see Box 13-4).

---

[6] Sterling Wortman and Ralph W. Cummings, Jr., *To Feed This World: The Challenge and the Strategy* (Baltimore, Md.: Johns Hopkins University Press, 1978).

BOX 13-4
LAND RESETTLEMENT AND REFORM IN ZIMBABWE

Prior to its invasion in 1890, land in Zimbabwe was held under communal tenure, and tribal leaders allocated rights to land. Members of Cecil Rhodes' Pioneer Columns were promised 3,000-acre holdings in exchange for assistance in colonizing the area. During the subsequent colonial period, Native Africans were relocated to low-rainfall, low-productivity land and were barred from landownership outside these tribal reserves. Large-scale resettlement without compensation continued through the early 1950s.

In 1965, the minority white government declared independence from Great Britain and continued enforcing restrictive conditions against Native Africans. The subsequent war of independence, which raged in the bush with increased intensity through the 1970s, used the "land question" as a uniting principle. Overcrowding and dwindling production on communal lands led to widespread rural poverty. The Lancaster House Agreements, which paved the way for independence in 1980, formed a basis for post-independence land reform. The British government allocated £44 million for purchase on a "willing seller, willing buyer basis." Robert Mugabe won the first free election and promised to resettle blacks on purchased white lands. Resettlement reduced civil conflict, provided opportunities for war victims and the landless, and relieved some population pressures on communal lands.

Between 1980 and 1990, the government obtained some 3 million hectares for resettlement and resettled roughly 54,000 households. The pace of resettlement slowed through the 1980s as attention moved toward providing agricultural services to communal areas. Farms in resettlement schemes had variable performance; many of those under private management did relatively well, while those that were managed under cooperative schemes performed less well. Critics of the resettlement program, however, noted that the most productive farm areas had been "resettled" by the President's political supporters.

The Lancaster House Agreements expired in 1990, effectively ending donor support of resettlement. Through the mid-1990s, few farms were resettled and government attention turned toward restoring macro economic balance and dealing with adverse consequences of severe drought. Following a major currency devaluation in late 1997, unrest grew and independence-war veterans began to demand access to land. The government responded by listing some large-scale commercial farms for "compulsory" resettlement. Although owners would be compensated, the method of compensation was unclear, and abandonment of the "willing buyer– willing seller" principle caused unease among owners of land and donors. "Listed" farms were said to be "underproductive," although government critics disputed many of these assertions. Between 1997 and 2000, resettlement became more contentious as rural interest groups increased

continued on next page

BOX 13-4, continued

pressure on the government through protests and forced seizure of white farms. In response to internal political pressure, the government began a "fast track" resettlement process and, between 2000 and 2003, acquired more than 75 percent of the nearly 4,500 white-owned commercial farms. By the end of 2003, fewer than 300 white farmers continued to farm.

Resettlement in Zimbabwe provides a number of lessons. Inequitable land access can create strong political forces, particularly following political change. The issue of compensation for lands remains complicated. While some white farmers obtained their lands through ancestral succession (their forebears had received land at no cost), many had purchased their land from others and were due fair compensation for their investments. As internal political pressure and the economic crisis grew, the Mugabe government felt increasing urgency for resettlement while resources to finance land purchase became scarcer. Land reforms of the type practiced in Zimbabwe can have huge unsettling effects. As commercial farmers were forced out, agricultural production and export earnings dropped by more than 50 percent, further exacerbating economic problems and contributing to near famine conditions. Neighboring countries absorbed many of the displaced white farmers, spreading social problems around the region. The politicization of the process damaged the credibility of government and undermined popular support (at home and among donors) for the program.

Source: William Masters, *Government and Agriculture in Zimbabwe* (Westport, CT: Praeger, 1994) (updated by press reports).

Political opposition to land reforms can be diffused by reforming policies that benefit owners of larger farms. Policies, such as tax exemptions, credit policies, and input subsidies that favor larger farms have the effect of increasing land prices, making compensation more expensive and providing economic advantages to the privileged class. Policies favoring large farms create incentives for opposition to land reforms and lower the probability of successful reform.

An administrative organization that coordinates national and local decision-making and implementation of the reform program is essential. This organization must implement a series of critical steps decisively and quickly. Speed is particularly important because if a reform is announced but not implemented quickly, capital will be removed from farms and productivity will suffer. For example, in Peru in the late 1960s a land reform was announced long before it was implemented. Landowners sold as many capital items as they could and halted any new investment.

Central authorities must act quickly to assemble land records that clearly identify the targeted land and its productivity. Criteria for acquiring land must be clear and simple, and rules for compensating former owners must be established. Former owners must not be allowed to reacquire the land following the reform.

Payments by new owners must be modest and should be integrated into a system of land taxes. One of the best ways for a country to raise revenues for local development is to tax owners of small landholdings. A land reform is incomplete if it is not associated with a tax reform that increases the tax burden on those who receive the land. Land reform increases the capacity of farmers to pay taxes, and the new landowners can identify the payment of taxes with the benefits they receive.

Land reform disrupts the institutions providing services such as credit, inputs, marketing, and technical information. Unless these systems are adjusted concurrently with the land reform, the reform will almost certainly fail. A marketing system that effectively serves small farms may be substantially different from the one that functioned before the land reform. New methods for distributing technical information may be needed, as the previous landowners may have been able to deal directly with the scientists in the agricultural research system. Group action such as cooperatives may be needed to coordinate services in a manner that is responsive to requirements of numerous owners of small farms.

Alternatives to Land Reform

The cost and political difficulty of attaining an effective land reform have led to alternatives to large-scale, centrally enforced redistribution. Examples of market-based land redistribution efforts include fortification of sales and rental markets, encouraging cooperatives to redistribute lands to their members, reduced government ownership, and creation of land banks.[7] These efforts generally require three complementary steps: (1) legal definition and assignment of property rights; (2) creating the legal framework for efficient functioning of the markets themselves; and (3) ensuring that complementary markets, particularly finance and insurance, function efficiently. These steps can be costly and difficult, and a market-based reform that does not address all will likely fail in its objectives — to redistribute land to more efficient users.

---

[7] These changes have been called "Phase III" of the process of land reform by Alain deJanvry, Marcel Fafchamps, and Elisabeth Sadoulet in "Peasant Household Behavior with Missing Markets: Some Paradoxes Explained," *Economic Journal*, vol. 101, no. 409 (1991), pp. 1400–17.

Definition and assignment of property rights involves surveying lands, titling them, and creating a land registry so that prospective participants can examine the land's history of transactions including liens and competing claims. The legal framework includes determining who can participate in transactions, means of contract enforcement, removal of implicit or explicit restrictions on rental and transfer, and so on. Implicit restrictions to rental, for example, may occur because without an adequate legal framework, squatters may possess strong claims to ownership. Land rentals may be discouraged due to fear of losing claims to ownership. Issues such as women's rights to own and transfer lands can have efficiency and equity effects on subsequent market processes.

Finally, while land markets have the potential to efficiently redistribute lands, in the presence of distortions in credit and insurance markets, creation of land markets alone may not solve the problem of inequitable distribution. For example, if banks or other creditors are unwilling to lend money in relatively small amounts due to transactions costs associated with such loans, then the poor may not be able to finance purchase or rental of the small amounts of land they seek. Unequal distribution of productive assets such as capital can exacerbate such problems because the poorest may not have collateral to support a loan. Insurance markets are important because the poorest of the poor may be less willing to risk their assets as collateral.

AGRICULTURAL LABOR MARKETS

Labor is often the most valuable resource the rural poor possess. This labor can be used to cultivate their own lands, to process and market products after harvest, to produce non-agricultural goods (some for own consumption and others to be sold in markets), to perform child-care, cooking, and other household activities. Alternatively, this labor may be sold or rented to others, both farm and non-farm employers. Agricultural labor markets exist because land markets alone can not balance out differences in land and labor endowments. Small-scale land owners or landless individuals supply labor to large-scale landowners, who need more than their own family labor to carry out their farming operations. Labor markets help allocate resources into their most valuable uses by transmitting signals about resource scarcity across space and time. As labor is an important input to production and a key asset held by the poor, the conduct and performance of labor markets are especially important for broad-based agricultural growth.

BOX 13-5
REFORM OF CAPITALIST AGRICULTURE IN COLOMBIA

Colombia presents an example of some of the pitfalls associated with land reforms in many countries. In the 1930s, there was a public outcry, mostly by urban consumers who desired cheaper foods, over the lack of productivity on the large landholdings of the rural elite. In 1936, Law 200 was passed which said that potentially productive but poorly cultivated or abandoned large holdings were to be expropriated by the government. This threat caused land productivity to rise for a short time, and virtually no land was confiscated. During the 1950s, a long period of civil conflict known as "La Violencia" hastened the destruction of traditional social relations and weakened the political powers of the old agrarian oligarchy.

Following a peace pact, a new phase of land reform began. Law 135 of 1961 set forth an ambitious reform package that included full compensation to existing landholders. The gradualist approach doomed the package from the start. Political pressure from landed groups allied with urban consumer interests successfully diverted inputs, often with substantial subsidies, to large-scale farms. Land values on favored farms increased dramatically, making compensation financially impossible. By 1972, only 1.5 percent of all land in large farms had been redistributed.

Law 4 in 1973 declared an end to this redistributive reform and returned the country to the principles of Law 200. At the same time, a political coalition between large-scale farmers, a small but substantial family-farm sector, and urban consumers formed and created pressure for a rural development program that favored the first two groups. Landless and marginal farmers were politically and economically excluded.

The conditions for a successful land reform never really existed in Colombia. Shifting alliances between urban and rural power groups diminished the political will. A lack of clear conviction for redistribution, combined with the slow pace of reform, further inhibited the efforts. Policies favoring large farms, largely intended to diffuse political opposition, had the effect of destroying any prospects for real reform.

Casual Versus Permanent Labor

Labor may be hired on a *casual* or temporary basis by the day or for some other short period of time such as for the harvest or weeding period. Alternatively, labor may be hired on a more permanent or longer-term basis, perhaps for months or years. Casual labor is usually paid in cash and in kind (for example, food; many day-labor wages include a meal for the worker). Laborers may be paid daily or on a piece-rate basis for certain tasks. Women are often paid less than men, even for the same task. Casual labor is characterized by strong *seasonality*; workers tend to be hired during planting and harvest, when agricultural labor demands are highest. Longer-term labor may have supervisory responsibilities or perform tasks that require special care such as applying farm chemicals. Formal or informal contracts may be developed to handle seasonal fluctuations and risks associated with agricultural production.

Transactions Costs, Asset Inequality, and Labor Markets

Labor markets in developing countries often contain imperfections due to power imbalances, imperfect information, and transactions costs. Power imbalances emerge when a single or small number of employers exist in an area. In such cases, the employers may exercise monopsony (single buyer) power over their employees and use fewer workers at lower wages than would exist in a competitive labor market. Large-scale plantations, such as those existing in Central America and in cocoa-producing areas in West Africa, may exhibit such power. Imperfect information and transactions costs also constitute major sources of labor market imperfections. Labor must be hired, with corresponding costs of search and contracting, and supervised. Supervision involves costs of monitoring and enforcement. Such costs may distort incentives for hiring and use of different types of labor.

Given information imperfections, employers may be unaware of the reliability of workers, some of whom shirk their duties. As a result, costly supervision or other contractual mechanisms must be undertaken to ensure the worker performs his or her duties as expected. Share cropping and piece-work contracts are two such mechanisms commonly found in less developed countries. One study of the effectiveness of such contractual arrangements conducted in the Philippines found that piece-rate and shared cultivation were associated with significantly higher worker effort than time-wage contracts.[8] Contracts that tie together labor, land-use agreements, credit, and other inputs often represent responses to imperfect information, transactions costs, and risk sharing.

Wages in agriculture, whether on a casual or a full-time basis, are relatively low. Throughout the developing world, people who rely on agricultural employment as their main source of income tend to be poor. Poverty rates are high among tobacco-estate workers in Zimbabwe and Malawi, among day laborers on the Indian subcontinent, and among coffee and other plantation workers in Central America. Government interventions into labor markets are often justified, based on this

---

[8] See A. Foster and M. Rosenzweig "A Test for Moral Hazard in the Labor Market: Contractual Arrangements, Effort, and Health," *Review of Economics and Statistics*, vol. 76 (1994), pp. 213–27.

observed poverty. Minimum wage legislation for farm workers has been tried in a number of countries including Zimbabwe, South Africa, and several Central American countries. Minimum wages for farm workers tend to be difficult to enforce, because of high transactions and enforcement costs and imperfect information endemic in rural areas of less developed countries. Other interventions include establishment of labor-enforcement standards, provision of labor-market information, investments in education and schooling to increase worker productivity, and promotion of non-agricultural job opportunities that compete with agricultural employment. As noted in Chapter 7, non-farm employment constitutes a large and growing share of the rural labor markets. As agriculture develops over time, it must compete with alternative employment opportunities in rural labor markets.

SUMMARY

Land tenure refers to the rights and patterns of control over the land resource. Land rights determine social and political status as well as the economic power of a large proportion of the population in developing countries. A land reform is an attempt to change the land tenure system through public policies. Land tenure systems vary in farm size and organization, affect incentives to produce and invest, and influence the distribution of benefits from agricultural growth. Family farms, corporate farms, state farms, and group farms are major types of farm ownership. Many types of tenancy or leasing arrangements also exist.

The post-feudal order has resulted in some large capitalist farms and a coexisting small farm sector in many countries. Socialist farms predominate in a few countries. No form of land tenure is universally efficient. Land reform is difficult to achieve because those holding the land rights have political power. Land reform is needed for improved economic efficiency, equity, and political and economic stability. Unless there is evidence that incentives have been created for farmers to undertake hard work and increase their capital investment, and, unless poverty has been reduced and social status improved for the rural poor, a successful land reform has not occurred. Government commitment, government power, and administrative organization are needed for a successful land reform. Land reform must take place quickly and must be accompanied by credit, marketing, and other services, and new land owners should be taxed to support development. Market-based land redistribution efforts include fortification of sales and rental markets, encouraging cooperatives to redistribute lands to their members, reduced government ownership, and creation of land banks.

Labor is often the most valuable resource the rural poor possess. Labor markets in developing countries often contain imperfections due to power imbalances, imperfect information, and transactions costs. Government interventions into labor markets are often justified based on this observed poverty.

IMPORTANT TERMS AND CONCEPTS

Capitalistic agriculture	Political stabilization
Compensation	Property rights
Corporate farms	Public capital formation
Entrepreneurial incentives	Semi-feudal land tenure
Family farms	Socialist agriculture
Group farms	State farms
Land reform	Successful land reform
Land tenure	Tenancy reform
Marketable surplus	Transactions costs
Casual labor	Permanent labor

Looking Ahead

In this chapter, we considered institutional changes related to land and labor. In the next chapter, we consider institutional changes related to input and credit policies. Governments often intervene in input and credit markets. We will examine the nature and advisability of these interventions.

QUESTIONS FOR DISCUSSION

1. What is land tenure?
2. What are the major ways farms are organized?
3. What are the major types of tenancy arrangements?
4. What is land reform?
5. How does an anti-feudal land reform differ from land reforms within a capitalist or socialist agrarian structure?
6. Why is a land reform often necessary?
7. Why is a land reform difficult to achieve?
8. How do you judge if a land reform has been successful?
9. Why are large landholdings in a densely populated country bad?
10. What are the requisites of a successful land reform?
11. What pressures might population growth or new technologies place on existing land tenure arrangements?
12. Why is it important for a land reform to take place quickly?
13. What services must accompany a land reform?

14. What distinguishes casual labor from permanent labor and why do both exist?
15. Why are transactions costs a problem in labor markets?

RECOMMENDED READING

Bell, Clive, "Reforming Property Rights in Land and Tenancy," *World Bank Research Observer,* vol. 5, no. 2 (July 1990), pp. 143–66.

de Janvry, Alain, "The Role of Land Reforms in Economic Development: Policies and Politics," *American Journal of Agricultural Economics*, vol. 63 (1981), pp. 384–92.

Deininger, Klaus, and Hans Binswanger, "The Evolution of the World Bank's Land Policy: Principles, Experience, and Future Challenges," *World Bank Research Observer*, vol. 14, no. 2 (August 1999), pp. 247–76.

Feder, Gershon, and Klaus Deininger, "Land Institutions and Land Markets," *World Bank Policy Research Working Paper Series*: 2014 (1999).

Ray, Debraj, *Development Economics* (Princeton, N.J.: Princeton University Press, 1998), chs 12-13.

Wortman, Sterling, and Ralph W. Cummings, Jr., *To Feed This World: The Challenge and the Strategy* (Baltimore, Md.: Johns Hopkins University Press, 1978), pp. 271–88.

CHAPTER 14

Input and Credit Markets

> To take maximum advantage of technological advances in farming systems, farmers must have access to recommended production inputs at the specific times and in the quantities and qualities needed; (and) access, if necessary, to outside sources of finance to purchase these inputs.
> —Sterling Wortman and Ralph W. Cummings[1]

This Chapter

1. Explains why it is important for farmers to have access to purchased inputs and to credit
2. Describes the nature of rural money markets and the determinants of rural interest rates
3. Discusses why governments tend to subsidize input prices and credit and why these subsidies are generally inadvisable.

IMPORTANCE OF NEW INPUTS

Successful agricultural development in most developing countries today requires increased output per hectare and per worker. This agricultural intensification depends in part on the availability and financing of new, often manufactured, inputs (see Box 14-1). Fertilizers and pesticides, new seeds, irrigation systems, mechanical power, and supplemental minerals and nutrients for animals are examples of these inputs. Governments must address a series of issues related to production, distribution, pricing, financing, and regulation of inputs, and to the identification and encouragement of optimal on-farm input usage.

Role of Manufactured Inputs

Manufactured inputs have an important role to play in agricultural development because the potential for expanding the land resource is limited in most countries. This scarcity or inelastic supply of land means

---

[1] Sterling Wortman and Ralph W. Cummings, Jr., *To Feed This World: The Challenge and the Strategy* (Baltimore, Md.: Johns Hopkins University Press, 1978), p. 343.

BOX 14-1
INPUT USE AND AGRICULTURAL OUTPUT IN INDIA

The contributions of modern inputs to increases in agricultural output are visible throughout Asia. India provides an interesting illustration. Production of foodgrains in India grew consistently from 1949 to 1984, averaging roughly 2.6 percent annually. Prior to 1967, few modern inputs were used by Indian farmers, and output growth was evenly attributable to increases in areas planted and increases in yields (see table). Following 1967, the growth rate of area planted fell dramatically, while growth in total output remained strong. Sustained increases in output during this latter period were created by widespread use of modern inputs, mostly irrigation water and fertilizer, and modern high-yielding varieties (HYVs). Yield-based growth contributed to more than 90 percent of the growth in production during the period from 1975/76 through 1983/84.

GROWTH RATES IN FOODGRAIN PRODUCTION AND INPUT USE IN INDIA, 1949/50–1983/84

	Annual growth rate (%)					
	Foodgrains			Inputs		
Period	Production	Acreage	Yield	Irrigation	Fertilizer	HYVs
1949/50– 1964/65	2.84	1.41	1.41	—	—	—
1967/68– 1975/76	1.91	.40	1.50	3.2	11.4	20.8
1975/76– 1983/84	2.48	.19	2.28	2.8	12.3	6.5

Source: J. S. Sarma and Vasant P. Gandhi, *Production and Consumption of Foodgrains in India: Implications of Accelerated Economic Growth and Poverty Alleviation*, International Food Policy Research Institute Report No. 91 (Washington, D.C., July 1990).

that its price tends to increase over time, both absolutely and relative to the price of labor. The induced-innovation theory described in Chapter 11 indicates that farmers will seek new agricultural technologies that will enable them to substitute lower cost inputs for those whose scarcity and price are rising. Agricultural research, described in detail in Chapter 12, will create the plant varieties that are responsive to these inputs. New, higher-productivity inputs include new seeds, fertilizer, irrigation, and pesticides.

Seeds, fertilizer, irrigation, and pesticides tend to be highly complementary inputs. To be more productive than traditional varieties, new varieties of wheat, rice, corn, and other food crops require more fertilizer and better water control than would be used under traditional practices. Water and fertilizer tend to induce lush plant growth and an environment favorable to weeds and other pests, thus raising the profitability of pesticides as well. If this package of inputs is available to farmers together with the necessary financing and information on usage, land and labor productivity can be raised. The result is an increase in output per hectare and per unit of labor applied, at least in those areas where the new inputs are suited and adopted. A description of these inputs will help better define their potential and limitations.

Seed. Seeds of high-yielding varieties are usually a relatively low-cost input. However, seed of superior varieties must be developed or identified, tested, produced and multiplied, monitored for quality, and distributed to farmers. The government often has a role to play in the development, testing, quality monitoring, and production of basic seed. Private firms can be involved in the multiplication of seeds and distribution to farmers. The exact roles of the public and private bodies in a particular country may change as the seed industry develops. As hybrid seeds continue to spread for crops such as maize, rice, and eggplant, the importance of the seed industry grows as farmers planting hybrids cannot save and use their seed from the previous crop if they expect to maintain productivity. One of the concerns with genetically modified crops is that one or a few seed companies may own the intellectual property rights associated with the new seeds and therefore may charge a significant seed premium (see Chapter 12 for further discussion). The government has a role to play in ensuring no undue exercise of monopoly power by seed companies.

Fertilizer. Higher-producing varieties require additional fertilizer, particularly nitrogen, phosphate, and potash. These nutrients can be obtained from natural fertility in the soil, animal and plant wastes, and leguminous plants that can fix nitrogen from the air. These natural sources often, but not always, must be supplemented by chemical (commercial) fertilizers to provide the necessary quantities and precise mixtures required. In areas where the supply of natural fertilizers is relatively inelastic, as commercial fertilizers become less expensive and are available in relatively elastic supply, their use can be expected to increase.

Unfortunately, for many remote, particularly upland, areas of developing countries, suitable high-yielding grain varieties have not been developed. Or the remoteness and lack of infrastructure hinders the distribution of chemical fertilizer to farmers. In these areas, traditional varieties and practices may continue to dominate and incomes remain low or even decline as higher production in more favored regions drives prices lower.

Water. Availability of irrigation water significantly influences the number of crops grown per land area per year, the inputs used, and hence production. Higher levels of fertilizer application require more and better-timed water input. Drainage is also important because few crops can tolerate excessive standing water or salinization. Several important factors complicate irrigation decisions. Irrigation infrastructure requires large financial investments, and governments often provide funds or encourage private entities to provide funds for such endeavors. Efficient water management requires proper pricing mechanisms: private users of water might over or under use irrigation water if it is not properly priced. Proper pricing is, however, complicated by difficulty in measuring the amount of water used. Water system management can have important direct effects on human health; malaria and schistosomiasis are common tropical diseases whose vectors thrive in standing water. Because of these considerations, development and management of irrigation and drainage systems often require a combination of public and private initiatives. Governments can seek to expand and modernize irrigation and drainage facilities. They can design rules of water pricing to encourage economically sound water use. Farmers and villages themselves can develop smaller, often well-based, systems and the necessary canals for distribution on farms along with rules for water distribution.

Pesticides. Farmers often find using pesticides (insecticides for insects, fungicides for diseases, and herbicides for weeds) highly profitable, as agricultural production intensifies through increased use of new seeds, fertilizer, and water. Sometimes these pesticides are applied as a preventative treatment and other times after a major pest problem develops.

Pesticides can have serious drawbacks, however. Some pesticides are toxic to humans and animals, and result in poisonings in the short run or chronic health problems in the longer term. Applications with improper equipment or inadequate clothing for protection exacerbate health problems. Improper storage and handling can create adverse health consequences. Chemical pollution can spread beyond the area where the pesticide is applied, with particularly deleterious effects on fisheries. Some pesticides kill insects that are beneficial to agriculture. Often, when pesticides are applied over a period of time, the target insects, diseases, or weeds develop resistance to the chemicals, making increased pesticide amounts necessary to maintain the same level of effectiveness.

Pesticides, despite these problems, will likely be needed for some time until new pest-resistant varieties, biological and cultural practices, and other substitute methods for pest control can further be developed. Several of these methods, called Integrated Pest Management or IPM, have already been developed and implemented for certain pests on certain crops in certain locations. Much additional research is needed, however, to make these practices more widely available in developing countries. Weed control is especially important to intensified production in Africa where labor for weeding is less abundant than in other regions.

Animal Inputs. As discussed in Chapter 7, livestock play an important role in farming systems in developing countries. Animal productivity is often low, and new inputs related to disease control, supplementary minerals and other feed supplements, improved shelter, and, in some

---

BOX 14-2
INTEGRATED PEST MANAGEMENT
IN ECUADORIAN HIGHLANDS

Potato producers in the Ecuadorian highlands face a number of important pests including Late Blight, Andean Weevil, and the Central American Tuber Moth. Late Blight is controlled through heavy applications of fungicides, while the latter two pests are controlled by spraying Carbofuran, a highly toxic pesticide. Farmers combine pesticides into a mixed "cocktail" containing as many as 12 chemical agents and spray their fields with up to 8 applications in a single season. Farmers complained about the high financial cost of these applications and health costs associated with chemical misuse. IPM techniques were developed through a USAID-sponsored research project (IPM-CRSP) and included identifying and screening Late Blight-resistant varieties, use of insect traps, more targeted spraying of a low-toxicity alternative, field sanitation and cultural techniques, and the use of biological control alternatives. Local farmers were invited to participate in Farmer Field Schools and field days to learn about the techniques. Pesticide applications were reduced by half, and experience on farmer fields showed that the IPM package yielded more than $600 per hectare in net benefit compared to alternative practices. This example shows that IPM technologies can be profitable at the same time that they reduce exposure to harmful chemicals.

Source: Alwang *et al.*, "Developing IPM Packages in Latin America," in *Globalizing Integrated Pest Management: A Participatory Process*, ed. G. W. Norton, E. A. Heinrichs, G. C. Luther, and M.E. Irwin (Ames, Iowa: Blackwell Publishing, 2005).

cases, better breeds can make a difference. Inputs for controlling diseases and parasites are perhaps the most important; the significance of feed supplements, shelter, and new breeds varies from country to country and by type of livestock. Because indigenous livestock have been adapted to their specific environments, the transfer in of new breeds is particularly complex, except perhaps for poultry.

Mechanical Inputs. Agricultural mechanization is frequently a controversial subject. Tilling, planting, cultivating, and harvesting are still done by hand in large parts of the developing world, particularly in Sub-Saharan Africa and in hilly regions on other continents. In many areas of Asia and Latin America, animals are an important source of power. Even in countries where farming is more mechanized, power tillers and tractors are often restricted to tillage and a few other operations.[2] The controversy arises because machinery usually substitutes for labor or animals. In many developing countries, labor is abundant and its cost is low. Alternative employment opportunities outside agriculture are limited, so that labor displacement is undesirable. Therefore, mechanization is most profitable in countries where land is abundant, labor is scarce, and capital is cheap; this situation would seem to exist in relatively few countries,

Does this mean that there is little role for agricultural mechanization? Not necessarily, but the types of mechanization should be different from what is observed in most Western developed countries. Highly productive cropping systems, whether on small or large farms, can often benefit from more precise planting depths and fertilizer placement, mechanically pumped irrigation water, mechanical threshing (but usually not harvesting unless labor is scarce), transport, power spraying of pesticides, and tilling when timing is critical for multiple cropping. Many of these mechanical devices, however, may be handheld (e.g., sprayers) or stationary (e.g., pumps and threshers). Even in areas where labor is usually abundant, shortages can occur in certain seasons, which, if relieved through mechanization, could increase the overall demand for labor.

Individual farmers will consider the private profitability when deciding whether to invest in a machine. If very large farms exist in countries with surplus labor in agriculture, operators of these farms may prefer labor-saving machinery because it allows them to deal with fewer employees, and, given the transactions costs and capital subsidies that may exist, it may be more privately profitable to follow

---

[2] See Hans Binswanger, "Agricultural Mechanization: A Comparative Historical Perspective," *World Book Research Observer*, vol. 1 (January 1986), pp. 27-56.

Treating cattle for parasites in Ethiopia.

large-scale mechanization even if society as a whole would be better off without it. Such behavior is one of the reasons why land reform is so important to many developing countries (see Chapter 13).

Governments and foreign assistance agencies must be careful not to encourage non-optimal mechanization (from society's viewpoint) through ill-advised subsidies or other means. Mechanization is inevitable over time, but the type of mechanization should be appropriate given the relative endowments of land, labor, and capital. Certain government policies, such as those influencing exchange rates, indirectly affect the prices of capital-intensive inputs such as machinery. Impacts on relative prices of inputs should be considered during policy formulation.

Input Markets

Developing countries often subsidize the purchase of seeds, fertilizers, irrigation water, pesticides, and occasionally mechanical inputs. Is this a good idea? Generally speaking, it is not. Such subsidies can lead to losses in economic efficiency for the country as a whole, can be costly to the government, can discourage private-sector competition in the provision of these inputs, and, particularly in the case of pesticides, may lead to environmental damages from over-application.

Governments frequently become involved in multiplying and selling improved seeds to farmers at or below cost. In some cases, scarce research resources are diverted to multiplying seeds rather than developing new varieties. In other cases, research systems are forced to focus on selling seeds to pay for operating costs for the system. As a result, private firms, unable to compete with the government treasury, do not take on the function of multiplying and selling seeds. Without development of these private firms, the government must continue to be responsible for this function.

Fertilizer subsidies can be used in selected situations in which governments desire to increase the adoption of new inputs by groups of farmers who might not otherwise adopt them. Unfortunately, in many countries these subsidies are necessitated by the artificially low prices imposed on agricultural outputs for the purpose of keeping food prices down for urban consumers. Input subsidies help compensate farmers for income losses from these policies. While this combination of policies can have the desired effect, at least in the short run, high costs to the government and potential fiscal problems result in the long run, making the policies non-sustainable. Also, the economic efficiency losses associated with these policies can be substantial. Finally, studies show that, from the farmer's perspective, access to inputs can be more important than their prices. Subsidy policies and government involvement in input markets often lead to shortages of inputs and their rationing, which are harmful to long-run growth.

Efficiency losses are also a problem for water and pesticide subsidies. The latter can create excessive use of toxic chemicals and can result in all the deleterious effects described earlier. In summary, input subsidies are generally inadvisable. The government can play a more constructive role by ensuring the availability of these inputs (including the improvement of rural roads), publishing price information to encourage competition, setting quality standards for seeds and fertilizers, requiring and enforcing labeling of input containers, and regulating use of toxic pesticides and transgenic seeds.

Role of Credit

Access to credit becomes important as a developing country moves from traditional to more modern agriculture. Credit helps farmers purchase inputs such as seeds, fertilizers, and chemicals. It facilitates purchase of durable productive inputs such as machinery, and helps households better manage their resources. Credit can be used for input purchases, investment, marketing, and consumption. Without credit, even high-return investments, long or short term, would be infeasible for many farmers. Loans enable farmers to better manage risks since they can borrow during bad years and pay back the loans during good years. Even within cropping seasons, short-term credit is used to smooth consumption and provide cash at times of acute needs.

BOX 14-3
MODERN INPUTS AND ECONOMIC GROWTH

New technologies and inputs help achieve increases in agricultural output and income in rural areas of developing countries. This income is spent by the households on goods and services, some of which are produced locally and others that are imported into the region. These expenditures induce income growth in the non-farm economy, the so-called *multiplier effects*. By far the largest portion of these multipliers is caused by household expenditures on consumer goods and services, though the effects resulting from increased use of farm inputs and in processing, marketing, and transportation of farm output are substantial contributors to regional growth.

Linkages between farms and suppliers of inputs also create spillovers into the local economy. Though seeds, agrichemicals, irrigation supplies, and farm machinery are usually not produced in agricultural regions, input supply services including technical advice, machinery repair, and a large proportion of irrigation construction and maintenance can be produced locally. These activities create opportunities for non-farm employment and income that is in turn spent locally. The creation and deepening of backward linkages from agriculture are important contributions to rural economic development.

Without widespread access to credit, inputs associated with improved technologies can be purchased only by wealthier farmers. Capital formation and improvements on smaller farms can be hampered. Fewer farmers are able to purchase or even rent land. In cases where produce marketing requires cash outlays, lack of credit can disrupt marketing activities. Well-functioning rural financial institutions are essential to improving economic efficiency, reducing income risk, and meeting income distribution goals.

NATURE OF RURAL MONEY-MARKETS AND DETERMINANTS OF RURAL INTEREST RATES

Finance in rural areas consists of three components: credit (borrowing and lending), saving, and insurance. These components frequently overlap, since savings can be used for capital purchases (and hence substitute for credit) or as a safety net or insurance substitute. Developing-country households use complex strategies to increase their productive capacity, share risk, and manage purchases of food and other goods over time. Access to finance helps determine the suitability of such strategies. Better understanding of rural financial markets requires consideration of the three components, and efforts to strengthen one component may be compromised by weaknesses in others.[3]

Credit facilitates the temporary transfer of purchasing power from one individual or organization to another. However, many types of lenders or *money-markets* exist, and credit institutions may or may not adequately serve the needs of a developing agriculture. Credit is often viewed as an oppressive or exploitive device in developing countries. We need to examine both the types of lending sources found in developing countries and the evidence of exploitive behavior associated with these money-markets.

Types of Money-markets

Rural money-markets consist of two broadly defined lending sources: organized (or formal) and informal. Private commercial banks, government-controlled banks, cooperative banks, and credit societies are called organized credit sources. Public or private, these lending sources are usually regulated by the government and are open to audit and inspection. In addition to credit, they may provide other financial services such as savings and certain forms of insurance. In general, formal money markets have historically not served well the needs of small- and medium-scale farmers in developing countries. As a result, over time many developing-country governments have intervened to promote better access to formal financial services, primarily credit. These programs have been largely unsuccessful for many reasons, one of which is that they did not recognize alternative, informal credit sources that are found throughout the developing world.

Informal or unorganized credit sources consist of moneylenders, merchants, pawnbrokers, landlords, friends, and relatives. Some credit sources — e.g., landlords and merchants — combine other economic activities with lending. Except for absentee landlords, the relationship between borrower and informal lender is marked by personal contact, simple accounting, and low administrative costs.

Informal lenders are important sources of funds in many rural areas. These lenders usually know the borrowers personally, require little collateral, make consumption as well as production loans, are accessible at all times, and are usually flexible in rescheduling loans. However, these informal lenders also tend to charge high rates of interest and are frequently accused of exploitive activities. In cases where lenders are landlords, merchants, or both, they have been accused of using

---

[3] See Manfred Zeller and Richard L. Meyer, *The Critical Triangle of Microfinance: from Vision to Reality* (Baltimore, Md.: Johns Hopkins University Press, 2003) for more details.

their position to tie borrowers to themselves by forcing their clients to rent from, borrow from, buy from, and sell to them. Thus, these agents are said to extract monopoly profits from their clients. Are borrowers being consistently exploited? It is important to examine this question because it has important implications for the role of more formal private and public credit institutions.

Do Informal Money-markets Exploit Borrowers?

The issue of borrower exploitation revolves around the existence of usury or monopoly profits earned by the lenders. Hence we need to consider the factors that determine the interest rates charged by these lenders. The major components of rates of interest on loans are: (1) administrative costs, (2) the opportunity cost of lending, (3) a risk premium due to the probability of default in repayment, and (4) monopoly profit.[4]

Administrative costs should not be too high for moneylenders, given simple contracting procedures and personal knowledge of clients. Many loans with small amounts of money per loan increase administrative costs, but these costs are probably not excessive. Opportunity costs of lending are low in rural areas because interest rates offered by organized money markets tend to be low. Therefore, the critical factor in determining whether interest rates are generating monopoly profits in the informal money market is the risk premium or the probability of default. The risk premium for loans to small-scale, particularly tenant, farmers can be high. These farmers are close to the margin of subsistence, and a streak of bad weather or a serious illness can spell disaster. Without formal collateral, the risk of default grows. Because weather tends to affect all farmers in a given area, a spell of bad weather creates potential for simultaneous default of many borrowers. Therefore, one would expect relatively high interest rates just to cover the risk factor. Exploitive situations do exist in which moneylenders extract monopolist gains. However, careful empirical studies seem to indicate that monopoly profits may not be as prevalent or large in informal credit markets as is often believed.[5] The reason is competition. The amounts of the loans are often small, and start-up costs required to become a moneylender are low. This ease of entry serves to keep interest rates at an appropriate level given the level of risk, administrative costs, and the opportunity cost of capital. If profit margins become large,

---

[4] See Subrata Ghatak and Ken Ingersent, *Agriculture and Economic Development* (Baltimore, Md.: Johns Hopkins University Press, 1984), p. 231.

[5] See, for example, P. Bardham and A. Rudra, "Interlinkage of Hand Labor and Capital Relations: An Analysis of Village Survey Data in East Asia," *Economic and Political Weekly* (1978), pp. 367–84.

incentives are created for new moneylenders to enter the business and compete away those profits.

High risks associated with loans to subsistence farmers, however, mean that lenders have incentives to maintain tight control over borrowers. Moneylenders who are also landlords or merchants have means of tying their clients to themselves through leases, consumer credit, and so forth. Other moneylenders may be hesitant to lend to someone who already owes substantial sums or who has defaulted to another. In summary, it appears that some exploitation by moneylenders does occur, particularly if the moneylenders control the land or the market. However, the magnitude of this exploitation may not be as great as is often believed. Evidence of high interest rates on rural loans is alone not sufficient to conclude that moneylenders are exploitive, since there are high costs associated with making these loans. Informal sources of credit serve a vital function in most developing countries because, without them, most small farmers would not have access to credit.

Organized Money-markets and Transactions Costs

Why are small farmers in developing countries not served more frequently by organized money-markets? Both private and public financial institutions find that transactions costs are high. Loans, savings, and insurance needs are small, and the paperwork and time spent evaluating potential clients, collecting payments, and supervising loans in order to reduce risks of default are costly. In many cases, the government regulates the maximum interest allowed, and that rate will fail to cover the administrative costs and risk. Thus, where private and public sources of finance exist, they tend to deal with larger-scale farmers to reduce administrative costs and the chances of default (see Box 14-4).

The magnitude of these transactions costs is illustrated by a relatively successful bank that provides credit to the rural poor in Bangladesh. The Grameen Bank of Bangladesh targets households that own less than 0.5 acres of cultivable land.[6] The bank organizes its clients into groups and associations, provides credit without collateral, and supervises utilization of the loans. A maximum amount (the equivalent of about $150) is lent to individuals within a group of five members. Nearly three-quarters of the borrowers are women. Peer pressure

[6] See Mahabub Hossain, *Credit for Alleviation of Rural Poverty: The Grameen Bank of Bangladesh*, International Food Policy Research Institute Research, Report No. 65 (Washington, D.C., February 1988), for an excellent discussion of the Grameen Bank; see also Mark M. Pitt and Shahidur Khandker, "The Impact of Group-based Credit Programs on Poor Households in Bangladesh: Does the Gender of Participants Matter?", *Journal of Political Economy* (1998), pp. 958–96, for information on the impacts of such programs.

BOX 14-4
ADMINISTRATIVE COSTS AND LOAN SIZE
The cost of lending to farmers includes relatively large fixed costs to pay for administration and bookkeeping. To cover these costs, interest rates must be higher for smaller loans, even if all borrowers have equal risk of default. For example, if the variable cost of capital is 10 percent but the bank incurs a fixed cost of $10 to administer each loan, for the bank to break even on each loan it must charge a total of 20 percent interest to those who want to borrow $100 for repayment after one year. In contrast, those who want to borrow $1,000 would have to pay only 11 percent.

---

together with close supervision ensures repayment rates of more than 90 percent. The interest rate charged is 16 percent a year and the default rate is less than 2 percent. The bank is subsidized, however, by the State Bank of Bangladesh and by the International Fund for Agricultural Development (IFAD). The interest on the loans would be around 5 to 10 percent higher than it is if the bank had, to break even, to borrow at the same rate as the other financial institutions in the country.[7] Because the default rate and the opportunity cost of capital are low, it is clear that most of the roughly 16 percent interest rate charged is administrative cost. The bank could lower this cost with less supervision, but the default rate would likely rise and offset the cost saving.

The Grameen Bank also lends very little money for activities associated with crop production. In general, this type of *micro-credit* lending serves mostly for livestock and poultry, for small-scale processing and manufacturing, and for trading and shopkeeping. These activities are less risky than crop production. The Grameen Bank model has been improved upon through sequential experimentation, and micro-finance organizations around the world are now flourishing without subsidization. These organizations and some of their organizing principles are discussed below.

One can see that small agricultural loans are costly and, as a result, most commercial lenders loan to larger farmers where the risk and administrative costs are lower. Government-supported credit programs often have subsidized interest rates in developing countries, but these rates tend to encourage loans to large farmers (many micro-finance institutions are exceptions).

Transactions costs for private or public loan transactions can also be high because of fraud, favoritism, or embezzlement of funds from within the system. This situation arises most frequently when loans

[7] Hossain, *Credit for Alleviation of Rural Poverty*, p. 11.

are subsidized, creating excess demand for credit and incentives for bribery.

Government-assisted Credit Programs

Many governments use credit programs as part of their development program, and many international donors support these programs. Government-supported credit is based on the notions that (1) credit is critical to the adoption of new technologies, (2) moneylenders exploit farmers and public credit can provide them with competition, (3) credit can be combined with supervision and education to increase the capacity of farmers to use modern inputs, (4) subsidized credit can offset disincentives to production created by other policies that discriminate against agriculture, and (5) government-supported credit programs can lessen inequities in the rural sector.[8]

Subsidized credit provides an easy vehicle for transferring public funds to the rural sector. Examples of subsidized credit programs abound in every region of the developing world. Dale Adams points to a number of studies of such programs in Honduras, Sudan, Jamaica, and elsewhere that demonstrate how ill-advised subsidized credit is from a development perspective.[9] Adams finds that while evaluations of subsidized credit programs often find favorable impacts on individual borrowers, a broader examination of the net effects of these programs usually finds few positive effects on economic development. They can also compromise the viability of the rural financial system.

Effects of Subsidized Credit

Subsidized credit creates excess demand for credit by lowering interest rates. In many cases, because interest rates are negative after controlling for inflation, the demand for credit is infinite. Subsidized credit erodes the capital available in financial markets and undermines rural financial institutions. Private banks cannot cover expenses (e.g., administrative costs, defaults) at low interest rates, yet they are forced to lower interest rates in order to remain competitive with public credit sources, even if not required to lower them by law. Thus, the survival of these private institutions is threatened. If they fail, additional government involvement and additional budget outlays will be needed. An equally important effect of subsidized loan rates is that they lower all interest

---

[8] See Yujiro Hayami and Vernon Ruttan, *Agricultural Development: An International Perspective* (Baltimore, Md.: Johns Hopkins University Press, 1985), pp. 398-403.

[9] See Dale W. Adams, "The Conundrum of Successful Credit Projects in Floundering Rural Financial Markets," *Economic Development and Cultural Change,* vol. 36 (January 1988), pp. 355–67.

Bangladesh families have benefited from Grameen
Bank loans.

rates and, hence, discourage private savings. If agricultural development is to be able to generate capital, then viable rural financial institutions are needed both to provide loans and to mobilize savings.

Because subsidized credit generates excess demand for loans, credit
is rationed and almost inevitably goes to the larger farms for which the
administrative costs are lower. The phenomenon of successful impacts
of subsidized credit on individual borrowers, yet negligible effects on
overall development, exists because rationed credit means that few
people are touched by the programs. Seldom are more than 5 percent of
the potential credit recipients reached. Because the subsidized loans
are valuable, the credit system can become politicized as large landowners offer favors to bank managers to obtain loans or financially support politicians to encourage continuation of the program. In addition
to these distributive effects, default rates on subsidized loans tend to
he high. Public sector lenders may be less familiar with the borrowers
than are the lenders in the private sector. Because there are pressures to
lend to larger borrowers, the productive potential of the loan may not
be considered, leading to high default tales.

Innovations in Rural Finance

Since the early 1970s, a gradual revolution in lending to the poor has
been occurring in a number of developing countries. As countries learn

from their failures with subsidized credit programs, and as experience grows with targeted small, group-loan programs such as the Grameen Bank in Bangladesh, the large-scale provision of small loans to low-income people has expanded. As of 2001, more than 1,500 micro-finance institutions (MFIs) with 54 million members existed in 85 developing countries.[10] The Grameen Bank, however, while demonstrating that poor people can be good credit risks given a credit program structured with proper incentives, has remained somewhat constrained by its inability to operate entirely without subsidy. Since the late 1980s, the *poverty lending* approach of banks such as the Grameen Bank has been challenged by a more commercially oriented *financial systems* micro-finance approach.[11] While both approaches focus on the poor, the latter emphasizes savings services to the poor as well as loans.[12] A financial systems approach enables banks to generate sufficient resources not only to be potentially sustainable but to provide opportunity for the economically active poor (as opposed to the extremely poor) to save and invest at a decent return during times when they are able to save. Examples of such micro-credit banking systems that have proven profitable are found in countries as diverse as Bolivia and Indonesia. Micro-financial institutions are also experimenting in offering the third component of the finance trinity: micro-insurance. These institutions have incentives to help their clients manage risks, since a poor outcome may lead to default on loans.

These new experiences in providing finance to small-scale and poor farmers have been built on a number of principles. The first principle is recognition that credit, savings, and insurance are interlinked, and efforts to provide one component should consider impacts on the others. Second, funds are fungible and can be used for things other than input purchases, such as consumption and other emergency needs. By better enabling risk management, access to financial services can improve income generation over time. A third principle is that, for sustainability, credit providers should charge what the loans cost, which is usually more than is charged to large commercial borrowers because of transactions costs on many small loans. Some of those costs arise from the necessity of screening credit applicants. The key innovation of micro-finance is the use of joint group liability; loans are made to groups and,

---

[10] Cecile Lapenu and Manfred Zeller, *Distribution, Growth and Performance of Microfinance Institutions in Africa, Asia, and Latin America*, IFPRI FCND Discussion Paper No. 114 (Washington, D.C., 2001), p. 111.

[11] See Marguerite S. Robinson, *The Micro-finance Revolution: Sustainable Finance for the Poor* (Washington, D.C.: World Bank, 2001), p. 7.

[12] Lapenu and Zeller, *Distribution, Growth and Performance*; note that of the 54 million members of MFIs worldwide, 44 million are savers and 23 million are borrowers.

if one member of the group defaults, the entire loan is considered to be in default. Using group liability, groups use local knowledge to screen members and moral suasion to enforce repayment. These factors reduce transactions costs and improve repayment rates. Commercially oriented micro-finance provides formal competition for informal moneylenders, and hence is most likely to succeed in precisely those areas where moneylender profits are excessive due to local monopoly power.

Lessons for Credit Policies

Several lessons emerge from the applied research on rural credit. First, adoption of new technologies often requires purchase of modern inputs. Consequently, credit availability has been found to be more important to development than the interest rate charged, and there is a tradeoff between credit availability and subsidized interest rates. Second, the viability of rural financial institutions is jeopardized by subsidized credit. This weakening of rural financial markets can constrict both the supply of and demand for credit. The rural poor are penalized on their deposits as well as their loans.[13] Third, credit is *fungible*; in other words, it may not be used for its intended purpose. It is easy for subsidized production credit to be used for consumption items or nonproductive assets. This fungibility is not necessarily bad unless it raises the default rate, but should at least be understood by policymakers. Fourth, a key to reducing market interest rates is to reduce agricultural risk (and hence defaults) and the transactions costs associated with lending and borrowing. Higher income levels associated with economic development may help reduce the risk of defaults as may certain crop insurance and other government policies discussed in Chapter 15. Improved roads and other means of communication, and in some cases, group borrowing and guarantee of loans, can help reduce transactions costs.

Because administrative costs per dollar lent to small farms are higher than to large farms, banks must either charge higher interest rates (or other hidden charges) to small farms than large, or give loans mainly to large farms. Thus, many countries need a land reform or the credit system will also work against the poorest farmers.

SUMMARY

Successful agricultural development requires increased output per hectare and per worker. This agricultural intensification depends on the availability of new, often manufactured, inputs. Seed, fertilizer, pesticides, irrigation, mechanical power, and supplementary minerals and

[13] See Adams, "The Conundrum of Successful Credit Projects," p. 366.

feeds are examples of these inputs. Manufactured inputs can substitute for inelastic supplies of land to increase production at a lower per unit cost. A variety of issues must be resolved by each country, however, with respect to externalities associated with certain inputs such as pesticides, the appropriate types of mechanization, and the role of the government in producing, distributing, and financing inputs. Governments often subsidize inputs. These subsidies can discourage private competition for input supply, can be costly to the government, and may encourage overuse of inputs such as pesticides that create externalities.

Credit is essential as a country moves from traditional to modern agriculture. Credit from informal sources such as moneylenders is often viewed as oppressive. However, risks and administrative costs of loans to small farms are high and, given the typical competition among moneylenders, monopoly profits may not be as prevalent or as high as is often portrayed. When moneylenders are also landlords or merchants, the chances of exploitation are greater. Formal private and public lenders do not serve a high proportion of farmers because risks and transactions costs are high. Because governments frequently subsidize interest rates, rationed credit tends to go to the larger farms. The subsidies erode the capital in the financial system, and, thus, the number of farms served. Low interest rates also discourage deposits and reduce the ability of formal private banks to compete. Credit, savings, and insurance are interlinked, and efforts to provide one component should consider impacts on the others.

IMPORTANT TERMS AND CONCEPTS

Administrative costs

Exploitation

Fertilizers and pesticides

Fungibility

Grameen Bank

Group lending

Informal credit sources

Input subsidies

Integrated pest management

Irrigation systems

Mechanical power

Micro-finance

Moneylenders

Money-markets

Monopoly power

New seeds

Opportunity costs of lending

Organized credit sources

Purchased inputs

Risk of default

Subsidized credit

Looking Ahead

Governments often intervene in agricultural markets to influence prices. In the next chapter we examine why governments intervene and the effects of those interventions. Efficient marketing systems are essential for agricultural development, and we consider the role that govern-ments can play in improving the marketing system.

QUESTIONS FOR DISCUSSION

1. Why are manufactured agricultural inputs usually necessary for ag-ricultural development?
2. What are some of the key manufactured inputs needed?
3. In what manner are agricultural inputs complementary in nature?
4. What are the advantages and disadvantages of pesticides?
5. Why is farm mechanization a controversial issue?
6. Why do governments subsidize the purchase of manufactured inputs?
7. Why is agricultural credit important to agricultural development?
8. How do organized and informal sources of credit differ?
9. Why might bankers be biased against small farmer loans in devel-oping countries?
10. What factors would you examine if you were trying to assess whether interest rates charged in informal money-markets were exploiting borrowers?
11. What are subsidized interest rates? Are they a good idea for getting agriculture moving?
12. What might be one problem associated with the fact that credit is fungible?
13. Why do governments support credit programs?
14. How might transactions costs associated with rural financial mar-kets be reduced?

RECOMMENDED READING

Adams, Dale W., "The Conundrum of Successful Credit Projects in Flounder-ing Rural Financial Markets," *Economic Development and Cultural Change*, vol. 36 (January 1988), pp. 355–67.

Adams, Dale W., Douglas H. Graham, and J. D. Von Pischke, eds, *Undermining Rural Development with Cheap Credit* (Boulder, Colo.: Westview Press, 1984).

Binswanger, Hans, "Agricultural Mechanization: A Comparative Historical Analysis," *World Bank Research Observer*, vol. 1 (January 1986), pp. 27–56.

Bouman, F. J. A., and R. Houtman, "Pawnbrokering as an Instrument of Rural Banking in the Third World," *Economic Development and Cultural Change*, vol. 37 (October 1988), pp. 69–99.

Ghatak, Subrata, and Ken Ingersent, *Agriculture and Economic Development* (Bal-timore, Md.: Johns Hopkins University Press, 1984), pp. 227–37.

Hayami, Yujiro, and Vernon W. Ruttan, *Agricultural Development: An Interna-tional Perspective* (Baltimore, Md.: Johns Hopkins University Press, 1985), pp. 398-403.

Hossain, Mahabub, *Credit for Alleviation of Rural Poverty: The Grameen Bank in Bangladesh*, International Food Policy Research Institute, Research Report No. 65 (Washington, D.C., February 1988).

Pingali, Prabhu, Yves Bigot, and Hans P. Binswanger, *Agricultural Mechanization and the Evolution of Farming Systems in Sub-Saharan Africa* (Baltimore, Md.: Johns Hopkins University Press, 1987).

Robinson, Marguerite, *The Micro Finance Revolution: Sustainable Finance for the Poor* (Washington, D.C.: World Bank, 2001).

Von Pischke, J. D., Dale W. Adams, and Gordon Donald, *Rural Financial Markets in Developing Countries: Their Use and Abuse* (Baltimore, Md.: Johns Hopkins University Press, 1983).

Wortman, Sterling, and Ralph W. Cummings, Jr., *To Feed This World: The Challenge and the Strategy* (Baltimore, Md.: Johns Hopkins University Press, 1978).

Yaron, Jacob, "What Makes Rural Financial Institutions Successful?" *World Bank Research Observer*, vol. 9, no. 1 (1994), pp. 49–70.

Zeller, Manfred, and Richard L. Meyer, *The Critical Triangle of Microfinance: From Vision to Reality* (Baltimore, Md.: Johns Hopkins University Press, 2003).

CHAPTER 15

Pricing Policies and Marketing Systems

The links between price polices and food marketing take the food
policy analyst to the very core of an economy and the most basic
issues concerning the consequences of market organization for
economic efficiency and income distribution.

— C. Peter Timmer[1]

This Chapter

1. Discusses the nature of markets, how and why governments tend
   to intervene in agricultural markets to affect prices, and the results
   of those interventions
2. Explains the importance of efficient marketing systems and describes how marketing systems have changed over time in developing countries
3. Considers the role that government can play in providing marketing infrastructure, market information, marketing services, and
   regulations.

PRICING POLICIES

Food and agricultural prices are major determinants of producer incentives and real incomes in developing countries. These prices are
influenced by government policies and by the efficiency of marketing
systems. Pricing policies and marketing systems have changed significantly over the past thirty years, especially in response to domestic
budgetary and global market pressures. The roles of government, processors, wholesalers, and retailers are changing. Governments in some
developing countries continue to adopt pricing policies that reduce food
prices for urban consumers even if farmers are forced to bear the costs.

---

[1] C. Peter Timmer, "The Relationship Between Price Policy and Food Marketing," in
Food Policy: Integrating Supply, Distribution, and Consumption, ed. J. Price Gittinger,
Joanne Leslie, and Caroline Hoisington (Baltimore, Md.: Johns Hopkins University
Press, 1987), p. 293.

In other developing countries, increased integration into global markets has resulted in freeing up of prices and in new approaches to processing, marketing, and regulating farm commodities and products. Ironically, in many developed countries where farmers are a much smaller proportion of the population, government price interventions continue to support agricultural prices, often at the expense of taxpayers and consumers, and in some cases with deleterious effects on developing countries. Why do we observe these policies? How are they implemented, and what are their short- and long-run effects? These questions are addressed below, followed in the next section by a discussion of the roles of agricultural marketing systems and how those systems have changed over time.

Reasons for Price Intervention

Governments intervene in agricultural price formation for two major reasons: to change the outcomes in agricultural markets themselves, and to raise revenue to pay for roads, police, and other public services. These interventions have a large influence on the welfare of both farmers and non-farmers. Sometimes government policies reflect the long-run interest of society as a whole, helping to stabilize and raise income for many people, but often they reflect more narrow or short-run political objectives.

The long-run interest of most societies calls for policies that (1) contribute to economic growth, (2) improve income distribution or at least meet minimum nutritional needs of citizens, and (3) provide a certain measure of food security or stability for the country over time. Governments vary widely in what they actually do, and their choice of policies helps explain the wide differences in economic outcomes across countries and over time. The choice of policy is much influenced by how governments respond to key interest groups. Urban consumers want lower food prices, particularly the poor who spend a large fraction of their income on food. Employers also prefer low food prices, which allow them to pay lower money wages. But low food prices hurt agricultural producers, and reduce investment in agriculture, thus lowering farm productivity over time.

In most developing countries, the balance of political power favors urban consumers and employers. Although farmers are in the majority, they are usually poorer, are often illiterate, and are geographically dispersed across the countryside. Thus, political power with respect to food prices is centered in urban-industrial areas.

As development proceeds and incomes grow, several factors may cause the balance of political power within countries to shift towards helping farmers, if necessary at the expense of consumers. First, food prices become less important in household budgets because the proportion of income spent on food declines with higher incomes. Second, the declining relative size of the agricultural sector makes it less costly for the government to succumb to pressures from farmers, while at the same time the reduced number and increased specialization of farms improves the ability of farmers to organize for collective action. Third, governments in richer countries have easier access to other sources of tax revenue outside the farm sector.

The form of government interventions into agricultural commodity markets also shifts as development proceeds. In the poorest countries, interventions often focus on international trade because that is easiest to control. Governments typically tax both imports and exports, and since poor countries often export farm goods and import manufactures, the result is a tax on farmers and protection for local industries. In somewhat higher-income developing countries, governments often introduce food-price subsidies, and increasingly try to support farm income as well. At the highest levels of economic development, perhaps the most important transition is toward increasingly well-targeted government programs that meet their political objectives with fewer side effects. For example, food price subsidies may be restricted to benefit only the poorest consumers, while farm subsidies may be made less distorting.

During early stages of development, agricultural policies are often highly inefficient, partly because governments have limited administrative capacity, but also because citizens who lose from bad policy may be unable to organize against them. Inefficiencies remain over time, but policies can improve as development proceeds, due partly to the structural transformation of the underlying economy but also to improvements in political accountability.

Methods of Price Intervention

Governments intervene to influence agricultural prices in several ways. They set price ceilings or floors and enforce them with commodity subsidies or taxes, manipulation of foreign exchange rates, commodity storage programs, restrictions on quantities traded, and/or other policy instruments. Let us examine how a few of these instruments work.

Suppose the government wants to lower the price of rice, an important food in the diet. The supply of rice must therefore be increased in the market relative to demand. Additional supplies can be created by increasing imports or by stimulating domestic production. In either case, government revenues must be used to bridge the gap between the

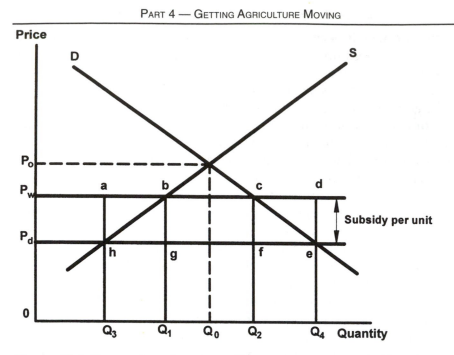

Figure 15-1. Economics of a price ceiling and consumer subsidy to lower agricultural prices.

initial price and the desired *price ceiling*. Figure 15-1 presents an illustration of how the price ceiling and subsidy might work.

The supply and demand schedules would intersect at price P_0 and quantity Q_0 if there were no trade in rice. However, the country in this example is assumed to be a rice importer, so the world price of rice, P_w, is below P_0. Initially, at P_w and without government intervention, quantity Q_1 is produced domestically, Q_2 is demanded by consumers, and the difference, $Q_2 - Q_1$, is met by imports. If the government desires to artificially create a domestic price for rice, P_d, below the world price, it must pay a subsidy per unit of rice equal to the difference between the world price and the desired domestic price $(P_w - P_d)$. This subsidy could be paid on a per ton basis to commercial importers to cover their losses for importing rice at a price below what they pay on the world market, or it could be paid to a government agency that imports rice.[2] In either case, the direct cost to the government of the subsidy is $(P_w - P_d)$ times $(Q_4 - Q_3)$, which equals area *adeh* in Figure 15-1. This kind of price ceiling program is common in many African and Asian countries. Consumers benefit but rice producers are hurt by the lower price of rice.

[2] C. Peter Timmer, *Getting Prices Right: The Scope and Limits of Agricultural Price Policy* (Ithaca, N.Y.: Cornell University Press, 1986), p. 36.

Sometimes the government prefers not to allow scarce foreign exchange to be spent on increased imports. In this case, farmers may be legally forced to sell their commodity to the government at a low price. For example, the government might force farmers to deliver Q_3 units of rice at P_d. Although nothing is imported, the demand for rice (Q_4) exceeds its supply (Q_3). The government must then ration rice to consumers. The shortage in the market provides incentives for farmers to sell their crop illegally on the *black market* for a higher price. Even if the government allows adequate imports to meet the projected demand at the lower price, if the price of the product is higher across the border, farmers will (usually illegally) sell in a neighboring country, thus further reducing domestic supplies.

One means to avoid reducing domestic production and illegal sales while at the same time supporting farm incomes is for the government to administer a two-price scheme in which producers are paid the world price but consumers pay only the subsidized price. This type of system is illustrated in Figure 15-1. Rather than paying *adeh* to importers, the government would pay $P_w bg P_d$, or a subsidy of ($P_w - P_d$) times ($Q_1 - 0$) to producers and a subsidy of *bdeg*, or ($P_w - P_d$) times ($Q_4 - Q_1$), to importers. Producers would still receive P_w, while consumers would face a price of P_d, thus the name *two-price scheme*. Of course an even higher subsidy could be paid to producers to further reduce imports and increase the producer price. The obvious difficulty with this scheme is its impact on the government budget. The subsidy costs have to be paid for by some means. Because of this cost, few major commodities are subsidized this way in developing countries, although related schemes are common in developed countries such as the United States and Japan, and in Europe. Two-price wheat programs have been operative, however, at various times in Brazil, Egypt, Mexico, and a few other low-income countries. Table 15-1 lists examples of past food subsidy programs in developing countries.

Developing countries often have food subsidy programs that are targeted toward the poor or to nutritionally vulnerable groups. These subsidies can be implemented through ration shops, ration cards, food stamps, or other means. Usually only the very poor are eligible, to keep the cost down, but in some cases ration shops, which sell basic grains and other staples, are located in poor neighborhoods under the theory that only the poor will frequent them. Alternatively, self-targeting can be achieved by subsidizing foods that the poor tend to buy, such as starchy staples or maize. Substantial savings can result from targeting: in Sri Lanka, targeting and program modification reduced outlays for

TABLE 15-1. EXAMPLES OF EXISTING OR PREVIOUS
CONSUMER PRICE SUBSIDY PROGRAMS
IN DEVELOPING COUNTRIES

Country	Principal foods subsidized	Type of program*	Food distribution	Actual coverage (implicit targeting)
Bangladesh	Wheat and rice	Price subsidy	Targeted and rationed	Mostly urban
Brazil	Wheat	Price sub*	General	Total population
China	Rice	Price sub*	General	Mostly urban
Colombia	Selected pro-cessed food	Food stamps	Targeted and rationed	Poor households with preschoolers or women who are pregnant or lactating
Egypt	Wheat	Price sub*	General	Total population
	Rice	Price sub*	Rationed	Mostly urban
	Sugar, tea, frozen meats, fish, and certain other foods	Price subsidy	Rationed	Total population
India	Wheat and rice	Price sub*	Rationed	Total population
Mexico	Maize and certain other foods	Price subsidy	General	Mostly urban
Morocco	Wheat	Price sub*	General	Total population
Pakistan	Wheat	Price sub*	Rationed	Mostly urban
Philippines	Rice and oil	Price subsidy	Targeted and rationed	All households in areas selected for high level of poverty
Sri Lanka (up to 1977)	Rice	Price sub*	Rationed	Total population
(from 1979)	Rice	Food stamps	Targeted and rationed	50 percent of population, biased toward the poor
Sudan	Wheat	Price sub*	General	Mostly urban
Thailand	Rice	Price sub*	General	Total population
Zambia	Maize	Price sub*	General	Mostly urban

* The abbreviation "sub" replaces the word "subsidy" for space reasons.

Source: Per Pinstrup-Andersen, *Food Subsidies in Developing Countries* (Baltimore, Md.: Johns Hopkins University Press, 1988), p. 6.

consumer food subsidies from 15 percent of total government expenditures to less than 3 percent.[3] The impact of targeted subsidies on agricultural prices and incentives depends on how they are financed, but food subsidies need not have adverse effects on agricultural incentives.

Another common price-policy instrument in developing countries is the export tax. The purpose of an export tax is to raise government revenues or reduce domestic commodity prices. The effects of the tax are illustrated in Figure 15-2. Because the country exports the commodity, the world price, P_w, is shown above the price, P_0, which would have prevailed domestically if there were no trade. If exports were freely allowed, this world price would prevail in the domestic market, and a total quantity of Q_2 would be produced domestically, Q_1 would be demanded by domestic consumers, and the difference $(Q_2 - Q_1)$ would be exported. Then, if an export tax equal to $P_w - P_d$ were imposed, the domestic price would fall to P_d, consumers would increase consumption to Q_3, producers would reduce the quantity supplied to Q_4, exports would decline to $Q_4 - Q_3$, and the government would earn an export tax revenue of $(P_w - P_d)$ times $(Q_4 - Q_3)$ or the area $bcfg$ in Figure 15-2. Poor countries may impose export taxes because they lack an alternative source of revenue. In the Figure 15-2 example, the country is unable to influence the world price, P_w, because it is a small producer in the world

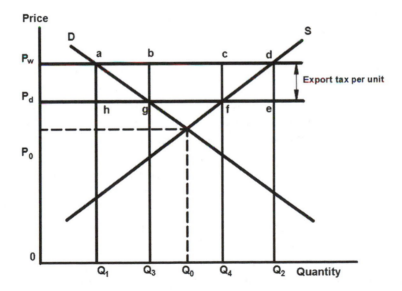

Figure 15-2. Economics of an export tax to raise revenue.

[3] Per Pinstrup-Andersen, "The Social and Economic Effects of Consumer-Oriented Food Subsidies: A Summary of Current Evidence," in *Food Subsidies in Developing Countries* (Baltimore, Md.: Johns Hopkins University Press, 1988), ch. 1, pp. 13-14.

market for the commodity. Domestic producers pay the cost of the tax through lower prices. If the country is a large producer, such as Brazil in the coffee market, its exports and any export tax influence the world market price. Therefore, part of the burden of the export tax can be passed on to consumers in other countries.

Governments follow many types of pricing policies; those described above are among the most common and direct pricing instruments employed in developing countries. Another common direct pricing policy is the attempt to stabilize commodity prices through a *buffer-stock* program. With such a program, supplies are purchased by the government if the price drops below a certain minimum floor level, and then dumped on the market if the price rises above a certain ceiling level. The purpose of the program is to stabilize short-run prices rather than alter the long-run price.

Perhaps the most common indirect pricing policy in developing countries is to overvalue the foreign exchange rate. The foreign exchange rate is the value of the country's currency in relation to the value of foreign currency: for example, the number of Mexican pesos that equal one U.S. dollar. If the official foreign exchange rate implies that the local currency is worth more than it actually is, and if exports occur at the official rate, then this overvalued exchange rate acts as an implicit export tax. However, it does not provide tax revenue to the government. More discussion of the trade effects of direct and indirect pricing policies is found in Chapters 16 and 18.

Interventions to shift either the supply of, or demand for, agricultural products also affect prices. Income transfer and employment programs are examples of policies to shift demands. Policies that steer investments into different sectors, credit programs, agricultural research, and land reforms all affect supplies. The net effect is to change equilibrium prices in markets. Governments tend to examine price trends and shifts and treat them as indicators of an underlying problem. For example, rapidly increasing food prices could be symptomatic of food demand increasing at a faster rate than food supply.

Prices provide important indicators of sector performance. However, policies that attack the symptom — in the above case, rapidly rising prices — by, perhaps, directly imposing price controls, can create long-run damage to economic growth. A preferred price intervention would be to address the causes of the problem by either investing in productivity-enhancing technologies or making more imports available. If demand lags behind supplies, then programs to stimulate demands, such as food stamps, might be contemplated. In general, it is preferable to directly address the causes of undesirable price trends rather than to directly intervene in the price process formation for reasons discussed below.

Short- and Long-run Effects of Pricing Policies

A few of the direct, short-run effects of food and agricultural pricing policies are illustrated in Figures 15-1 and 15-2. As producer and consumer prices are raised or lowered, changes in production and consumption occur. Producer incomes, foreign exchange earnings, price stability, and government revenues are also directly influenced by price policies. These and other direct and indirect, short- and long-run effects of pricing policies are summarized in Table 15-2 and illustrated in Box 15-1.

An important short-term effect of many price policies is to transfer income from producers to consumers. Within consumer groups, the poor tend to be the most sensitive to food prices, since they spend proportionately more income on food. The poor are usually targeted either indirectly because a food they eat is subsidized, or directly by being given food stamps or access to ration shops. However, studies show that even well-targeted price subsidy programs are associated with large "leakages" to the non-poor. These leakages imply higher program costs to the government and create distortions.

TABLE 15-2. SUMMARY OF PRICE POLICY EFFECTS

Direct short-run effects of price policies

1. Changes in consumer and producer prices
2. Changes in quantities produced and consumed
3. Changes in exports, imports, and foreign exchange earnings
4. Income transfers between and among consumer and producer groups
5. Government budget effects
6. Price stability effects
7. Changes in marketing margins and their effects on efficiency of resource allocation

Indirect and long-run effects of price policies

1. Employment changes
2. Incentives for capital investment
3. Incentives for technical change
4. Changes in health and nutrition
5. Long-run changes in allocation of resources in production, storage, transportation, and processing

---

BOX 15-1
FOOD PRICE POLICY IN EGYPT

The Egyptian government for many years operated a complex set of policies that distorted agricultural output and input prices. These policies initially were imposed to transfer resources out of agriculture for industrialization while maintaining low prices for consumers. These policies slowly evolved so that the tax burden on agriculture, resulting from the policies, was no different from that on other productive sectors of the economy, while consumer price subsidies increased dramatically. This food subsidy bill accounted for more than 15 percent of total government expenditures in the 1980s.

Consumer prices for numerous goods were subsidized below free-market levels. Every family received a ration card entitling it to a certain quantity of goods at these fixed prices. Additional goods had to be purchased at substantially higher prices in the black market. The government imported large quantities of goods to maintain these low prices. The government was also heavily involved in marketing both agricultural products and agricultural inputs. Input prices were subsidized, partly to compensate producers for the implicit tax of other policies. Livestock producers were protected and received subsidized feeds.

All of these policies created substantial distortions in resource allocation. Important incentives for the production of livestock were created, while wheat, rice, and cotton production were discriminated against. The budgetary cost of the subsidies was enormous.

Source: Joachim von Braun and Hartwig de Haen, *The Effects of Food Price and Subsidy Policies on Egyptian Agriculture*, International Food Policy Research Institute, Research Report No. 42 (Washington, D.C., November 1983).

A major feature of both direct and indirect effects of many price policies is the influence of those policies on efficiency of resource allocation, depending on the program. In the short run, resources are diverted to less productive uses because of the subsidy or tax. Additional indirect or long-run misallocation of resources can result as investments and structural changes occur that expand less efficient sectors of the economy at the expense of more efficient ones. In addition, efficiency losses occur due to the resource costs associated with collecting taxes or administering the policy. Food stamp and ration shop programs have fewer distortionary impacts because they shift food demands among recipient groups rather than working through price signals.

Distortions in the normal price differences for a commodity across locations, between points in time, and at different levels of processing can influence storage, transportation, and processing of the commodity. For example, urban prices are normally expected to be higher than rural prices for the same food commodity because of transportation costs. If the government sets a ceiling price that is equal in both rural and urban areas, transporting the good from the rural to the urban area may no longer be profitable. In fact, in some cases governments have been known to set urban food prices lower than rural prices, with the result that food, supplied by imports, is transported from urban areas to rural areas.

Likewise, ceiling prices can discourage the normal seasonal storage of a crop if prices are not allowed to rise to cover storage costs. Also, if a government reduces the price margin allowed between farm and retail levels, processors and marketers can be forced out of business.

Pricing policies may be implemented through government procurement agencies with *monopsonistic* (single buyer) power. Thus, opportunities are created for illegal garnering of rents by government employees, and inefficiencies can arise that may force additional reductions in farm prices. These often unintended results of pricing policies can be particularly severe in countries with poor communications and underdeveloped legal systems.

Other indirect effects of pricing policies include employment changes, incentives to develop and adopt new technologies, and changes in health and nutrition. If total revenues for one sector or commodity are raised through pricing policies, more people may be employed. Also, producer incentives to press private firms or public research agencies for new technologies as well as incentives to adopt technologies may he enhanced. Consumer price subsidies can have important impacts on health and nutrition. In cases where they are financed through government tax revenues and not by depressing producer prices, they can be an effective means of transferring income to targeted groups.

Once price policies are instituted, they are difficult to repeal. Urban consumers in numerous countries have reacted in negative and sometimes violent manners to government attempts to lower subsidies. In summary, price–policy effects are pervasive and influence the efficiency of the production and marketing systems.

MARKETING FUNCTIONS AND DEFICIENCIES

Marketing transforms products over time, space, and form through storage, transportation, and processing. Through marketing, goods are exchanged and prices set. Markets communicate signals to producers, processors, input suppliers, and consumers about the costs of buying, selling, storing, processing, and transporting. These major marketing functions and their linkages to price policies are summarized in Figure 15-3.

In the earliest stages of development and in remote areas, a high proportion of the population lives on farms and is relatively self-sufficient. The demand for agricultural marketing services is limited. As development proceeds, with resulting increased living standards and urbanization, the size and efficiency of the marketing system become more important. Unless marketing services are improved concurrently with the development and spread of new technologies, improvements in education and credit, and the other factors discussed in this section of the book, economic development will be hindered. An inefficient

FUNCTIONS OF MARKETS AND MARKETING

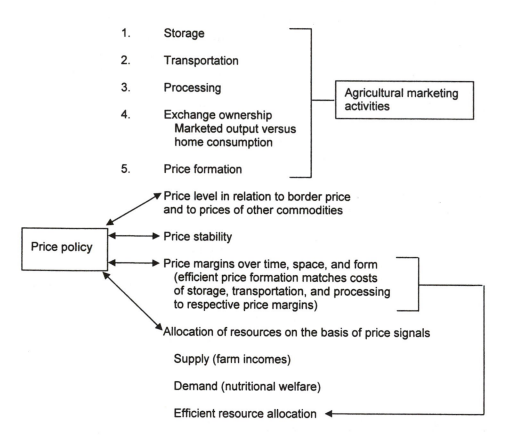

Figure 15-3. Links between agricultural price policy and agricultural marketing (Source: C. Peter Timmer, "The Relationship Between Price Policy and Food Marketing," in *Food Policy: Integrating Supply, Distribution, and Consumption*, ed. J. Price Gittinger, Joanne Leslie, and Caroline Hoisington (Baltimore, Md.: Johns Hopkins University Press, 1987), p. 294.

marketing system can absorb substantial private and public resources and result in low farm-level and high retail-level prices.

Marketing System Deficiencies in Developing Countries

Private marketing systems in many developing countries operate relatively well, in that prices are influenced by underlying supply and demand conditions. Products are stored, transported, processed, and exchanged in roughly the amounts expected given prevailing costs, except where governments have intervened with price policies. Price rigging by opportunistic marketing agents is generally not a serious problem. However, because marketing costs can be high and some price distortions do occur, marketing system deficiencies may retard the rate of agricultural growth and influence the distribution of the benefits of that growth. Let us consider the nature of these deficiencies before turning in the following section to the possible public role in solving them.

The principal weaknesses in marketing systems in developing countries are: (1) infrastructure deficiencies that raise the cost of transport, (2) producers' lack of information, (3) the weak bargaining position of producers of certain commodities, and (4) government-induced market distortions. The magnitude of each of these deficiencies differs across regions and by country, and is changing rapidly for the better in many countries, but severe problems are found in some countries, particularly in Sub-Saharan Africa. The most visible effect of these weaknesses is to create a large spread between the prices producers are paid for their products and the retail prices. Marketing system deficiencies also create wide variations in producer prices within countries and within years. Examples of producer/retail price spreads and of intra-country price variations are presented in Table 15-3 for selected countries in Africa and Asia. The Sub-Saharan African countries have larger price spreads than do the Asian countries, indicating more deficient marketing systems.

Good communications (e.g., roads, railroads. telephones, postal services) and storage infrastructure are crucial to a well-functioning agricultural marketing system. The availability and quality of rural roads, in particular, have a strong influence on marketing costs and on the willingness of farmers to adopt new technologies and sell any surplus production. A farmer who has only a few hectares may still have to market several tons of output to generate revenue needed to apply new seeds, fertilizers, and other modern inputs. Telephones, postal services, radio stations, and so on, increase access to information. Modern storage facilities are important, to minimize rodent, insect, and water damage while commodities are being held. Most storage occurs on the

TABLE 15-3. PRODUCER/CONSUMER AND REGIONAL PRICE SPREADS, SELECTED AFRICAN AND ASIAN COUNTRIES

Country	Commodity	Producer/consumer[a] price spread	Regional[b] spread
Nigeria	Maize	54.5	35.6
	Rice	57.0	72.9
	Sorghum	59.8	45.9
Malawi	Maize	48.2	21.9
	Rice	55.1	68.2
Tanzania	Maize	38.2	25.7
	Rice	56.6	61.3
	Sorghum	48.1	35.5
Kenya	Maize	42.0	30.0
Sudan	Sorghum	61.2	48.2
	Wheat		52.1
Indonesia	Rice	84.0	71.9
India	Rice	82.0	68.9
	Wheat	79.5	65.9
	Sorghum	80.0	63.5
Bangladesh	Rice	79.0	75.0
Philippines	Rice	87.8	82.
	Maize	71.5	64.2

[a]Producer price/retail price X 100   [b]Lowest price/highest price X 100

Source: Raisuddin Ahmed, "Pricing Principles and Public Intervention in Domestic Markets," ch. 4 in *Agricultural Price Policy for Developing Countries*, ed. John W. Mellor and Raisuddin Ahmed (Baltimore, Md.: Johns Hopkins University Press, 1988), p. 67.

farm or at facilities owned by private traders. Storage may also be provided by the government for buffer stocks and food distribution programs.

Producers require information to improve market efficiency and reduce transactions costs, as discussed in Chapter 11. Unequal access to information can give a competitive advantage to particular groups of farmers or traders who have more information. When roads, basic telecommunications, and news services are lacking or are available to only a few, those with better information on market prices, crop prospects, prospective changes in international forces, and so on, can earn higher profits, and in some cases, gain political power as well. Thus, access to information is of fundamental importance for agricultural development. The wireless communications revolution is having a profound effect on information availability and, hence, marketing efficiency. Lower-cost cell-phones are widely available in developing countries, even in relatively remote rural areas. Vegetable producers from China to Brazil now receive price information through cell-phones. Coffee producers in the Guatemalan highlands use beeper technology to receive up-to-the-minute price information. These innovations lower the cost of obtaining information, enable farmers to retain more value on their farms, and enhance planning for deliveries, thus lowering waste and improving efficiency.

The structure of agricultural markets is usually such that the number of middle agents is smaller than the number of producers. Economists hold differing views on whether relatively fewer such intermediaries result in monopolistic power on the part of the intermediary and an unfair bargaining advantage. One needs to be cautious in drawing conclusions. Because the more efficient traders and processors tend to deal in large volumes, there are naturally fewer of these people than there are producers. On the other hand, in most countries with private marketing systems, ease of entry is such that there are still enough processors and other middle agents to provide competition for each other. Examples of collusion and monopolistic power, however, undoubtedly exist for certain products, particularly in isolated areas, where information is costly, and where social and cultural factors play a contributing role. This form of market power is undergoing pressure from the telecommunications revolution described above.

A common marketing problem for producers in some developing countries is a situation in which government-controlled marketing organizations (often called *parastatals*) are given monopolistic power and legal authority to purchase all of a product while setting its price as well (Box 15-2). As discussed in the price-policy section, these tightly controlled markets can have negative effects on producer incentives and market efficiency. Agricultural economic systems are inherently complex. A large amount of information is transmitted through market signals, and decisions made by central marketing boards and parastatal agencies can create serious market distortions. If these types of government agencies are a cause of, rather than a solution to, marketing problems in developing countries, how might the public sector improve marketing efficiency? This issue is addressed in the following section.

BOX 15-2
COMMODITY MARKETING BOARDS
IN SUB-SAHARAN AFRICA

In some Sub-Saharan African states, publicly sanctioned monopolies still purchase and export agricultural goods. These marketing boards serve as the sole buyers of major exports, purchase crops at administratively determined prices, and sell them at prevailing world market prices. These state marketing agencies are vestiges of the colonial period, and their origins and histories vary considerably. Many were established during the Great Depression of the 1930s or World War II. Their official mandates were almost invariably to benefit producers by reinvesting revenues in agriculture and, especially, stabilizing producer prices.

As the colonial governments were confronted with growing needs for revenues, they quickly found ways of diverting marketing board funds away from agricultural development and into general revenue coffers. Following independence, African governments continued to use the commodity marketing boards as extensions of their normal revenue-generating arms, and the initial purposes of the boards were ignored. Examples are found in Ghana and Nigeria immediately following independence.

Since colonial times, these boards have been used to transfer resources from agriculture into "modernizing" and mostly urban development. They have served political objectives by raising revenues, increasing employment of favored groups, and keeping primary commodity prices low to benefit urban and industrial concerns. The boards never really fulfilled their mandate to improve and stabilize conditions in agriculture. In combination with other policy distortions, they contributed to the stagnation and decline of agriculture in many African countries.

THE ROLE OF THE PUBLIC SECTOR IN AGRICULTURAL MARKETING

The primary role of the government is to provide the infrastructure required for an efficient marketing system, particularly roads; a market information system; a commodity grading system; and regulations to ensure the rights of all participants. The underlying rationale for government involvement is the presence of public goods and market failures creating externalities. Public goods provide benefits to society as a whole but would be supplied in less than the socially desirable amounts by the private sector alone. Externalities involve often unintended positive or negative effects of the actions of one person (firm) or persons (firms) on other people.

Roadside market in Bangladesh.

Provision of Infrastructure

The private sector can be expected to build many of the required storage facilities, processing plants, and so on, but investments in roads, seaports, airports, and, in most cases, telecommunications, will require government involvement. One firm, or even a small group of firms, will lack the incentives to build sufficient roads, not just because of their high cost but because of the difficulty of excluding others from, or charging for, their use. Roads are a public good that serve all industries, consumers, and national defense.

Several studies have estimated the economic importance of roads to agriculture in developing countries. For example, Spriggs estimated a benefit/cost ratio of 8 for surfaced roads in the eastern rice regions of India.[4] Ahmed and Hossain estimated that incomes were roughly one-third higher for villages with better infrastructure, compared to those with poor infrastructure, in Bangladesh.[5] Fan and Chan-Kang found

[4] John Spriggs, "Benefit–Cost Analysis of Surfaced Roads in the Eastern Rice Region of India," *American Journal of Agricultural Economics*, vol. 59 (May 1997), pp. 375–9.
[5] Raisuddin Ahmed and Mahabub Hossain, *Developmental Impact of Rural Infrastructure in Bangladesh*, International Food Policy Research Institute, Research Report No. 83 (Washington, D.C., October 1990), p. 70.

that even low-quality rural roads in China are excellent investments, with a 5-1 benefit–cost ratio.[6] The evidence in numerous countries suggests that investments in infrastructure have greatly narrowed farm-retail margins.

Provision of Information

Provision of accurate crop and livestock reports requires investments in data collection and dissemination. Production and consumption data may be poor quality, but accurate data on marketed quantities, qualities, and prices can give essential information for formulating agricultural policies and for decisions by individual economic agents.

To ensure equal access to information, data need to be collected in all-important markets and disseminated on a regular basis. Information on current market prices, crop prospects, and factors influencing demand can be spread through radio broadcasts and newspapers once the government reports are released. An efficient, competitive market requires widespread access to information. Otherwise, a small group of large-scale farmers, traders, or processors can gain market power at the expense of small farmers, particularly those in remote areas. These agents can then use the resulting profits to influence political and economic policy to favor themselves. The result is both efficiency losses (reduced economic growth) and distributional inequities.

In economies highly oriented toward subsistence production, markets offer few premiums for higher-quality products. As interregional communication, and particularly as export trade, develops, quality standards increase in importance because buyers need to compare the products of many different sellers, often without seeing the product before the sale. In markets using modern technology, purchases are often made electronically or over the phone, something that can happen only with a recognized system of grades and standards.

Threshing, drying, cleaning, storage, and processing practices for crops and feeding, slaughtering, storage, and other practices for livestock influence the quality of the final product. Unless grades and standards are established with corresponding price differentials, then producers and processors have little incentive to incur the costs of producing higher-quality goods.

---

[6] See Shenggen Fan and Connie Chan-Kang, *Road Development, Economic Growth and Poverty Reduction in China*, International Food Policy Research Institute, Research Report No. 138 (Washington, D.C., 2005). Chapter 4 in this research report contains an extensive review of research findings about infrastructure and income growth in developing countries.

Regulations

Market regulations related to factors affecting health and safety, but also to weighing practices and other legal codes that influence the enforceability of contracts, are important to a well-functioning marketing system. The purposes of many of these regulations are to ensure basic honesty and reduce transactions costs in marketing. As discussed in Chapter 11, development brings with it a reduction in personal exchange and associated social and cultural constraints on behavior. Increased impersonal exchange requires new institutional arrangements to substitute for the rules of behavior that had been imposed previously by a more personal society.

The importance of market regulations does not imply a need for heavy involvement of government marketing boards or other public trading agencies. Banning private marketing activities does not improve the welfare of either farmers or consumers. While there is a role for the government in the activities discussed above and perhaps in implementing a price stabilization scheme, more extensive public monopolization of domestic marketing functions tends to produce high marketing costs and large market distortions.

THE CHANGING STRUCTURE OF WHOLESALE AND RETAIL FOOD MARKETS

At the retail level in developing countries, there has been a restructuring in urban areas, which began in earnest during the 1990s, toward increased involvement of large wholesalers and supermarkets in food marketing. In a few cases, the supermarkets are owned by multinational companies and, in most cases, the result has been higher-quality products and more efficient (lower-cost) marketing. These markets are, however, increasingly forcing small-scale retailers out of business, just as they did in many developed countries. Before the advent of supermarkets, local brokers or small-scale wholesalers brought relatively undifferentiated commodities from the rural areas to small shops or central markets in the urban areas. In most of the developing world, this structure still predominates. However increasingly, large and often specialized wholesalers bring products from the rural areas to larger processors, supermarkets, and food service chains in urban areas.[7]

---

[7] See Thomas Reardon and C. Peter Timmer, "Transformation of Markets for Agricultural Output in Developing Countries Since 1950: How Has Thinking Changed?" Chapter 13 in *Handbook of Agricultural Economics, Volume 3: Agricultural Development: Farmers, Farm Production, and Food Markets*, ed. R. E. Evenson, P. Pingali, and T. P. Schultz (Amsterdam: Elsevier, 2006).

As market structures change, they do so unevenly in the developing world, with urban retail markets changing before rural markets, and certain geographic areas undergoing a more rapid transformation than others. For example, according to Reardon and Timmer, the degree of transformation is greatest in South America, East Asia outside China, and North-central Europe. The second wave of market change is occurring in Central America and Mexico, Southeast Asia, South-central Europe, and South Africa, and the third wave is just beginning in South Asia, China, Eastern Europe, and parts of Africa.[8]

The market transformation tends to include five sets of changes: (1) a shift from raw commodities to more specialized products; (2) rapid organizational change involving consolidation in the processing and retail segments of the food system with the rise of supermarkets; (3) institutional change in the markets with the rise of contracts and private grades and standards for food quality and safety; (4) rapid technological and managerial change among suppliers, wholesalers, and retailers; and (5) distributional and technological impacts of the wholesale and retail market changes back on farmers. Efforts are needed to prepare poor and small-scale producers to access these new marketing channels; improve quality, adhere to size and other standards, and develop organizational and contracting skills. Off the farm, impacts of this retail revolution on participants in traditional supply and retailing channels are not well understood, but may be substantial. Reardon and Timmer provide a detailed synopsis of what has occurred in food markets in developing countries since the 1950s, and the reasons for those changes.[9]

SUMMARY

Food and agricultural prices are major determinants of producer incentives and of real incomes in developing countries. Governments in those countries often adopt pricing policies to reduce food prices for urban consumers at the expense of producers. Political leaders devise policies to meet society's objectives and the demands of interest groups, to generate revenue, and, in some cases, to line their own pockets. Governments can influence agricultural prices by netting price ceilings or floors and enforcing them with subsidies, taxes, manipulation of exchange rates, storage programs, quantity restrictions, and other policy instruments. These interventions influence producer and consumer prices and incomes, production and consumption, foreign exchange earnings, price stability, government revenues, the efficiency of resource allocation,

[8] Ibid.
[9] Ibid.

employment, capital investment, technical change, health and nutrition, and marketing margins.

Marketing refers to the process of changing products in time, space, and form through storage, transportation, and processing. Goods are exchanged and prices determined in markets. The importance of these functions increases as markets become more commercialized. Developing countries often have marketing systems characterized by deficient infrastructure, inadequate information, weak bargaining position for producers for certain commodities, and government-induced distortions. The government can help solve certain marketing deficiencies, particularly the lack of roads and information. The public sector can provide a system of grades and standards as well as other regulations. These contributions can help reduce transactions costs that rise as markets become less personal. Governments should avoid the larger parastatal marketing agencies that tend to introduce marketing distortions.

Private marketing systems have gradually evolved over the past 50 years in developing countries, with many countries currently experiencing a shift from raw commodities being sold in small shops to more differentiated food products being assembled and processed by larger wholesalers. Supermarkets are increasingly opening in the urban areas of the richer developing countries. This market consolidation is likely to continue at a fast pace in the future, and will have profound impacts on producers, consumers, and middle agents.

IMPORTANT TERMS AND CONCEPTS

Buffer-stock programs	Marketing margin
Competitive market	Middle agents
Export tax	Monopsony
Externalities	Parastatal
Foreign exchange rate	Price ceiling
Grading system	Price distortions
Infrastructure	Price floor
Interest groups	Price formation
Market information	Pricing policies
Market regulations	Public good
Market transformation	Resource allocation efficiency
Marketing board	Two-price programs
Marketing functions	Time, space, and form

Looking Ahead

This chapter concludes the discussion of technical and institutional factors that can influence development of the agricultural sector. The following set of chapters moves beyond the agricultural sector and considers international trade, macroeconomic forces, international capital flows, and other policies that feed back on agricultural development. We begin in the next chapter by considering the importance of international trade. Problems faced by developing countries with respect to agricultural trade, and potential solutions to those problems, are explored.

QUESTIONS FOR DISCUSSION

1. Why do developing-country governments frequently set agricultural prices below market levels?
2. Why do governments get involved in stabilizing prices?
3. What are the direct short-run effects of price policies in agriculture?
4. What are the indirect and long-run effects of price policies in agriculture?
5. Draw a graph to illustrate the effects on supply and demand of a price ceiling set above the market equilibrium price.
6. Draw a graph to illustrate the effect of a price support to farmers set above the market equilibrium price.
7. What are the major food marketing functions? Why are these functions necessary to get agriculture moving is developing countries?
8. What are the major deficiencies in agricultural marketing systems in developing countries?
9. What role might the government play in improving an agricultural marketing system?
10. Discuss the potential rote of buffer stocks in an agricultural development program in a developing country.
11. Why might government marketing boards and parastatals create inefficiencies in resource use?
12. Why do governments in developing countries use export taxes on agricultural commodities more frequently than do governments in more developed countries?
13. Why does the increasing impersonal exchange that accompanies development imply a need for increased government regulation?
14. Why are marketing grades and standards important?
15. Why does increased market information improve marketing efficiency?
16. What has happened to the growth of supermarkets in developing countries over the past few years and why?

RECOMMENDED READING

Anderson, Kym, and Yujiro Hayami, *The Political Economy of Agricultural Protection* (Sydney: Allen & Unwin, 1986).

Byerlee, Derek, and Gustavo Sain, "Food Pricing Policy in Developing Countries: Bias Against Agriculture or for Urban Consumers?" *American Journal of Agricultural Economics,* vol. 68 (November 1986), pp. 961–9.

Fan, Shenggen, and Connie Chan-Kang, *Road Development, Economic Growth and Poverty Reduction in China,* International Food Policy Research Institute, Research Report No. 138 (Washington, D.C., 2005).

Gittinger, J.Price, Joanne Leslie, and Caroline Hoisington, eds, *Food Policy: Integrating Supply, Distribution, and Consumption* (Baltimore, Md.: Johns Hopkins University Press, 1987).

Krueger, Anne O., Maurice Schiff, and Alberto Valdes, "Agricultural Incentives in Developing Countries: Measuring the Effects of Sectoral and Economywide Policies," *World Bank Economic Review,* vol. 2 (September 1988), pp. 255–71.

Pinstrup-Andersen, Per, ed., *Food Subsidies in Developing Countries* (Baltimore, Md.: Johns Hopkins University Press, 1988).

Reardon, Thomas, and C. Peter Timmer, "Transformation of Markets for Agricultural Output in Developing Countries Since 1950: How Has Thinking Changed?" Chapter 13 in *Handbook of Agricultural Economics,* vol. 3: *Agricultural Development: Farmers, Farm Production, and Food Markets,* ed. R. E. Evenson, P. Pingali, and T. P. Schultz (Amsterdam: Elsevier, 2006).

Streeten, Paul, *What Price Food? Agricultural Price Policies in Developing Countries* (Ithaca, N.Y.: Cornell University Press, 1987).

Timmer, C. Peter, *Getting Prices Right: The Scope and Limits of Agricultural Price Policy* (Ithaca, N.Y.: Cornell University Press, 1986).

Timmer, C. Peter, Walter P. Falcon, and Scott R. Pearson, *Food Policy Analysis* (Baltimore, Md.: Johns Hopkins University Press, 1983), ch. 4.

Agricultural Development in an Interdependent World

Wheat being loaded on a ship.

CHAPTER 16

Agriculture
and International Trade

The evidence over the past four decades is suggestive ... that improved trade opportunities for developing countries ... could make an important contribution to growth and hence poverty reduction over time.
                                                          — William R. Cline[1]

This Chapter

1. Explains why countries trade
2. Describes the recent experience of developing countries with trade and why trade patterns change as economic development occurs
3. Discusses problems that impede developing countries from realizing their trade potential with respect to agriculture.

WHY COUNTRIES TRADE

The role of international trade in economic development has attracted the attention of economists and policymakers for almost 200 years (Box 16-1). Despite some views to the contrary, most people now agree that relatively open trade is helpful for successful economic development, as countries with more open trade often (though not always) have faster and more sustained economic growth. In this chapter we ask why this is, and also why so many governments choose to restrict trade despite its potential economic benefits.

Need for Imports and Exports

Trade facilitates development because it permits more efficient use of resources. A country can benefit if it exports what it can produce relatively cheaply, and imports items that would require more resources to produce locally. Most countries import and export the same goods year

---

[1] William R. Cline, *Trade Policy and Global Poverty* (Washington, D.C.: Institute for International Economics, 2004), p. 45.

BOX 16-1
HISTORICAL ROOTS OF INTERNATIONAL TRADE DEBATE

Trade among countries has existed for thousands of years, most of that time in a very loosely structured system. By the sixteenth and seventeenth centuries, money, goods, and credit markets had developed to facilitate trade and colonial expansion. An economic doctrine known as *mercantilism* encouraged exports but discouraged imports. The preferred form of payment was gold rather than goods. A wide range of restrictive trade policies was implemented including tariffs, licenses, export subsidies, and general state control of international commerce. As the Industrial Revolution spread in the late 1700s, mercantilist ideas were increasingly questioned. Raw materials for expanding factory output were imported, and markets for the output were sought abroad. Technological advances in transportation and communications further stimulated trade.

A strong movement toward economic liberalization began in the early 1800s. Perhaps the most important factor in the movement was the unilateral removal of trade restrictions in the United Kingdom. The world's leading economic power at the time, the United Kingdom repealed its Corn Laws in 1846, ending the world's first major price-support program for agricultural commodities. Britain then sought worldwide trade liberalization, with some success. World trade was relatively free until World War I, although several countries, including the United States and Germany, followed selective protectionist policies. World War I changed the trading environment. Industries, including agriculture, which had expanded during the war, suffered slack demand and falling prices afterward. Governments attempted to protect these industries by introducing protectionist policies during the 1920s and 1930s that the world is still struggling to remove today. Persistent protectionist policies for agricultural products are especially evident.

after year, but especially in agriculture there can be wide fluctuations in the quantities traded due to temporary shortages or surpluses, especially for food products. Annual or seasonal production surpluses or deficits can be smoothed out by trade, thereby reducing price swings that might have occurred without it. Most countries also run persistent trade surpluses or deficits from year to year, corresponding to offsetting capital flows. Net inflows of foreign investment are matched by trade deficits, and net outflows are linked to trade surpluses, and when investment flows change there may be a sudden need to alter trade patterns accordingly.

Comparative Advantage

Surprisingly, the rationale for trade does not depend on absolute cost differences between countries. Absolute cost differences determine a country's wealth, not its pattern of trade. Trade is driven by *relative* cost differences among goods within each country, as countries export goods whose cost is relatively low in terms of other goods. This *principle of comparative advantage*, first articulated by David Ricardo in 1817, states that it is best for each country to export those goods for which it has the greatest relative cost advantage and to import goods which are relatively more costly.

The comparative advantage principle implies that one country could produce all goods at lower cost than other countries, yet it would still raise its standard of living through trade, exporting what it produces relatively best (Box 16-2). What counts is the *opportunity cost* in terms of *other goods*.

Some have taken a more negative view of trade and argued that as countries become more integrated into the world economy, they open themselves up to increased exploitation by the more developed countries, by multinational corporations and other actors in international markets, and by the wealthy elites within their own countries. They have argued that the terms of trade, or the prices received for exports from developing countries compared to the prices paid for imports, tend to decline over time. Prices for developed-country products are also said to be high because of monopolistic elements in the production of developed-country products that are imported by developing countries, and protectionist measures in the more developed countries. Some have also argued that dependence on international markets for food endangers national security since international markets are volatile and unpredictable. Finally, they have argued that at very early stages of development "infant industries" may need to be protected from international competition in order to survive. The solutions to these perceived problems are import-substitution policies. Examples of these policies are direct import restrictions, setting of foreign exchange rates above the market equilibrium (which discourages exports for reasons discussed below), and export taxes that discourage exports and stimulate substitution of domestically produced, often industrial, goods for imports.

Although most economists favor freer trade, there are strong debates within economics regarding (1) the degree to which any gains from trade will, in fact, be retained in the developing country and be relatively broadly distributed, and (2) the magnitude of the efficiency losses resulting from attempts to become relatively self-sufficient

---

BOX 16-2
ILLUSTRATION OF THE PRINCIPLE
OF COMPARATIVE ADVANTAGE

To illustrate the principle of comparative advantage, we will consider two countries, each of which produces two stylized outputs: manufacturing (MFG) and agriculture (AGR). Assuming factors of production are mobile within each country, we can specify a production possibility frontier (PPF) for manufacturing and agriculture for each country. The PPF shows the maximum (or total) amount of MFG and AGR that the resources in each country can produce. For example, suppose we have the linear PPF for each country shown below. Country A can produce 30 units of MFG and no AGR, 30 units of AGR and no MFG, or a combination in between such as 15 MFG and 15 AGR.

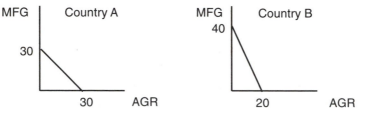

For country A, the internal terms of trade equals 1. It must give up 1 unit of AGR to produce 1 additional unit of MFG, and vice versa. For country B, the internal terms of trade between MFG and AGR is 2. It must give up 1 unit of AGR to produce 2 units of MFG, or 2 units of MFG to produce 1 unit of AGR.

If country A gave up 4 units of AGR, it could increase production of MFG by 4 units. But if it could sell 2 units of AGR to country B, country B could exchange those for 4 units of MFG (i.e., 1 unit of AGR costs 2 MFG in country B). Therefore, country A now has 2 units of MFG more than it would have without trade. Likewise, country B could give up 4 units of MFG to country A and receive up to 4 units of AGR (i.e., 1 unit of MFG costs 1 unit of AGR in country A). Trade would leave country B better off by 2 units of AGR than before. Country A will specialize in AGR and country B in MFG, and the international terms of trade (price of AGR in terms of MFG) will be somewhere between the slopes of the two PPFs (perhaps 1.5 to 1). Exactly where will depend on the demands for MFG and AGR within each country.

---

through import-substitution policies. Few dispute the *potential* for gains from trade, most desire that any gains be broadly distributed, and most agree that increased trade can result in both gainers and losers even if the gainers outnumber the losers.

Almost all developing countries do in fact trade, and most of them also follow some restrictive trade policies. In addition, many LDC exports come initially from agriculture because agriculture is the largest sector. However, the preponderance of empirical evidence over the past 40 years supports the view that a relatively open trading environment is more conducive to economic development than a highly restrictive one. In fact, when the world or a group of countries want to punish a nation (for example, Iraq after the 1991 Gulf War), the first step taken is often to refuse to trade with it.

Gainers, Losers, and the Politics of Trade Policy

Despite the potential for significant economic gains from trade and accompanying specialization, some of the more bitter disputes in economic policy involve international trade policies. Policy-makers, farm groups, consumer advocates, labor leaders, and environmental groups constantly debate the benefits and costs of trade policies that affect exports, imports, the balance of payments, prices, jobs, and the environment. The reason for such contentiousness is twofold: (1) some people do lose from freer trade even as others gain, and (2) the losers are usually more concentrated than the winners. Much of the tendency for governments to restrict trade stems from the fact that trade restrictions can generate highly concentrated and easily visible gains, while spreading their cost broadly among the population over time. For example, protecting a particular industry generates immediate high-wage employment and other benefits in that sector, at a cost that is spread over many other activities in the country. Advocates for the protection can readily identify the winners and tell their story, whereas the losses can be seen only through abstract reasoning and aggregate statistics.

The groups that are best able to act collectively and to lobby policy-makers tend to see trade policies enacted in their favor, even though doing so may impose even greater costs on other, less influential groups. Agricultural lobbies are particularly strong in Europe and Japan, and have obtained relatively large income transfers from other sectors. Within the United States, commodity groups such as those for sugar, cotton, and dairy have been particularly successful in securing government benefits over time. Sometimes, commodity groups align with environmental and labor groups to achieve specific trade policy objectives.

An important fact about trade policy is that, while the debate often focuses on foreign countries, the actual effect of a policy change occurs mainly within the country concerned. Any trade restriction may have some impact on world prices and hence economic conditions in foreign countries, but most of its effect is on domestic prices.

DEVELOPING COUNTRY EXPERIENCE WITH TRADE

During the 1950s and 1960s, import-substitution policies predominated in many developing countries. These inward-oriented policies helped produce a decline in the ratio of exports to GDP in many developing countries until the early 1970s. Since that time, the ratio of exports to GDP has generally increased, paralleling an overall expansion in world trade. However, many developing countries still pursue import-substitution policies. Countries that followed these policies for several years — for example, Argentina, India, and Egypt — tended to grow more slowly than those that followed more open trading regimes, for example, Malaysia, South Korea, and Botswana. While it is difficult to generalize based on a few cases, studies that have examined the overall statistical significance of trade restrictions have generally found a negative impact on economic growth.

It is often difficult to classify a country as relatively open or relatively restricted because policies change over time. Brazil, for example, followed an import-substitution policy from 1955 to 1965 but a more open trading regime from 1966 to 1976.[2] The Mexican economy was quite closed until 1985, but has been relatively open since then. Even South Korea, which is often cited as an example of a successful export-oriented economy, has imposed substantial restrictions on trade from time to time. Trade intervention is usually a matter of degree.

Changing Structure of Trade

Total trade has grown for developing countries over the past 30 years, but the share of agricultural exports in LDC trade has declined steadily from about 60 percent of total exports in 1955 to about 20 percent in recent years. This lower share partly reflects the import-substitution policies mentioned above, but it mainly reflects the shifts in comparative advantage of developing countries from agricultural to manufactured products, demand changes in the more developed countries, increased domestic demand for food in developing countries, and the increased importance of petroleum-based fuels in LDC exports and imports. Nevertheless, several developing countries still depend on a few agricultural exports for a major share of their foreign exchange earnings.

---

[2] See Anne O. Krueger, *Trade and Employment in Developing Countries* (Chicago, Ill.: University of Chicago Press, 1983), p. 44.

The change in comparative advantage is best illustrated by Southeast Asian countries such as Singapore, South Korea, Taiwan, the Philippines, Malaysia, and Indonesia. As these countries invested in human and physical capital, their comparative advantage shifted from natural resource-based and low-skilled, labor-based production activities, to more skill-intensive and, for some countries, capital-intensive products. Their export mix partly reflects this changing comparative advantage (Table 16-1). Agricultural trade is often important in early stages of development but other nonagricultural products assume more importance as development proceeds due to changing comparative advantage.

Developing countries have had, and will continue to have, a comparative advantage in several tropical or subtropical commodities such as coffee, cocoa, tea, rubber, bananas, and sugar. It appears that their comparative advantages in citrus and soybeans have increased over time as well.

Increased demand by more developed countries for many of the agricultural exports of developing countries are limited due to relatively small income elasticities of demand for those commodities and, in some cases, to the development of synthetic substitutes (e.g., for rubber, jute, sisal, cotton). On the other hand, domestic demand for food within the developing countries often increases rapidly with development. Not only are populations growing, but a high proportion of any income increases are spent on food. Also, the mix of foods demanded shifts toward more expensive foods (often meats and vegetables, wheat, and certain other grains rather than roots) as incomes grow (Fig. 16-1). As

TABLE 16-1. STRUCTURE OF EXPORTS FOR SOUTHEAST ASIAN COUNTRIES, 1965, 1989, AND 2003

| | Percentage share of total merchandise exports | | | | | |
| | Primary and other products | | | Manufacturing | | |
Country	1965	1989	2003	1965	1989	2003
Indonesia	96	68	48	4	32	52
Malaysia	94	56	23	6	44	77
Philippines	95	38	10	6	62	90
Singapore	65	27	15	34	73	85
South Korea	40	7	7	59	93	93
Taiwan	30	8	NA	69	93	NA

Note: Numbers may not add up to 100 because of rounding.
Source: World Bank, *World Development Report* (New York: Oxford University Press, 1991, 2005).

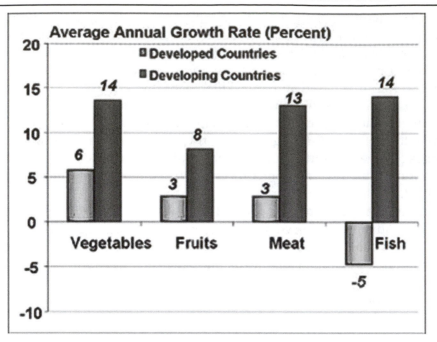

Figure 16-1. Average annual growth rate of consumption of high value food products, 2000–2004 (Source: FAOSTAT as presented at the CGIAR Annual General Meeting, Marrakesh, Morrocco, December 6, 2005).

a result, the more rapidly growing, middle-income countries have actually become less self-sufficient in food production over the past two decades, even as their agricultural production and incomes have risen. Food trade between developing countries grows rapidly with income growth.

Some countries have reacted to the increased domestic demand for food by setting artificially low prices for food commodities and overvaluing their exchange rates to tax exports and lower the prices of imports. These policies tend to be counterproductive, as they discourage production. The effects of exchange rate manipulation are discussed in more detail in Chapter 18.

Trade, Employment, and Capital Interactions

Employment growth is crucial for economic development. While few people are totally idle, there is clearly underemployment in most developing countries. By underemployment we mean people working only part-time or in very low-productivity jobs. Several possible linkages exist between trade and employment. One such linkage is the effect of trade on overall growth through more efficient resource allocation, assuming faster growth entails more employment. A second linkage is that export industries in countries in early stages of development tend to be labor-intensive, consistent with the Factor Endowment Theory of Trade (Box 16-3). Thus, increased exports might lead to greater employment. A third possible linkage is that trade policies might influence the degree of labor intensity in all industries. For example, trade policies might encourage capital-intensive industries through subsidized capital-goods imports.

Empirical evidence suggests that increased exports from developing countries, including agricultural exports, have positive employment implications. Those countries that have followed import-substitution policies (e.g., India) have suffered greater employment problems than more open economies. Research in several countries by the International Food Policy Research Institute (IFPRI) indicates that an export-oriented agriculture increases the demand for hired labor, raises family incomes, and benefits both landowners and landless laborers.[3]

BOX 16-3
FACTOR ENDOWMENT THEORY OF TRADE[a]

The Factor Endowment Theory of Trade (often called the Heckscher-Ohlin-Samuelson Theory because it is derived from their work) argues that because countries have different factor endowments, they adopt different production techniques, and the result is profitable trade. A country with relatively abundant labor (compared to land and capital), will have a low wage rate relative to land prices, rents, and interest on capital-borrowing. Such a country will find it optimal to adopt labor-intensive rather than capital-intensive technologies. The opposite would be true for capital-abundant countries. Without trade, the price ratio of labor-intensive goods to capital-intensive goods will be lower in the labor-abundant country than in the capital-abundant country. Opening the country up to trade would mean that the labor-abundant country would export labor-intensive goods in exchange for capital-intensive goods. Trade will have the effect of increasing the demand for the abundant factor, thus bidding up its price, and increasing the supply of the scarce factor (in the form of imported goods), thereby reducing its price. Trade is expected to reduce factor price differences between countries.

[a]This discussion is drawn from David Colman and Trevor Young, *Principles of Agricultural Economics: Markets and Prices in Less Developed Countries* (Cambridge: Cambridge University Press, 1989), pp. 232-4.

---

[3] Several studies conducted by Joachim Von Braun and others at the International Food Policy Research Institute involved farm and household surveys in Guatemala, The Gambia, Rwanda, and elsewhere.

Small-scale farmers who produce sugar cane, non-traditional vegetables, and other cash crops for export usually maintain some production of subsistence crops as insurance against market and production risk, but these farmers also benefit from the additional income from the cash crops.

The Role of Trade in Agricultural Development

Agriculture has many roles to play in economic development, and trade can affect the relative importance of these different roles. In fact, an outward-looking trade orientation helps solidify the role of agriculture in development, especially if the outward orientation is accompanied by an agriculture- and employment-based growth strategy. Removal of impediments to trade will facilitate exports, and thus will enhance the foreign exchange contribution of agriculture. An open trading regime helps provide accurate signals of relative resource scarcity to producers and investors; the abundance of labor usually found in most developing countries signals the need for employment-intensive investment. With no bias in favor of capital-intensive industries, demands for capital-intensive manufacturing processes can be met through imports, increasing the importance of agriculture's labor contribution.

The food and fiber contribution of agriculture under an outward-looking strategy is usually of most concern to policymakers. Fear of excessive reliance on imports to meet domestic food needs can lead to protectionist policies. But protection raises the cost of food, and combining freer trade with more investment in domestic agricultural production usually results in faster and more stable economic growth. Of course, if growth in demand exceeds domestic food production, then imports may be needed to fill the gap, but these imports should be viewed as evidence of success in generating employment and income growth. Income growth will enhance food security, and open trade will reduce reliance on often unstable domestic food production.

TRADE IMPEDIMENTS

The variety of agricultural trade strategies that exists in developing countries reflects differences in resource endowments, history, food security, sources of government revenues, balance of payments, and so on. This variety also indicates differences in perceptions about the ability of markets to generate prices consistent with desired income distributions. Virtually no country in the world operates with a completely free-trade regime. Most developing countries employ trade policies that discriminate against the agricultural sector, as discussed in Chapter 15. Domestic trade policies, however, are just one of the impediments to agricultural trade. In this section we discuss the major constraints to trade and in Chapter 17 we suggest potential solutions to trade problems. Impediments to agricultural trade for developing countries can be classified into three major categories: (1) external demand constraints, (2) restrictive trade policies at home, and (3) market instability.

External Demand Constraints

Developing countries have long been concerned that as producers of primary products they face relatively inelastic demands in more-developed countries. With inelastic demands, additional exports may result in a fall in world prices for the commodities. While individual countries face relatively elastic export demands, when several countries that export the same products (e.g., cocoa, coffee, bananas) all try to increase exports simultaneously, prices might fall by a higher percentage than export quantities increase. Thus, their collective export revenues could decline, even as export quantities grow. An important recent example is the coffee crisis which began in the late 1990s as Vietnam, Indonesia, and other relative newcomers to coffee production began to expand their exports. World prices of coffee fell to around $.50/lb compared to average prices of $1.20/lb during the 1980s. Prices during the early 2000s were below the cost of production in many Central American countries, and an estimated 540,000 workers in Central America lost their jobs as coffee farms discontinued harvesting.[4] Since 2004, coffee prices have begun to recover, but the example illustrates high short-run costs associated with international price fluctuations.

The products that developing countries import may have more elastic demands, thus creating the terms-of-trade problem mentioned earlier. Historical evidence suggests that the terms of trade for commodities traded by developing countries may have declined over time, as their output growth outpaced any demand increases.[5]

Trade Restrictions in More Developed Countries

The demand for certain LDC agricultural exports is affected by trade restrictions in more developed countries (MDCs). The MDCs are more protectionist of their agricultural than of their industrial products. Whereas LDCs often discriminate against agriculture, MDCs often

---

[4] See Panos Varangis, Paul Siegel, Daniele Giovannucci, and Bryan Lewin, *Dealing with the Coffee Crisis in Central America; Impacts and Strategies*, World Bank Development Research Group Policy Research Working Paper No. 2993 (March 2003).

[5] *Gross* terms of trade do not take into account differences in costs of production between the products. However, it is difficult to draw a firm conclusion about the *net* terms of trade because improved technologies have reduced the cost of producing the exports as well. It is possible for the *gross* terms of trade to decline but the *net* terms of trade and comparative advantage for agricultural products to improve.

support farm prices above market equilibrium levels in the hope of supporting farm incomes (see Chapter 15). Thus, MDCs have to restrict imports to avoid supporting the whole world price structure. Restrictions particularly affect exports from temperate and subtropical areas of LDCs that compete with MDC agricultural products: commodities such as beef, certain fruits and vegetables, and sugar. The European Union, Japan, and the United States have particularly high average levels of producer support (Table 16-2).

Raw tropical products such as cocoa and coffee face few restrictions because they do not compete with more developed country production. However, semi-processed products, such as cocoa paste and certain fibers such as cotton, do face restrictions. Developing countries would like to export more processed commodities because those products have a higher unit value and provide more employment.

Quotas and tariffs are two of the more common import restrictions placed on agricultural commodities by MDCs. An example of how an import tariff works to increase price in the country imposing it and to reduce imports from the exporting countries is illustrated in Figure 16-2. The tariff increases domestic supply and reduces domestic demand. In Figure 16-2, the country imposing the tariff is small in the world market; therefore the tariff does not reduce the world price. However, if the country is large in the world market, such as the United

TABLE 16-2. TOTAL SUPPORT TO AGRICULTURAL PRODUCERS IN SELECTED COUNTRIES (2001 US$ MILLIONS)

Country/region	Total producer support including border restrictions	Total domestic producer support, not including support due to border restrictions such as tariffs and quotas
EU15	87.7	39.6
Switzerland	4.4	1.9
Canada	4.0	2.1
United States	51.4	31.9
Mexico	7.3	2.6
Japan	45.4	4.6
Republic of Korea	16.7	1.0
Australia and New Zealand	0.8	0.8

Source: Thomas Hertel and Roman Keeney, "What is at Stake: The Relative Importance of Import Barriers, Export Subsidies, and Domestic Support," in *Agricultural Trade Reform and the Doha Development Agenda*, ed. Kym Anderson and Will Martin (Washington, D.C.: World Bank, 2005), p. 42.

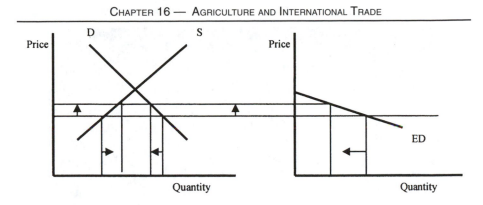

Figure 16-2. Effects of an import tariff.

States with sugar, a tariff (or a quota, which would act just like the tariff in its effects on the market) would depress the world price as well.

It is estimated that if the more developed countries removed all barriers to market access for agricultural products from other countries, the world would gain about $44 billion (in 2001 dollars), about a quarter of which would accrue to developing countries.[6] If all tariffs and subsidies were removed by more developed countries, the largest individual country winners would be Brazil, Argentina, and India.[7]

Subsidized agricultural prices in the more developed countries encourage increased production in those countries while high prices discourage consumption. If production exceeds consumption, stocks accumulate unless they are exported at subsidized prices. The additional volume of exports can depress world prices, making it difficult for developing countries to compete. Dairy products and wheat are examples of subsidized exports of high-income countries. Urban consumers in developing countries can benefit from these policies, at least in the short run, due to lower prices, but farmers in those countries are faced with production disincentives and lower incomes. These price distortions, though benefiting MDC farmers, are globally inefficient. They create conditions for lower growth worldwide. One of the purposes of negotiations under the auspices of the World Trade Organization (discussed in Chapter 17) is to reduce these trade restrictions.

---

[6] Thomas Hertel and Roman Keeney, "What is at Stake: The Relative Importance of Import Barriers, Export Subsidies, and Domestic Support," in *Agricultural Trade Reform and the Doha Development Agenda,* ed. Kym Anderson and Will Martin (Washington, D.C.: World Bank, 2005), p. 49.

[7] Ibid., p. 52.

Developed countries protect their producers at the expense of producers in developing countries.

Restrictive Trade Policies at Home

Many developing countries proclaim food self-sufficiency as an objective, but employ direct and indirect policies that, on net, tax farmers, subsidize consumers, and increase dependence on food imports. Examples of direct policies that influence agricultural trade are export taxes and subsidies, import tariffs, export and import quotas, import or export licenses, and government-controlled marketing margins. Multiple and overvalued exchange rates and high rates of industrial protection are the principal indirect means of discriminating against agriculture.

Agricultural export taxes are one of the oldest and most common trade interventions in developing countries. Export taxes tend to raise the prices of the products to foreign buyers and reduce the prices received by domestic producers. Producers of cocoa in Ghana, cotton in Mali, coffee in Togo, tobacco in Tanzania, and tea in India — to name just a few products and countries — typically receive much less than the border prices for their products. Some of this difference is due to marketing system inadequacies (see Chapter 15), but a significant portion is caused by export taxes.

Some taxation of export crops involves direct taxation of products as they move through ports. Alternatively, public marketing agencies are established that control marketing margins or set farm prices lower than market equilibrium. These agencies, often called marketing boards or parastatal marketing agencies, were discussed in Chapter 15. They are granted monopoly power for buying and selling the commodity, and they may set quotas for exports or imports.

Export taxes are prevalent in the developing countries because they are a relatively easy tax to institute and collect compared to alternatives such as income or land taxes. Export taxes generate government revenues and, in some cases, reduce exports and encourage the shifting of production from exports to domestic food crops.

Occasionally, developing countries impose export taxes in attempts to exploit monopoly power that they believe they hold in world markets. If a country is a large enough exporter in the world market to affect the world price, it can use a tax to raise the world price. Although the volume of trade would be lower following the imposition of the tax, the hope is that additional income is earned at the expense of purchasing countries because the price is higher. Ghana has used this rationale for its export tax on cocoa, Brazil for a tax on coffee, and Bangladesh for a tax on jute. Although some world price increase is possible, the ability of individual developing countries to exploit monopoly power for particular commodities is quite limited. Higher prices create incentives for increased production in other countries as well as for the development of substitute products.

Developing countries sometimes use export quotas to partially or totally restrict exports. These restrictions force the sale of the products in domestic markets, thereby reducing prices to consumers. The result, however, is to discourage domestic production and to generate profits for those holding the quota rights.

Import tariffs and quotas are also used on agricultural products in developing countries, and are commonly employed on industrial products as well. When an import tariff or quota is imposed on industrial goods, the prices of the goods are raised relative to those of agricultural goods, creating an indirect tax on agriculture. Another significant source of indirect taxation is exchange rate misalignments that result from both macroeconomic policies and direct industrial protection policies. When fiscal and monetary policies (see Chapter 18) lead to a higher rate of inflation at home than the rate prevailing abroad, the value of the local currency falls. If governments fail to adjust the official exchange rate downward, the currency becomes *overvalued*. An overvalued currency makes exports from a country more expensive and imports into it cheaper. Thus, fewer goods are exported and more imported. The additional supply of agricultural products on the domestic market reduces farm and consumer prices. Exchange rate overvaluation is common in developing countries and historically has been particularly severe in several African countries such as Nigeria, Ghana, and Tanzania.

Countries sometimes establish a *multiple exchange rate system.* With this system, different commodities are traded at different rates. For example, the government allows one rate of exchange for a commodity it wants to keep inexpensive in the country and another for a commodity it wants to make expensive. Multiple exchange rate systems often discriminate against the agricultural sector.

Historically, indirect taxation of agriculture in developing countries has often been more than twice as large as direct taxation.[8] Examples for several countries are provided in Table 16-3. Ghana, for example, experienced a taxation rate for agriculture of 59.5 percent from 1958 to 1976, more than half of which was indirect. The Philippines had a 27.4 percent rate of taxation on exportables from 1960 to 1986, most of which was indirect.

The differences between direct protection of importable commodities and of exportable commodities in Table 16-3 indicate a strong historical anti-trade bias. Importable goods have generally received positive protection — that is, imports have been taxed either directly or indirectly. This taxation reflected a desire to achieve a certain level of self-sufficiency in food production. Exportable goods have been taxed, reflecting policies designed to raise government revenues. It is not infrequent for direct price intervention in agriculture to account for 5 to 10 percent or more of total government revenue in developing countries.

Farmers in developing countries respond strongly to prices. The output responses to price changes for different commodities in Africa and the rest of the world are indicated in Table 16-4. Production of individual crops has tended to be very responsive to changes in price. Total agricultural output has been less responsive to changes in the average price of agricultural products, but substantial shifts have occurred among commodities produced as prices changed. The types and amounts of commodities produced and the technologies adopted depend on the prices farmers receive. Thus, government policies that affect these prices influence production and resource allocation. In a study of 18 developing countries over the period 1960 to 1989, Schiff and Valdes found that removal of taxation on agriculture would have increased the annual rate of agricultural growth in these countries from 2.5 to 3.1 percent.[9]

---

[8] See Anne O. Krueger, Maurice Schiff, and Alberto Valdes, "Agricultural Incentives in Developing Countries: Measuring the Effect of Sectoral and Economy-wide Policies," *World Bank Economic Review*, vol. 2 (September 1988), p. 262.

[9] See Maurice Schiff and Alberto Valdes, *A Synthesis of the Economics of Price Interventions in Developing Countries* (Baltimore, Md.: Johns Hopkins University Press, 1991).

TABLE 16-3. DIRECT, INDIRECT, AND TOTAL NOMINAL PROTECTION RATES FOR AGRICULTURE
(Average, Percent)

Country	Period	Indirect nominal protection rate	Direct nominal protection rate[a]			Total nominal protection rate
			Importable commodities	Exportable commodities	All commodities	
Cote d'Ivoire	1960-1982	-23.3	26.2	-28.7	-25.7	-49.0
Ghana	1958-1976	-32.6	42.9	-29.8	-26.9	-59.5
Zambia	1966-1984	-29.9	-16.4	-3.1	-16.1	-46.2
Egypt	1964-1984	-19.6	-5.1	-32.8	-24.8	-44.4
Morocco	1963-1984	-17.4	-8.2	-18.5	-15.0	-32.4
Pakistan	1960-1986	-33.1	-6.9	-5.6	-6.4	-39.5
Sri Lanka	1960-1985	-31.1	39.0	-18.4	-9.0	-40.1
Malaysia	1960-1983	-8.2	23.6	-12.7	-9.4	-17.6
Philippines	1960-1986	-23.3	17.9	-11.2	-4.1	-27.4
Thailand	1962-1984	-15.0	n.a.	-25.1	-25.1	-40.1
Argentina	1960-1984	-21.3	n.a.	-17.8	-17.8	-39.1
Brazil	1969-1983	-18.4	20.2	5.4	10.1	-8.3
Chile	1960-1983	-20.4	-1.2	13.5	-1.2	-21.6
Colombia	1960-1983	-25.2	14.5	-8.5	-4.8	-30.0
Dominican Republic	1966-1985	-21.3	19.0	-24.8	-18.6	-39.9
Total Average		-22.5	14.4	-12.6	-7.9	-30.3

n.a. = not available. [a]The direct nominal protection rate reflects the degree of price support, on a percentage basis, compared to a situation with no direct price interventions. A negative value indicates taxation, or negative protection. The indirect nominal protection rate reflects the degree of price support (taxation), on a percentage basis, that results from indirect government intervention. The total nominal protection rate is the sum of the direct and indirect interventions (Source: Alberto Valdes, "Agricultural Trade and Pricing Policies in Developing Countries: Implications for Trade," invited paper presented at the XXI Conference of the International Association of Agricultural Economists, Tokyo, August 1991).

TABLE 16-4. SUMMARY OF OUTPUT RESPONSES TO PRICE CHANGES

| Crop | Percentage change in output with a 10% increase in price[a] | |
	African	Other developing countries
Wheat	3.1–6.5	1.0–10.0
Maize	2.3–24.3	1.0–3.0
Sorghum	1.0–7.0	1.0–3.0
Groundnuts	2.4–16.2	1.0–3.0
Cotton	2.3–6.7	1.0–16.2
Tobacco	4.8–8.2	0.5–10.0
Cocoa	1.5–18.0	1.2–9.5
Coffee	1.4–15.5	0.8–10.0
Rubber	1.4–9.4	0.4–4.0
Palm Oil	2.0–8.1	

[a]The lower end of the range shows short-term supply responses, and the upper end shows long-term responses.
Source: World Bank, *World Development Report 1986* (New York: Oxford University Press, 1986), p. 68.

Not only is agricultural growth retarded by trade restrictions, but real incomes in agriculture are substantially reduced. Lower real incomes hasten the exodus of people from rural areas, creating social costs in urban areas, as sewer, water, health systems, and other infrastructure are stretched to their limits. Just as important, lower incomes in agriculture reduce farmers' incentives to invest in land improvements such as irrigation and farm buildings, to adopt new technologies, and to support rural schools with local resources.

Arguments against a relatively free trade regime are often based on anticipated effects of trade on income distribution. The basic concern is that the benefits of trade may accrue to the wealthiest segments of society. While there is reason for concern that a disproportionate amount of economic gains from trade might go to the wealthiest, historical evidence suggests that a high proportion of the benefits from trade *restrictions* also accrues to them. Trade restrictions provide a fertile environment for powerful domestic interest groups to pressure for advantages. The benefits of quota rights, export and import licenses, and subsidized inputs provide economic incentives for people to lobby for these privileges. Visible corruption often emerges as well. It is naive to assume that governments are simply selfless protectors of social welfare. They are politicians and civil servants who respond to pressures from private individuals and interest groups.

While many government employees act with the overall public good in mind, they may also be just as concerned with their own self-interest as people are in the private sector. Self-interest can encompass monetary gain, re-election, promotion, or other rewards. In addition, even when there are no conflicts of public and private interests, administrative complexities associated with trade restrictions can lead to waste, costly time delays in marketing, and other types of inefficiency.

Market Instability

Government officials in developing countries often perceive food insecurity and national income risks to be associated with a relatively open agricultural trade orientation. Price instability in international commodity markets contributes to these risks. The result is a set of policies to restrict food exports to promote domestic food production and diversify into nonfood agricultural exports. The concern over food security arises due to both production and price risks. Agricultural production is sensitive to weather and pest risks (e.g., droughts, floods, typhoons, locusts) Because world prices for agricultural products are highly variable, countries worry that they may not be able to afford imports in years of poor production.

Why are agricultural prices so variable? First, world demand for most primary commodities is relatively inelastic. As supply shifts back and forth against an inelastic demand curve, prices vary substantially for small changes in quantity supplied (Fig. 16-3). If the demand curve in Figure 16-3 were elastic (flatter), then the price changes would be much less pronounced. Second, many MDCs reduce the linkage between their domestic prices of food commodities and world prices. They do this through import quotas, variable levies, and other methods of price fixing. Because producers and consumers in the developed countries bear less of the price risks as a result of these policies, the price variability is greater in the rest of the world.

The concern over national income risk is especially acute in those countries that rely on one or very few export commodities for a large portion of their foreign exchange earnings, employment, and national income. Prices of agricultural commodities often fluctuate more than 10 percent per year. Hence national income, foreign exchange earnings, inflation, and employment in countries that depend on only a few exports are highly volatile. To reduce this volatility, and to insulate themselves from variability of prices and food imports, many countries attempt to become more self-sufficient in food and to diversify their

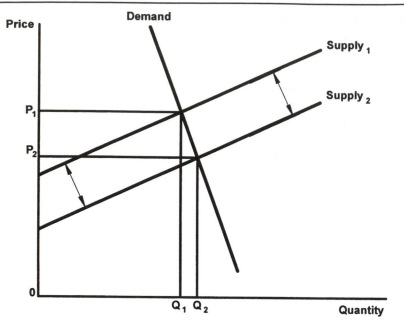

Figure 16-3. Small changes in the supply of agricultural products can result in large changes in price.

exports even if they have to sacrifice average income gains. However, many means exist for promoting diversification, with different associated costs. Diversification can be achieved by stimulating domestic production of new products. Small-scale farmers in Central America, for example, are diversifying into production of non-traditional exports, such as fruits and vegetables. Governments in the region have invested resources in pest-reducing research, provision of market and production information, infrastructure, and organizational skills. The diversification is associated with higher levels of exports, but less overall exposure to single market risks.

Another means of managing international market risks is by depressing trade through the many restrictions described above. Risk is like a cost. If policy-makers are averse to risk, it may seem optimal for them to raise the price of imports and to export less. They should, however, recognize that alternative, less costly, options exist.

SUMMARY

Proponents of a negative view of trade argue that as countries become more integrated into the world economy, they open themselves up for exploitation by more developed countries. Proponents of the more positive view of trade argue that trade facilitates development because it permits more efficient use of resources. It gives countries access to goods and services that otherwise would be unavailable or more expensive. Well-motivated efforts to restrict trade often serve merely to benefit the wealthy. Most developing countries do trade and also follow some restrictive trade policies. Many developing country exports come from agriculture. The preponderance of evidence supports the view that a relatively open trading environment is more conducive to economic development than is a highly restrictive one.

Developing countries have a comparative advantage in several agricultural products, particularly tropical ones. They often become less self-sufficient in food in the middle stages of development. Trade tends to have favorable employment implications.

External demand constraints, market instability, and internal direct and indirect trade restrictions all impede exports from and imports into developing countries. Lack of access to developed country markets is probably the most severe external problem. Governments impose internal trade restrictions to raise revenue, to distribute income to particular groups in response to pressures from interest groups, to exploit monopoly power for certain export crops, and for reasons of food security. Indirect restrictions such as overvalued exchange rates are often greater sources of discrimination against agriculture than are direct restrictions such as export taxes and quotas.

IMPORTANT TERMS AND CONCEPTS

Mercantilism	International trade
Comparative advantage	Multiple exchange rates
Export taxes	Foreign exchange rates
Free trade	Overvalued exchange rate
Protectionism	Quotas
Import substitution	Tariffs
International commodity agreement	Terms of trade

Looking Ahead

A variety of steps can be taken to enhance international trade. The next chapter considers those steps, including the role of regional groupings of countries, multilateral trade negotiations, and other changes in domestic and international policies.

OUESTIONS FOR DISCUSSION

1. Why do countries trade?
2. Why do some argue that the terms of trade turn against developing countries over time?
3. What is comparative advantage?
4. Has agriculture as a percent of total cap on earnings increased or declined for developing countries over the past 30 to 40 years?
5. Why might a country's comparative advantage for particular products change over time?
6. Identify the possible linkages between trade and employment.
7. What are the major external trade impediments facing developing countries?
8. Why is world price instability a problem for developing countries?
9. What are the major direct and indirect agricultural trade restrictions employed by developing countries?
10. Why do developing countries impose trade restrictions?
11. Why does an overvalued exchange rate hurt agricultural exports from a country?

RECOMMENDED READING

Anderson, Kym, and Yujiro Hayami, *The Political Economy of Agricultural Protection* (Sidney: Allen & Unwin, 1986).

Anderson, Kym, and Will Martin, eds, *Agricultural Trade Reform and the Doha Development Agenda* (Washington, D.C.: World Bank, 2005).

Colman, David, and Trevor Young, *Principles of Agricultural Economics: Markets and Prices in Less Developed Countries* (Cambridge: Cambridge University Press, 1989), ch. 11.

Dornbusch, Rudiger, "The Case for Trade Liberalization in Developing Countries," *Journal of Economic Perspectives*, vol. 6 (Winter 1992), pp. 69-85.

Johnson, D. Gale, *World Agriculture in Disarray*, 2nd edn (New York: St. Martin's Press, 1991).

CHAPTER 17

Trade Policies, Negotiations, and Agreements

[T]he WTO at least provides a system of rules for world trade....
The rules may not be perfect, but they are certainly better than no
rules at all. — Eugenio Dias Bonilla and Sherman Robinson[1]

This Chapter
1. Explores solutions to internal constraints to trade
2. Discusses trade negotiations, regional cooperation, and other international solutions to trade problems
3. Considers means of reducing price, production, and income instability problems associated with trade.

REDUCING INTERNAL BARRIERS TO INTERNATIONAL TRADE

Barriers to expanded international trade in agricultural products are both self-imposed by developing countries and the result of protectionist policies of more developed countries. We begin by considering what developing countries can do internally to solve their trade problems. Trade restrictions are imposed within developing countries in attempts to distribute benefits to particular groups, to generate government revenues, and to offset economic instability and food insecurity. Removal of these restrictions requires enhanced information to reduce transactions costs, develop alternative revenue sources, and design institutional arrangements that offset the short-term losses associated with long-term gains.

Enhanced Information to Reduce Transactions Costs

Transactions costs, particularly unequal access to information, when combined with collective action (especially informal lobbying and

---

[1] Eugenio Dias Bonilla and Sherman Robinson, "The WTO Can Help World's Poor Farmers," *International Herald Tribune*, March 28, 2001.

protesting) enable particular groups to pressure the government for trade policies that distribute income streams in their favor at the expense of others. Collective action is not necessarily undesirable. It may help focus government behavior to remedy market failures. However, many internal trade restrictions in developing countries merely enhance the already favored positions of wealthy individuals and reduce economic efficiency. Collective action and political power can present obstacles for reform of these policies. A concerted and sustained effort by governments is needed to reform policies benefiting powerful groups.

The development of basic infrastructure, communications systems, education, and other factors that enhance information flows is essential for reducing transactions costs. Information is needed to help in the design of institutions to constrain unscrupulous behavior. In essence, policy prescriptions mentioned in earlier chapters with respect to land tenure, environmental policy, price policy, and research policy are also relevant for trade policy.

One external means of encouraging internal policy reforms is for an organization such as the International Monetary Fund (IMF) to require certain policy changes as conditions for loans. This type of activity is viewed by many as meddling in the internal affairs of developing countries. And to a certain extent it is. It certainly places a high burden on an institution such as the IMF to get its policy prescription right, lest it force unnecessary hardship on a country. Sometimes, however, such international actions provide cover for a government to undertake reforms it feels are necessary but could not otherwise accomplish because of political opposition.

Alternative Revenue Sources

Export taxes and import tariffs are among the easiest mechanisms for raising government revenues in developing countries. Improved information systems are essential if less distorting income taxes are to replace export and import taxes. Some form of property tax may be possible in a few countries. Converting quantitative restrictions (quotas) to export taxes or import tariffs as an intermediate step to their removal, while not removing the distortion, would at least provide more revenues to the government rather than to private individuals.

Foreign debt reduction would reduce the pressure on LDC governments to generate revenues. The nature of LDC debt problems and potential solutions are discussed in Chapter 18. Most of these solutions require action on the part of both more and less developed countries. Several developing countries are realigning their exchange rates to encourage more exports, but this solution will be insufficient for most countries without additional assistance from more developed countries.

Institutional Changes to Reduce Short-term Losses

Developing countries often find it difficult to undertake necessary long-term policy reform because the short-term consequences are so severe. Devaluation of an overvalued exchange rate raises the cost of imports and reduces the cost of exports. While these cost changes improve the foreign exchange balance and may improve economic efficiency, they also mean that fewer goods are in the domestic market and that there may be severe price increases in the short run. Food prices may rise, real incomes fall, and a disproportionate burden may be placed on the poorest members of society.

International organizations can play a role in providing financial assistance to help offset short-term *cost-of-adjustment* problems associated with policies or structural adjustment programs. In fact, multilateral donors, led by the World Bank, instituted social funds as a means of providing some protection to the poor during structural adjustment programs.[2] Since the losers from policy reform can block changes that would help many more people in the long run, assistance aimed at facilitating adjustment can have significant benefits over time.

REMOVING EXTERNAL CONSTRAINTS TO INTERNATIONAL TRADE

The primary methods that have been suggested as potential solutions to external trade constraints include trade negotiations and special preferences, regional cooperation, and product diversification. Countervailing trade restrictions to offset the external constraints have also been suggested, but they can generate their own set of problems.

Trade Negotiations and Special Preferences

Bilateral and multilateral negotiations have provided opportunities for liberalizing external restrictions on LDC trade. Bilateral negotiations occur when one country negotiates preferential trade arrangements with a second country either for specific goods or for whole categories of goods and services. For example, nation A might grant nation B preferential access to its sugar market — that is, reduce or remove restrictions to sugar imports from nation B — in exchange for special access to nation B's wheat market. Or, nation A, a more developed country, might simply grant a special preference to nation B, a less developed country.

---

[2] See Carol Graham, *Safety Nets, Politics, and the Poor* (Washington, D.C.: The Brookings Institution, 1994).

Numerous variations of bilateral trade negotiations and special preferences are found.

Since World War II, the primary focus for trade negotiations has been multilateral rather than bilateral under the auspices of the General Agreement on Tariffs and Trade (GATT) and, since 1994, the World Trade Organization (WTO). The GATT, signed in 1947, replaced a series of bilateral agreements that segmented world trade before the war. More than 100 countries were signatories to the GATT, and currently about 150 countries are members of the WTO, its successor organization. The GATT and WTO have attempted to foster adherence to the principle that countries should not discriminate in the application of tariffs.[3] Nondiscrimination implies that bilateral preferential agreements are not allowed. The rules allow for exceptions for developing countries. Several developed countries maintain preferential trading arrangements with particular groups of developing countries for certain categories of products. For example, the United States instituted a Caribbean Basin Initiative that eliminated tariffs and quantitative restrictions for many agricultural products from Caribbean countries. Several countries in West Africa have had special preferences with France. Some developing countries have called for more generalized preferences to be granted to countries with incomes below a particular level.

The GATT contained provisions related to consultation and negotiation to avoid disputes, rules concerning non-tariff as well as tariff barriers, and agreements to periodic multilateral negotiations to lower trade barriers. Over time, success in reducing tariff barriers increased the importance of non-tariff barriers. Non-tariff influences on trade include, but are not limited to, certain types of health and safety regulations (see Box 17-1), domestic content restrictions, complex customs formalities and reporting requirements, and rules on intellectual properties.

Eight rounds of multilateral trade negotiations took place under the GATT. Most of the early rounds involved negotiations on tariffs and on rules for trading blocs such as the European Community (EC). The middle rounds focused increasingly on non-tariff issues. Agricultural trade restrictions received relatively little attention until the Uruguay Round from 1986 to 1994.[4] They are at the heart of the most recent

---

[3] Nondiscrimination has been called the *most-favored nation principle,* that a country should apply to other countries the same tariff levels that it applies to its most-favored nations.

[4] Tariff rounds are frequently named after individuals or after locations where the initial discussions in the latest round have taken place. The Uruguay Round began with a meeting in Punta del Este, Uruguay, in 1986.

BOX 17-1
ENVIRONMENTAL, HEALTH AND SAFETY REGULATIONS

Environmental or health and safety regulations can have a significant effect on trade. The United States prohibits the importation of products that have certain pesticide residues. Fresh or frozen beef is prohibited from countries that have a history of foot-and-mouth disease. Clearly, governments are wise to regulate trade in products potentially injurious to public health. More developed countries usually have tighter environmental and food-safety regulations than those of less developed countries. These regulations raise the cost of production so that without corresponding restrictions on trade, not only might there be environmental or health threats, but developed country producers might be placed at a competitive disadvantage. However, environmental or health and safety restrictions appear sometimes to be used arbitrarily to protect the economic health of an industry when the true human health hazard is seriously in doubt. As a result, recent multilateral trade negotiations have included tighter rules on when such restrictions may be applied.

Doha Round negotiations under the WTO, also called the Development Round.

Developing countries have felt that trade negotiations have focused too little on developed country trade restrictions that affect developing countries. Since 1964, they have met periodically under the auspices of the United Nations Conference on Trade and Development (UNCTAD), a permanent organization within the United Nations, to develop proposals for trade arrangements more favorable to developing countries. These discussions led to calls for a *new international economic order* (NIEO). The NIEO contains provisions for improved access to developed country markets through a generalized system of trade preferences and for a set of mechanisms aimed at reducing price and foreign exchange earnings instability.

Aside from some compensatory financing schemes for stabilizing foreign exchange (discussed below), some specific trade preferences, and a few other measures, UNCTAD proposals for a NIEO went largely unheeded. The Uruguay Round of the GATT produced the first serious attempt to address agricultural trade restrictions, including some of particular concern to developing countries. The reason for finally considering agricultural restrictions had little to do with agricultural development problems per se. By the mid-1980s, budget costs, shrinking foreign demand, and world surpluses that threatened a global trade war forced agricultural issues to the top of the GATT agenda. The Uruguay Round negotiations highlighted the divisions among more developed countries and between more developed and less developed countries with respect to trade policy, and also illustrated the diversity of interests among less developed countries. Net-exporting developing countries were very concerned about market access and effects of developed country export subsidies. Net-importing developing countries, while concerned about market access, were also concerned about possible rising prices in world markets, particularly for food grains.

The Uruguay Round ended with a very modest reduction in trade barriers, but success in reorienting the trade debate in several respects. Prior to the Uruguay Round, trade in many agricultural products was unaffected by the tariff cuts that had been made for industrial products in previous rounds. In the Uruguay Round, there was agreement to convert all non-tariff agricultural trade barriers to tariffs. These tariffs were subject to bindings that limit countries' ability to increase them. The Round also contributed to a shift in domestic support for agriculture away from those policies with the largest potential to affect production, and therefore, to affect trade flows. Countries accepted commitments to reduce expenditures on export subsidies and not to apply new subsidies to unsubsidized commodities. Because the base periods chosen had generally high protection, the way non-tariff barriers were converted to tariffs, and the modest percentage reductions agreed to, the overall reduction in trade barriers was small.[5] However, the base was established to build on in future negotiations.

The Uruguay Round resulted in separate agreements on (a) sanitary and phyto-sanitary (SPS) measures to protect humans, animals, and plants from foreign pests, diseases, and contaminants; and (b) intellectual property rights to protect patents, copyrights, and other such rights from infringement abroad. Both of these measures have been difficult for developing countries to accept. The SPS rules can be credited with increasing transparency of countries' SPS regulations and providing a means for settling disputes. Still, the rules can be manipulated to some extent to create barriers to trade that may not be related to SPS concerns. The rules state that science should be the deciding factor as to whether an imported good poses a threat, but science can still be debated. Intellectual property rights are monopoly rights that are granted

---

[5] Developed countries committed to reducing tariffs by 36 percent from the levels in the late 1980s, developing countries 15 percent. The Uruguay Round allowed countries to institute "tariff-rate quotas." A tariff-rate quota applies a lower tariff to imports below a certain quantitative limit (quota), and permits a higher tariff on imported goods after the quota has been reached. The purpose was to ensure that historical trade levels could be maintained, while creating some new trade opportunities. However, the effect has been to slow down the rate of trade liberalization.

to create incentives for private individuals and firms to innovate. However, they can also lead to companies charging high prices to poor countries for drugs and production inputs.

World Trade Organization

The WTO was created in 1994 to replace the GATT, and to strengthen the enforcement of international trade rules and the settlement of trade disputes. For example, a single country can no longer block the formation of a dispute resolution panel, or veto an adverse ruling by blocking the adoption of a panel report. However, it can still be difficult to get countries whose practices have been successfully ruled against to change their behavior, because the only sanction which the WTO can impose when a member government is found to have violated its WTO commitments is to grant to other governments the permission to impose limited, specific retaliatory sanctions.

The WTO matters mainly as a framework for negotiation. Despite concerns of developing countries that the WTO is dominated by more developed countries, the WTO does give developing countries more say than they would have outside it. It is also more open than the GATT. Partly for this reason, more developing countries have joined the WTO than were members of the GATT. Because developing countries can vote as a bloc or blocs, they can force issues more strongly than before. Future negotiations under the WTO are likely to succeed only with some concessions to developing country concerns. In the 2001 meeting in Doha, Qatar, developed countries agreed to place export subsidies higher on the agenda. They also agreed to some relief on intellectual properties, such as for drugs to fight AIDS. The Doha Round has been called the Development Round, to indicate international commitment to addressing concerns of developing countries. In the 2003 Ministerial-level WTO meeting in Cancun, Mexico, a group of 21 developing countries (which together represented about two-thirds of all the world's farmers) called for tighter domestic support restrictions for developed countries and more flexibility for special and differential treatment for developing countries. Their strong position was one of the reasons leading to a breakdown of that meeting, but it marked a negotiating milestone in that, for the first time, several developing countries negotiated as a block and were able to affect the outcome.

The WTO faces significant obstacles in its role as an international forum for trade negotiations. Some poor countries fear that developed countries will use labor standards as a protectionist tool. Others are concerned that little progress will be made to strengthen anti-dumping rules, and to continue to remove protectionist policies on textiles and apparel. The Europeans want stronger environmental rules than either the United States or developing countries would like, the latter preferring environmental issues to come under separate, non-trade agreements.

Most would argue that the WTO is at least potentially more of a friend than a foe to developing countries. It has been estimated that global free trade would confer income gains of about $150 to 200 billion annually to developing countries and reduce the number of extremely poor people.[6] About half of those gains would arise from removing restrictions (e.g., tariffs and quotas) on exports from developing country products to developed country markets, especially in agricultural goods, textiles, and apparel. The gains would be roughly twice the amount that developing countries currently receive through foreign development assistance. However, in the Doha Round negotiations neither developed nor developing countries have sought the degree of trade liberalization that would come close to generating this level of benefits. Almost half of what developing countries could gain from free trade would come from their own tariff reductions, because about one-third of their exports are to other developing countries and because their tariffs are higher than those of the developed countries.[7]

Regional Trade Agreements

International trading relations are increasingly influenced by regional organizations and trading blocs. The economic union in Europe and the North American Free Trade Area (NAFTA) are examples, but so too are the more loosely integrated free-trade areas that have been established in the Asian-Pacific countries, the Andean countries, and the Southern cone countries in Latin America, in Southern Africa, and elsewhere. *Free-trade areas* are trading blocs whose member nations agree to lower or eliminate tariffs and perhaps other trade barriers among themselves, but each country maintains its own independent trade policy toward nonmember nations. Free movement of production factors, such as labor, is usually not included.[8]

---

[6] William Cline, *Trade Policy and Global Poverty* (Washington, D.C.: Institute for International Economics, 2004).

[7] Kym Anderson and Will Martin, eds, *Agricultural Trade Reform and the Doha Development Agenda* (Washington, D.C.: World Bank, 2005), p. 12.

[8] Free movement of factors is allowed in a tighter form of economic integration such as a Common Market or an Economic Federation or Economic Union. One type of regional economic integration that is tighter than a free-trade area but looser than a Common Market is a "Customs Union" in which member countries agree to a common trade policy against all outside countries.

One of the recommendations in the NIEO proposed by UNCTAD was for increased *collective self-reliance* among developing countries. Reduced trade restrictions among a group of those countries could allow for increased specialization, economies of scale (particularly for manufacturers), and competition that reduces costs of production and improves economic efficiency. Occasionally a group of countries can gain some market power through closer economic integration. However, exercise of that power usually creates incentives for one member country to undercut another in terms of production or prices. The power is then eroded and the cohesion of the group jeopardized.

Regional economic groupings can be helpful to developing countries, but their usefulness is limited somewhat by the fact that gains from trade among themselves are often constrained by the similarity of products produced among different countries in a region. For this reason, there has been interest in developing countries to link to more developed countries in these groupings. NAFTA is a good example, whereby Mexico is linked to the United States and Canada. Trade liberalization under NAFTA has been accompanied by substantially larger volumes of trade of agricultural commodities among the three countries. NAFTA eliminated many tariffs and quantitative restrictions among the participants, beginning in 1994, and provides for progressive elimination of tariffs and other trade barriers between the countries over a 15-year period. Both exports and imports are higher in each of the three countries than would have been otherwise. The result has been gains from trade as well as resource adjustments within individual commodity sectors. Despite these gains, extending NAFTA to include first Central America and the Caribbean and then all of South America has been controversial in developing countries and more developed countries alike. Fears for loss of jobs on both sides are strong, and there is little question that expanded regional trade would force many cost adjustments. Some fear that the signing of regional trade agreements will lessen incentives for countries to enter into meaningful multilateral negotiations at the global level.

REDUCING INSTABILITY PROBLEMS

The primary solutions suggested for instability in prices of traded goods and of foreign exchange earnings include: diversification, commodity agreements, compensatory financing, and enhanced use of market information.

Product Diversification

Many countries that receive a high proportion of their export earnings from one or two commodities could likely moderate the effects of external trade restrictions by some diversification of exports. The terms of trade can turn against any single product as substitutes are developed (e.g., for jute and sisal) or new technologies shift supply out against a relatively inelastic and slowly shifting world demand (e.g., peanuts). In addition, even if progress is made through negotiations in opening up market access for commodities such as sugar or cotton or reducing explicit or implicit export subsidies for commodities such as peanuts, total removal of developed country policy distortions is unlikely. Diversifying the production of export and food crops can help not only to reduce the terms-of-trade problems arising from external constraints, but may reduce risks associated with price, production, and foreign exchange variability.

The difficulty for developing countries is in deciding how much to diversify away from a commodity for which it has a strong comparative advantage. Diversification out of agriculture is a natural consequence of economic development that may eventually increase exchange earnings stability, but too much diversification within agriculture can be a costly means of achieving stability.

Commodity Agreements

One possible solution for reducing price variability for individual traded commodities is to develop international commodity agreements. Several of these agreements have been concluded over the past 30 years for commodities such as wheat, sugar, coffee, and cocoa. However, few of these agreements have been effective for very long. Some, like the international wheat agreement, have attempted to balance interests of producers and consumers, setting no limits on production. Others, such as the coffee and sugar agreements, have attempted not only to stabilize prices but also to prevent competitive price-cutting by setting export and import quotas. However, when production varies, these quotas can actually serve to destabilize world prices.

A third type of commodity agreement involves *buffer stocks*. With a buffer-stock scheme, when supplies are high, the commodity is bought up and stored. These buffer stocks provide protection against a time when supply of the commodity drops for some reason. If there is a shortage, stocks are released on to the market to keep prices down. The agreement might specify a minimum and a maximum price, a buffer stock of say 15 percent of world production, a tax on imports or exports to build up the stocks, and perhaps some quotas for producing countries.

An international agreement was in effect for coffee a few years ago.

With most commodity agreements, exporters and importers have difficulty agreeing on an appropriate target price range. The agreements have also proven expensive to administer, especially buffer-stock programs with their high costs of storage.

Compensatory Financing

Schemes aimed at stabilizing foreign exchange earnings rather than directly intervening in commodity markets have met with some success. The basic idea behind compensatory financing schemes (CFS) is illustrated in Figure 17-1. A reference line is set for each country for its total export earnings or earnings from particular commodities. Upper and lower acceptable bounds are set around the reference line. When earnings go below the lower bound, the CFS fills in the shortfall by providing cash or credit to the particular country. When earnings are in excess of the upper bound, developing countries may pay back what was previously taken out.

One CFS operated by the International Monetary Fund's (IMF) Compensatory Financing Facility (CFF) was established in 1963 to provide financial assistance to member countries experiencing temporary export shortfalls. To use the CFF, the IMF must be convinced that the country will seek means to correct its balance of payments problem in the case that export earnings shortfalls are caused by structural problems. Countries can also borrow against the CFF when adverse weather and other circumstances beyond their control result in high

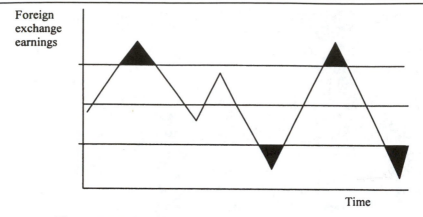

Figure 17.1. Example of a compensatory finance scheme.

cereal import costs. This component of the CFF, called the *cereal import facility*, was set up in 1981, but has seldom been used.

A second compensatory finance scheme is run by the European Community (EC) and is called the STABEX. The STABEX was established in 1975 under the Lomé Convention.[9] This scheme was restricted to African, Caribbean, and Pacific countries and was aimed at stabilizing export earnings for 48 agricultural products. Usually only exports to the EC were covered. A reference line was set for each commodity based on the average value of exports for the products in the preceding four years. To qualify for compensation, export earnings had to fall at least 6.5 percent below the reference line. All loans were interest-free and the least developed countries repaid nothing. The major commodities supported were cotton, sisal, coffee, cocoa, and peanuts. Major beneficiaries were Senegal, Sudan, Cote d'Ivoire, Mauritania, and Tanzania.

Enhanced Use of Market Information

A country can attempt to minimize the negative effects of commodity price fluctuations through use of market information. Futures markets exist in London, New York, Sydney, and elsewhere; the largest futures market "exchange" is in Chicago. With futures markets, commodities can be bought and sold for delivery at a future date. Farmers or export-

[9] The EC's economic arrangement with African, Caribbean, and Pacific countries, which replaced former colonial preference schemes, are spelled out in the Lomé Convention. Other arrangements include free access for many African, Caribbean, and Pacific products to EC markets and the European Development Fund, which administers foreign aid to these countries.

ers can fix a price for goods to be sold later, thus reducing the risk. This activity is called *hedging*. Alternatively, sellers can insure against extremely low prices and buyers against extremely high prices by trading in options on futures contracts. Farmers or exporters can insure against low prices by purchasing an option to sell if prices fall to a specified level. If prices fall below that level, they can exercise their option to sell at that price. If prices rise above it, they lose what was paid for the option, but they can sell the products for the higher price.

The usefulness of international futures and options markets is limited for developing countries because internal commodity prices may not follow the same pattern as commodity prices in Chicago, New York, and so on. However, if trade becomes more liberalized in the future, these markets may become more useful.

SUMMARY

External demand constraints, market instability, and internal direct and indirect trade restrictions all impede exports from and imports into developing countries. Lack of access to developed country markets is probably the most severe external problem. Governments impose internal trade restrictions to raise revenue, to distribute income to particular groups in response to pressures from interest groups, to exploit monopoly power for certain export crops, and for reasons of food security. Indirect restrictions such as overvalued exchange rates are often more significant sources of discrimination against agriculture than are direct restrictions such as export taxes and quotas.

Trade negotiations were undertaken under the GATT beginning in 1947, but have only recently addressed in any substantial way the restrictions on agricultural products that are so important to developing countries. The WTO was formed during the Uruguay Round of negotiations to replace the GATT; the WTO currently has roughly 150 countries as members. Developing countries have more say in the WTO than they had under GATT. Regional economic groupings of countries such as NAFTA have also become more prevalent and have increased regional trade, although their effects on total trade are less certain. International commodity agreements, compensatory financing, product diversification, and enhanced use of market information may help developing countries deal with economic instability.

IMPORTANT TERMS AND CONCEPTS

Compensatory finance

Doha Round

Free-trade area

GATT

International commodity agreement

International trade

Multilateral trade negotiations

NAFTA

Protectionism

Quotas

Tariffs

Terms of trade

Trade preferences

Uruguay Round

World Trade Organization
(WTO)

Looking Ahead

The macroeconomic environment strongly influences agricultural production incentives, agricultural trade, and employment. Domestic macroeconomic policies affect key prices in the economy, including exchange rates, interest rates, wages, food prices, and land prices. Government revenues, taxation, borrowing, and inflation all influence agriculture. In the next chapter we will consider the effects of both domestic macroeconomic policies and world macroeconomic relationships. Particular attention is devoted to world capital markets and the debt crisis facing many developing countries today.

QUESTIONS FOR DISCUSSION

1. Why do developing countries impose trade restrictions?
2. What is the GATT, and why have developing countries felt that it has focused too little on their problems?
3. What is the WTO, and why was it created?
4. What is the difference between multilateral and bilateral trade negotiations?
5. What are components of the new international economic order (NIEO) called for by developing countries under UNCTAD?
6. What is the purpose of a compensatory finance scheme and how might one work?
7. Why might product diversification be helpful to developing countries?
8. What is a free-trade area? Give an example.
9. How do buffer stocks relate in international commodity agreements?
10. How might enhanced information help reduce internal trade restrictions in LDCs?
11. Why does an overvalued exchange rate hurt agricultural exports from a country?

RECOMMENDED READING

Anderson, Kym, and Will Martin, eds, *Agricultural Trade Reform and the Doha Development Agenda* (Washington, D.C.: World Bank, 2005).

Cline, William, *Trade Policy and Global Poverty* (Washington, D.C.: Institute for International Economics, 2004).

Colman, David, and Trevor Young, *Principles of Agricultural Economics: Markets and Prices in Less Developed Countries* (Cambridge: Cambridge University Press, 1989), ch. 11.

Houck, James P., *Elements of Agricultural Trade Policies* (New York: MacMillan, 1986).

International Agricultural Trade Research Consortium Webpage: http://www.iatrcweb.org/

International Monetary Fund webpage on the Compensatory Financing Facility: http://www.imf.org/external/np/ccffbsff/review/index.htm#i

Johnson, D. Gale, *World Agriculture in Disarray*, 2nd edn (New York: St. Martin's Press, 1991).

U.S. Department of Agriculture, *NAFTA*, ERS WRS-99-1 (Washington, D.C., 1999).

World Trade Organization webpage: http://www.wto.org/

CHAPTER 18

Macroeconomic Policies and Agricultural Development

In the long run, macroeconomic forces are too pervasive and too powerful for micro-sectoral strategies to overcome. When they work at cross-purposes, as they do in many developing countries, an unfavorable macroeconomic environment will ultimately erode even the best plans for consumption, production, or marketing.
— C. Peter Timmer, Walter P. Falcon, and Scott R. Pearson[1]

This Chapter

1. Discusses the importance of government policies associated with taxation, spending, borrowing, interest rates, wage rates, the money supply, and exchange rates in influencing the performance of the agricultural sector
2. Examines why governments in less developed countries tend to pursue specific types of macroeconomic policies
3. Describes the significance of the interrelationships among macroeconomic policies across countries; international capital, labor, and product markets; and domestic agricultural markets.

MACROECONOMIC POLICIES AND AGRICULTURE

Macroeconomic policies have a strong influence on output prices, factor prices, marketing margins, and, hence, on incentives for agricultural producers, consumers, and marketing agents. Foreign exchange rates, for example, affect export and import prices and quantities and, thus, output and input prices. Interest rates determine the cost of investments in machinery and equipment and, when combined with wage rates, the capital intensity of production. Interest rates also influence the cost of storage.

---

[1] C. Peter Timmer, Walter P. Falcon, and Scott R. Pearson, *Food Policy Analysis* (Baltimore, Md.: Johns Hopkins University Press, 1983), p. 215. This chapter draws on updates/ideas in *Food Policy Analysis*, especially in the first section.

The macroeconomic environment conditions the rate and structure of agricultural and urban-industrial growth. Job creation and income growth and distribution are as much a function of macroeconomic policies as are policies and projects targeted at specific sectors. The short-run effects of macro policies on employment and income distribution can be quite different from their long-term effects. Real incomes of urban consumers can be sharply reduced in the wake of macroeconomic policy adjustments aimed at reducing public debt or controlling inflation. Policymakers often seek means of softening the short-run income and nutritional consequences of policy changes needed for long-term growth.

Understanding the effects of macroeconomic variables on food and agriculture is important for designing economically and politically viable short- and long-run policies. When macro-policies create distortions such as overvalued exchange rates, heavily subsidized interest rates, and inflationary fiscal and monetary policies, agriculture is usually discriminated against and long-term prospects for development are compromised. Pressures build for major macro-policy reforms that, even if unintentionally, usually help the rural sector by increasing farm incomes and rural employment. Price increases and lower subsidies, however, necessitate painful adjustments by urban consumers. The pervasive nature of these macro-policy effects makes it imperative for those interested in agricultural development to understand how the macro-economy works.

Describing a Macroeconomy

The "macroeconomy" is the aggregate of all economic activity in the country. It is the sum of all individual goods and services, at the prevailing "macro-prices" for foreign currency, capital, and labor that cut across all sectors. The value of the activity at current exchange rates, interest rates, and wage rates can be added up in terms of demand, supply, or income (see Fig. 18-1). A country's gross domestic product (GDP), a measure of its domestically produced national income, will, in theory, be identical regardless of whether it is calculated by summing demands, supplies, or incomes. In practice, differences in measurement errors lead to different measures of income, depending on the adding-up technique used. Macroeconomic policies in developed countries often focus on managing the demand side of the economy. Governments implement policies to stimulate private consumption or investment, use public expenditures to create demand, and closely manage trade. Policies in developing countries frequently are more concerned with managing aggregate supply. Governments in developing

Demand description	Supply description	Income description
Consumption + Private investment + Government expenditures + Excess of exports over imports ↓ Gross domestic product (GDP) + Net income transfers abroad ↓ Gross national product (GNP)	Agricultural production + Industrial production + Production of services + Government production ↓ Gross domestic product	Wages + Interest + Rents + Profits ↓ Gross domestic product

Figure 18-1. Three descriptions of a macroeconomy.

countries tend to use the types of policies described in Chapter 15 to manage agricultural supply; similar policies affect the other productive sectors. Numerous developing countries have attempted to stimulate supply by involving the government directly in the production of goods and services.

Demand equals supply when the components in Figure 18-1 are expressed in real terms (inflation is netted out). The basic factors of production (land, labor, and capital), together with management, earn incomes when they produce goods and services. These incomes are spent on the components of aggregate demand; hence, total income equals GDP. Developing countries are often very concerned about the distribution of total income among wages, interest, rents, and profits, and undertake policies to manage this distribution.

The prices of goods and services are generally expressed in the country's currency units. The monetary value of a good or service can change due to inflation even when its real value has not changed. Policies that create inflation can change real values as well, though often indirectly. The causes of inflation are discussed below, but many of inflation's effects are, in a sense, unintended results of fiscal and monetary policies. We turn our attention to these policies first, highlighting their effects on agriculture. Then we consider the effects of macro-price policies, particularly those policies related to exchange rates, interest rates, and wage rates. Finally, we consider the effects of macro-policies on rural-urban terms of trade and land prices. The major macro-economic and agricultural policy connections are summarized in Figure 18-2; these connections are described below.

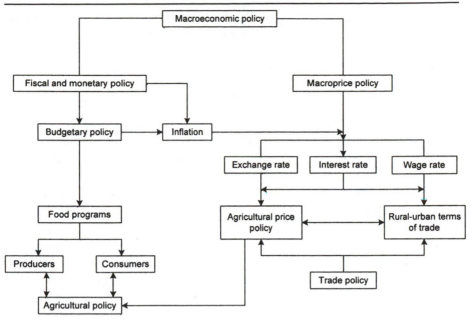

Figure 18-2. Major connections between macroeconomic policy and food policy (Source: Based on Figure 5-1 in *Food Policy Analysis*, ed. C. Peter Timmer, Walter P. Falcon, and Scott R. Pearson (Baltimore, Md.: Johns Hopkins University Press, 1983), p. 223).

Fiscal and Monetary Policy

Fiscal policy is the use of taxes and spending by government to influence employment, income growth and distribution, and other objectives. *Monetary policy* is the use of the money supply and the interest rate to influence these things. The two kinds of policy are closely related. In particular, since the government can print money and never goes bankrupt, expanding the money supply or borrowing from foreigners can be tempting alternatives to raising taxes. Governments differ substantially in their ability and willingness to run budget deficits, and in the way these deficits are financed.

Governments in developing countries often go into debt because of their many pressing needs and limited tax revenues. Tax collection, particularly income-tax collection, is difficult and costly, and taxes are easy to evade in countries with poor information systems. Consequently, developing countries raise large proportions of their tax revenues from export taxes, import tariffs, and sales taxes, as these taxes tend to be easier to collect than others.

Because agriculture is usually the largest sector in the economy in developing countries, it generally provides more revenue to the government than it receives in return in the form of government programs. However, there are usually substantial budget allocations to the agricultural sector. Programs for producers include items such as irrigation systems, roads, agricultural research and extension, market information, and certain output or input subsidies. Programs for consumers include items such as targeted and non-targeted food price subsidies. Many of the investments in agricultural research and extension, irrigation, roads, and so on, also benefit consumers.

Foreign aid can ease some of these revenue needs, as discussed in Chapter 19. A few countries have petroleum and other mineral resources they can export so that foreign consumers help provide revenues for government spending. However, given the limitations to raising taxes, obtaining foreign aid, or exporting petroleum or minerals, most developing countries incur sizable budget deficits. They meet these deficits by borrowing, often from abroad, or by increasing the money supply (that is, by printing more money).

Currently, several developing countries are heavily burdened by debts incurred through previous borrowing abroad. This debt problem, its causes, its effects, and potential solutions to it, are discussed later in the chapter. The debt incurred in previous borrowing currently constrains the ability of many countries to take on additional debt. Consequently, domestic money supply and budget finance policies become that much more important. The size of the money supply must match the needs for operating capital in the productive sectors of the economy. However, when a country prints money to finance a large budget deficit, inflation is the usual result (see Box 18-1).

Inflation can be linked to increases in particular prices; for example, if a country devalues its exchange rate so the prices of all traded goods rise, or it keeps a fixed exchange rate and sees foreign prices rise. But in such cases, the rising prices of those items translate into economy-wide inflation only if the whole money supply rises accordingly. Otherwise, the prices of other things would fall, and only relative prices would change.

Whatever its source, inflation does not usually imply a change in all prices by the same amount, and so it creates some winners and some losers. Indeed, it often hurts agriculture because the prices of inputs usually rise by more than the prices of farm outputs. When inflation occurs, the foreign exchange rate should change to reflect the reduced value of the currency. Many developing countries do not allow this adjustment to take place completely. The resulting overvalued exchange

---

BOX 18-1
EFFECT ON INFLATION OF A GOVERNMENT BUDGET
DEFICIT FINANCED BY EXPANDING THE MONEY SUPPLY

Inflation is a sustained rise in the general price level for a country's goods and services. It is usually measured by a price index. The following example illustrates why expansion of the money supply to finance government budget deficits creates inflation. The aggregate supply of goods and services produced must equal the aggregate demand from total expenditures or, $Y = P \times Q = C + I + G + X - M$, where:

Y = monetary value of national output or income
P = price index for all goods and services produced
Q = quantity index for all goods and services produced
C = national consumption expenditures in private sector
I = national investment expenditures in private sector
G = government expenditures on consumption and investment
X = total value of exports
M = total value of imports

If government demand for goods and services (G) increases because the government prints money to pay for a budget deficit, either the quantity produced of goods and services (Q) must increase, imports (M) must increase, or prices (P) will rise. Most developing countries do not have enought idle resources to meet this demand with enough Q. Changes in imports require foreign exchange. Thus, the usual result is an increase in prices.

Source: C. Peter Timmer, Walter P. Falcon, and Scott R. Pearson, *Food Policy Analysis* (Baltimore, Md.: Johns Hopkins University Press, 1983), pp. 227-8.

---

rate increases the price of agricultural exports (thus reducing export demand) and makes food imports cheaper. The resulting increased supply of agricultural products on the domestic market reduces farm product prices. The foreign exchange rate policy is just one of the macro-price policies that have significant impacts on agriculture.

Macro-prices and Agriculture

Governments use macroeconomic policies to influence inflation, provide incentives, and distribute income. Three prices — foreign exchange rates, interest rates, and wage rates — have major effects on the macro-economy and can be manipulated by the government. These *macro-prices* are all, in fact, determined by supply-and-demand conditions in their respective markets, so that if the government decides to set them by fiat, conditions of excess supply or demand can result. Two of these prices, interest rates and wage rates, signal the scarcity of basic factors of production, capital, and labor. Governments are often tempted to set wage rates artificially high to directly raise incomes of workers. They are tempted to set interest rates low to encourage borrowing and investment. Wages set above the free market, equilibrium value determined by supply and demand conditions will lead to excess supply of labor and, hence, unemployment. Interest rates set below equilibrium values will create excess demand for credit, which will then have to be rationed. Government policy can be used to affect those macro-prices indirectly by intervening to change the underlying supply and/or demand conditions. Public works projects, for example, stimulate demand for labor and could be used to raise wages.

The foreign exchange rate is relatively easy to control, and governments often do control it. Two other prices with major effects on the macroeconomy, food prices and land prices, are influenced indirectly through exchange rate manipulations. These prices can also be affected directly by imposing tariffs, or by government interventions in their respective markets.

Exchange Rates. An exchange rate is the number of units of one currency that it takes to buy a unit of another currency, or the price of one currency in terms of another. For many relatively developed countries, the foreign exchange rate is determined in international money markets by the supply of and demand for a country's currency. For example, there is a demand for U.S. dollars in Japan in order to pay for agricultural products imported from the United States. Similarly, there is a supply of dollars in Japan coming from the purchase of Japanese cars by U.S. consumers. The balance of payments of any country summarizes all economic transactions between it and the rest of the world. The current account, largely reflecting trade balances in goods and services, is balanced by the capital account, which reflects changes in ownership of assets between countries. A country with a trade deficit (i.e., it currently imports more than it exports), by the nature of the accounting relationship must run a capital account surplus (i.e., it is selling more of its assets) to foreign investors. Thus, the supply of, and demand for, dollars are affected by international trade and capital flows for investment or other purposes.

These same supply-and-demand factors exist in developing countries, but the exchange rates in these countries are frequently set by governments rather than determined in currency markets. A developing country may fix or "peg" the value of its currency to that of a major trading partner such as the United States. For example, Honduras for many years fixed its currency, the Lempira, to the dollar at a rate of 2

Lempira equals 1 dollar. The Lempira then followed the fate of the dollar in foreign exchange markets. It declined in value when the dollar declined against third countries, and rose when the dollar rose.

A government can set a new official exchange rate to raise or lower the value of its currency. For example, Honduras eventually devalued its currency relative to the dollar and set it at a ratio of 4 to 1. This devaluation made imports into Honduras more expensive and its exports cheaper. In recent years, countries as diverse as Thailand, Indonesia, South Korea, Russia, Brazil, Argentina, and Turkey have used pegged exchange rates, at least for a period of time. In some cases, such as Ecuador, the country has even done away with its currency and just used the dollar as its currency. In other cases, countries have used what is called a *crawling* or *soft peg*, where the currency is allowed to shift gradually over time or move within a pre-specified range with respect to another currency.

Many countries overvalue their exchange rates for long periods of time. Overvalued exchange rates usually result from differences in inflation rates between a country and its major trading partners. Domestic inflation in the presence of fixed exchange rates means that imports seem cheaper relative to domestically produced goods. At the same time, exports from the country become more expensive abroad. But the market for foreign exchange in the country will not balance unless capital flows in; thus the value of the currency is driven up. Any policy that creates inflationary pressures, such as government budget deficits or expansion of the money supply, will, when combined with fixed exchange rates, lead to overvaluation. Countries maintain overvalued exchange rates by controlling the movement of foreign exchange and foreign investment (see Box 18-2).

Countries overvalue exchange rates in part to keep domestic prices down. More imports and fewer exports mean more goods in the domestic market. The greater the domestic supply of goods relative to demand, the lower the price. The result of an overvaluation is that the prices of traded goods produced in the country, such as many agricultural goods, are depressed relative to those of non-traded goods and services. Thus rural incomes tend to be lowered compared to urban incomes.

Devaluation can correct the problem, at least temporarily, but unless fiscal and monetary policies are changed to reduce either government expenditures or aggregate demand, inflation will rather quickly result in a recurrence of the overvalued exchange rate. Devaluation can also cause hardship for those who produce non-tradable goods and services and consume tradable goods, for example, civil servants and

---

BOX 18-2
HOW A GOVERNMENT MAINTAINS
AN OVERVALUED EXCHANGE RATE

Since supply and demand factors determine exchange rates, if a government wishes to fix the official rate at a level other than its equilibrium, then it must intervene in the foreign exchange market. It can support an overvalued rate by selling foreign exchange reserves (dollars or some other currency) and purchasing its own currency, thus supporting its value. Overvaluation thus diminishes foreign reserves and cannot be sustained for long periods of time. In the absence of significant reserves, a government can restrict access to foreign currency at the official rate, and thus effectively ration the commodity (foreign exchange) for which excess demand exists. This rationing is usually implemented by imposing direct currency controls, by controlled allocations of foreign exchange to preferred importers, and by tariffs and other barriers to imports.

---

certain groups of factory workers. Food prices generally rise in response to currency devaluation, helping farmers and hurting urban consumers. Policies are often needed to protect the welfare of the very poor when a devaluation occurs, especially if the currency has been allowed to become substantially overvalued and a large adjustment is needed.

Over time, countries that are open to international capital flows have found that either a fixed exchange rate or a flexible exchange rate that is allowed to float against other currencies is more sustainable than an exchange rate that is managed by the government so that it adjusts gradually. Countries with a history of sharp monetary instability, or that are closely tied in trade and capital flows to another country, seem to be the ones who choose the fixed rate system.

Interest Rates. The price of capital investment is represented by the interest rate. The interest rate reflects, in part, the productivity of capital or the opportunity cost of using capital for one purpose rather than another. Interest rates also reflect risk and the value of currency as opposed to future consumption. Interest rates are determined by the interaction of the supply of investment funds, basically household savings, and the demand for these funds.

Governments can influence interest rates by setting them for public credit sources and by imposing regulations such as reserve requirements on private financial sources. In addition, the method by which the government finances a fiscal deficit affects interest rates. If a deficit is financed by domestic borrowing, then interest rates may rise in response to the increased demand for funds. The alternative means of financing deficits is to print money, a policy that is inflationary. Thus, higher interest rates in the presence of budget deficits can help keep inflation down. Macroeconomic policy with respect to interest rates often represents an attempt to balance the value of capital in increasing production with the valuation of future, relative to current, consumption.

Governments may set a maximum interest rate that can be charged by lenders in the country. If the rate is set too low, excess demand for credit is created because demand for credit will exceed its supply. Under these circumstances, credit has to be rationed to borrowers who are fortunate to have access to the funds, and private lenders will have incentives not to lend or to circumvent the regulations. Formal lending institutions may be forced out of business. Moneylenders and other informal credit sources not under the control of the government find it easier to charge higher rates.

When interest rates are controlled, they may even be set below the inflation rate. When this happens, the real interest rate is, in fact, negative.[2] Negative real interest rates create credit crises since they spur demand for borrowing far above the supply of savings. Even less extreme interventions can have negative effects, however, as they encourage use of government credit for those who can obtain it and drive out private credit institutions.

Wage Rates. The primary source of income for most people in the world is returns to their labor. Hence creating jobs at decent wages is essential to reductions in poverty and hunger. Governments recognize the importance of labor remuneration and often set minimum wages in an attempt to raise people out of poverty. Unfortunately, in low-income countries where most people are self-employed, minimum-wage legislation is a relatively impotent tool for raising returns to labor and can have unintended effects that hurt labor.

Labor markets are complex because they are segmented by skill levels, occupations, and locations. In rural areas, labor arrangements may include payment in kind (e.g., food or other goods), may involve conditional access to a piece of land, or may depend on other special relationships between employers and workers that are determined by local customs or institutions. Wages for unskilled workers in these areas may be close to the average product of labor rather than the marginal product (Chapter 6). This level, in turn, is close to a basic subsistence level. Minimum-wage legislation is virtually unenforceable in rural areas in developing countries.

---

[2] The *real* interest rate is equal to the nominal interest rate minus the rate of inflation.

In urban areas, minimum wage legislation has been successful in large industries and government organizations. People who are able to obtain jobs at or above the minimum wage clearly benefit. Unfortunately, by raising the price of labor, minimum wage legislation reduces the demand for labor by these industries and organizations. Thus unemployment (or excess supply of labor) may result in the short run. In the long run, the industries may adapt more capital-intensive technologies, further displacing labor, or close their doors and move to a country with lower and more flexible wages. The possibility of higher wages in the formal sector may attract more migrants to the urban area, even if jobs are scarce. This influx of migrants will also swell the informal sector. Consequently, minimum-wage legislation in the formal sector may, over time, depress wages in the informal sector. In summary, wages are an important macro-price, especially to the poor, but governments have little ability to raise people out of poverty by legislating wage levels.

Prices of Agricultural Products and Land. Agricultural prices are influenced by government interventions in output and input markets, as discussed in Chapters 15 and 16. Price supports, input subsidies, export taxes, and so on, directly influence the terms of trade between the agricultural and nonagricultural sectors. Fiscal and monetary policies and macro-price, however, usually have even larger effects on the terms of trade between the sectors than do the more direct price policies. For example, the agricultural sector produces a high proportion of tradable commodities. Thus, an overvalued exchange rate that encourages imports and discourages exports typically has a strong negative effect on the agricultural sector.

When macro-policies and prices discriminate against the agricultural sector so that agricultural prices are depressed, downward pressures are placed on land prices as well. Incentives are reduced for improving the land base or for developing technologies to utilize land more efficiently.

In summary, macro-prices reflect basic economic conditions in an economy. Unless agricultural productivity is increased, simply distorting these prices through government policies is likely to hinder the development process and create distributional effects that hurt the rural poor.

362

WHY GOVERNMENTS PURSUE PARTICULAR MACROECONOMIC POLICIES

Why do governments in developing countries often follow macroeconomic policies that discriminate against rural producers in favor of urban consumers? Why do they sometimes change course and introduce structural adjustment programs that may partially reverse this discrimination? Let us examine the motivation for these public interventions and the reasons why governments often use indirect macro-policy and macro-price interventions as much as policies targeted directly at particular sectors.

Political leadership and individual personalities play an important role in any individual government's specific choices, but looking across a large number of governments over time, certain patterns become clear. At the simplest level, governments follow policies that respond to the balance of political power within their countries. They distribute income in particular ways to help certain sectors, to correct past problems such as external debts, to reduce inflation, and to react to changing world conditions. Because food is a wage good (i.e., food is a high proportion of consumer budgets in developing countries), the interests of urban consumers coincide with owners of industrial firms. Consumers view lower-priced food as higher real wages, while industrialists see it as serving to decrease upward pressure on nominal wages. Thus, an overvalued exchange rate, for example, is a tempting quick fix for stimulating industrial growth, distributing income toward politically influential urban consumers and industrialists, and reducing inflationary pressures.

The growth stimulus of macroeconomic intervention is often short-lived. Discrimination against agriculture reduces both agricultural growth and investment and foreign exchange earnings from agricultural exports. A severely overvalued exchange rate can turn a food exporter into a food importer. Rural opposition to the macro-policies increases over time, inflation worsens due to higher food prices, and unemployment grows. Then, because pressures from urban groups continue, governments may subsidize agricultural inputs, raise output prices through subsidized market margins for food staples, and undertake other measures to reduce prices to consumers. In other words, they pursue partially offsetting policies.

Why do governments institute such complex sets of policies? The basic reason is political expediency. Urban consumers and industrialists are potent pressure groups that demand low food prices and relatively more public goods for urban compared to rural areas.

Transactions Costs and Collective Action

Both macroeconomic interventions and sectoral policies provide benefits to politically favored groups. Individuals may belong to several different groups, and may be simultaneously helped and harmed by different policies. The net benefit obtained from policy, often called political rents, is rarely clear. Macro-policy interventions are particularly difficult to observe. Thus, governments may provide direct subsidies to agricultural producers that are more than offset by overvalued exchange rates and still appear to be helping farmers. Food prices are kept low in urban areas, at least in the short run, and urban industrialists and civil servants, with better information than most farmers, press for the continuation of exchange rate distortions and other forms of protection that benefit the urban sector.

Rural and urban households can form coalitions and lobby collectively for their interests. The policy preferences of politicians and other government officials are affected by the relative strength of these rural and urban lobby groups. The urban lobby is often quite strong because it may represent a coalition of households, students, civil servants, military factions, labor unions, and industrialists.

It is not the sheer size of the urban lobby that gives it power to influence policy. The rural lobby is even larger in many developing countries. However, the urban lobby is much more concentrated geographically, and this concentration facilitates its ability to organize. Students are concentrated near universities, civil servants in government offices, and labor unions and industrialists in a relatively small, concentrated formal sector. The military is highly organized. If people decide to protest rising food prices, the costs of organizing and coming together for this purpose are relatively small in the urban sector.

Because the urban lobby is made up of several relatively small but homogeneous groups, members of these groups see the benefits of organizing collectively to press for their interests. Rural interest groups, particularly small-scale farmers, are so dispersed that individual members often see few benefits to themselves. Communication is difficult so that, even if collective benefits are perceived, the costs of organization and action are prohibitively high. Ironically, as development proceeds and the agricultural sector declines in relative and absolute size, its ability to organize and lobby often increases. Also, the cost to the government of subsidizing a small agricultural sector is lower than for a larger sector. Therefore, once a country is relatively well-developed, it usually reduces its discrimination against agriculture.

Sometimes government policies are motivated by corruption among politicians and other officials. Policy distortion creates gains for certain groups, and some of these gains are appropriated by individuals in public service as payment for instituting the policies.

Historical Factors, Structural Adjustment, and External Forces

A government at any particular point in time is constrained by the accumulated effects of past policy choices, interacting with worldwide economic conditions. One of the most dramatic examples of history colliding with external forces involves government debt. It is natural for developing countries to go into debt to some degree, but, at the start of the 1980s, there was a simultaneous increase in international real interest rates and decline in world commodity prices that sharply increased the difficulty of repayment. Many countries, particularly in Latin America and in Africa, had no choice but to devalue their currencies and cut consumption expenditures in an effort to generate more foreign exchange. Similarly painful "structural adjustment" programs were forced on other countries in the 1990s, when their sources of capital suddenly disappeared. The term *structural adjustment* is often associated with policies aimed at repaying government debt, usually foreign debt. This adjustment typically involves a devaluation of the foreign exchange rate to increase exports and reduce imports, a reduction in government spending and increase in tax collection, sale of government assets, and removal of restrictions on economic activity. The devaluation, privatization, and various types of liberalization are important because external debts cannot be reduced without earning or saving foreign exchange. Reduced government spending and increased efficiency in tax collection are needed to bring spending more in line with revenues. The removal of policy distortions is needed to stimulate economic growth.

The economic growth effects of structural adjustment programs may take several years. Inflation may decline relatively quickly after an initial increase in prices caused by the devaluation, but at a cost of increased unemployment in urban areas. Food prices will usually rise, hurting the urban poor. Public sector employees can be laid off, and standard services curtailed. Higher agricultural prices help farmers and, in the long run, increase employment and output in agriculture.

Governments undertake structural adjustment programs because the economy has deteriorated to the extent that existing policies are unable to control inflation, the country is shut off from additional borrowing, and political support has deteriorated, even from urban sources. External sources of funding such as the International Monetary Fund (IMF) of the World Bank may require structural adjustment as a precondition for additional assistance. Thus, external pressures become another source of policy change. There is no question, however, that structural adjustment programs are painful. For example, in Latin America, the net capital transfer out of the region to repay previous debts was approximately $30 billion per year during the 1980s. Imports were slashed by about one-third. Sharp cutbacks were realized in public services such as water and sewage systems, schools, and hospitals. Social funds and other safety-net programs are often instituted to protect the most vulnerable members of society, but pain inevitably occurs.

Some policy changes are made necessary by changing world economic conditions. A recession in the industrialized countries, for example, can reduce the demand for products from developing countries. High interest rates elsewhere in the world can exacerbate debt problems for developing countries. A shock to the oil market can strain exchange reserves for countries without petroleum. Consequently, some policy changes are necessitated just to react to these external forces. In the next section we examine how these world macroeconomic linkages occur and how they affect developing countries.

WORLD MACROECONOMIC RELATIONSHIPS[3]

Starting from the end of World War II when there was virtually no international capital market, the international monetary system has grown to the point that transfers of capital between countries dwarf the values of international trade in goods. International capital flows are now roughly 25 times the amount of trade flows. These flows ensure a close link between interest rates and exchange rates across countries, and heavily influence countries' trade and their fiscal and monetary policy options.

A second major change in the structure of the international economy was the shift beginning in 1973 from a system of fixed exchange rates to one of bloc-floating exchange rates. With the fixed system, currencies around the world were fixed for long periods of time against the dollar.[4] With the bloc-floating system, the values of major

---

[3] Parts of this section draw on and update material from G. Edward Schuh, "The Changing Context of Food and Agricultural Development Policy," in *Food Policy: Integrating Supply, Distribution, and Consumption,* ed. J. Price Gittinger, Joanne Leslie, and Caroline Hoisington (Baltimore, Md.: Johns Hopkins University Press, 1987), pp. 78-87.

[4] The fixed exchange rate system had been established at the Bretton-Woods Conference in 1944. Trade expanded rapidly under this system, but the system eventually became unworkable when certain currencies, particularly the U.S. dollar, became seriously overvalued and others, particularly the German Deutschmark and the Japanese yen, became severely undervalued.

currencies are allowed to change rapidly against each other in response to market conditions. Several developing country currencies, however, remain fixed to the major floating currencies such as the U.S. dollar.

Implications of Well-integrated Capital Markets and Bloc-floating Exchange Rates

The emergence of a well-integrated international capital market and bloc-floating exchange rates meant that interest rates, capital movements, exchange rates, and trade became interconnected. It meant that fiscal and monetary policies in each country were tied into a single global macroeconomy, with a common core rate of inflation and interest rates. For example, when the United States issues bonds at higher interest rates to pay for a government budget deficit, the capital to buy those bonds comes from a wide range of foreign as well as domestic sources. This foreign purchase of U.S. bonds increases the demand for dollars, driving up the value of the dollar. It also reduces the capital available for other purposes, raising worldwide interest rates. The higher interest rate makes it harder for developing countries to pay off their remaining foreign debt, forcing them to reduce consumption more than they otherwise would.[5]

The higher value of the dollar also makes U.S. exports more expensive abroad and encourages imports into the United States. Developing countries with currencies that are pegged to the dollar will also find it harder to export and easier to import. Then tradable goods sectors in those countries, such as agriculture, suffer from downward pressure on prices.

Governments often try to partially isolate their domestic agricultural sectors from changes in international markets, but any such isolation would mean loss of gains from trade and from access to foreign capital to facilitate development. Consequently, developing countries usually choose to absorb a certain amount of instability in interest rates, exchange rates, and so on, caused by world macroeconomic forces in order to benefit from international goods and capital markets. These countries, however, may need to: (1) protect the poorest of the poor through targeted food subsidies or other means of ensuring basic food security, and (2) take full advantage of international schemes aimed at

---

[5] Capital flows can also keep interest rates low in some cases. For example, in 2004 to 2005, China purchased many U.S. bonds even though interest rates were low, which meant that the United States did not have to raise interest rates to sell bonds to finance its budget deficit. These low rates kept the demand for home mortgages and other loans strong and stimulated the U.S economy because people were willing to borrow and spend.

The poorest of the poor may need to be protected by targeted food subsidies during structural adjustment.

stabilizing foreign exchange such as the compensatory finance arrangements discussed in Chapter 17. The IMF does play a role in trying to help stabilize LDC economies during times of financial crises. In a sense, the IMF is the closest thing the world has to an international central bank. However, the relatively small resource base of that institution, and lack of explicit mandate, keeps it role circumscribed, as discussed below.

Changes in International Comparative and Competitive Advantage

Comparative advantage increasingly is less influenced by physical resource endowments and more by human capital endowments. Government spending on education and agricultural research and the rapid international diffusion of certain technologies, particularly biotechnology, has the potential to influence human capital accumulation in many developing countries by improving education, nutrition, and incomes. These changes may eventually lead to restructured trade patterns.

Government macroeconomic and sectoral protectionist policies, however, can suppress underlying comparative advantage and distort a national economy away from what the physical and human resource base would seem to dictate. As exchange rates swing, so too does competitive advantage, in directions discussed previously. For example, a long decline in the value of the U.S. dollar can mislead U.S. producers and producers in other countries about their long-term ability to compete. A sustained rise in the value of the dollar can send opposite but still misleading signals. These movements can be induced by U.S. and foreign government macroeconomic policies that do not reflect any changes in fundamental comparative advantage.

The External Debt Problem: Causes, Effects, and Potential Solutions

It is natural for the governments of developing countries to borrow to finance productive investment. As long as a country has investment opportunities in the public or private sector that yield returns comfortably above the cost of funds in the world market, then such investments should be made even if external borrowing is required. The country will grow more rapidly and can export to repay the loan in the future. A country may also borrow at times to finance consumption, a policy that would be appropriate, for example, if a natural disaster or a short-run economic shock such as a sharp oil-price change makes it reasonable to sustain consumption even though current income is lower.

Borrowing is imprudent, however, when the debt is increased to cover longer-run consumption, questionable investments, large government deficits, or capital flight out of the country.[6] Imprudent, large-scale borrowing by the government occurred in many developing countries during the 1970s, particularly in Latin America and Sub-Saharan Africa. The result was a *debt crisis* that began in the early 1980s and has only slowly receded over the past two decades. When a country has a debt crisis, it lacks the foreign exchange needed to make payment of interest and principal on its loans. Why did the crisis occur in so many countries, what were its effects, and how might debt problems be solved? The answers to these questions are important for understanding why several developing countries experienced a "lost decade" during the 1980s and how these countries can be extricated from economic predicaments that continue in several of them to this day.

Causes of the Debt Crisis. When a country makes more payments to the rest of the world than it receives in payments, it has a *current account deficit* (see Box 18-3). It has to sell off assets or borrow to finance the deficit. Developing countries began running abnormally large current account deficits beginning in 1973 with the first OPEC oil shock.

---

[6] Capital flight occurs when capital leaves a country due to perceived risk at home. Capital flight, however, is difficult to distinguish from normal capital flows. It often occurs when the government borrows foreign exchange and makes it available to residents at a subsidized price. People acquire this foreign exchange, if they can, and move it to banks or other investments abroad.

BOX 18-3
CURRENT ACCOUNT DEFICIT

The current account deficit represents the excess of spending on imports and interest payments on the external debt over export revenues. In other words, it equals the trade deficit plus interest payments. The current account deficit in a particular year also represents the increase in the net debt for a country. Unless the trade surplus is large enough, the mere existence of an external debt means that interest on that debt will cause the debt to keep growing.

Prior to that time, the total accumulated external debt for all developing countries was less than $100 billion. During the 1970s, commercial banks received a flood of dollars from the oil-producing countries. The banks loaned these dollars to developing countries to finance their current account deficits. Banks were happy to invest in developing countries because they received high interest rates and the loans appeared to be guaranteed by governments. Several Latin American and Asian countries seemed to be good risks because they had grown rapidly for several years. In Africa, growth had, for the most part, not occurred, but countries there borrowed from official sources for development purposes or to meet consumption shortfalls caused by production problems.

By 1980, many developing countries were heavily in debt, which became suddenly much harder to repay when worldwide interest rates rose sharply due to tight monetary policy in the United States and Britain. Many of the commercial loans to developing countries had been made at adjustable interest rates, and borrowers found it hard enough to pay interest let alone repay principal. As a result, the stock of LDC debt kept rising, from $400 billion by 1980 to $660 billion by 1982.

Repayment of debt became even harder in 1982 when a world recession struck, depressing demands for LDC exports. Even the demand for oil declined, resulting in a drying up of money for new loans. The first reaction of countries seriously in debt was to refinance the loans and spread them out over a longer period of time. Several countries, however, found it extremely difficult to service their debts (make scheduled interest and principal payments) or to acquire new funds. For Latin America, debt servicing exceeded 50 percent of the value of the region's exports during the early 1980s, and much of the debt was owed on short-term loans at variable interest rates that were rising.

The first of the large debtors to announce it could no longer service its debts was Mexico in 1982. Mexico was a net oil exporter, but it had borrowed heavily against anticipated future oil revenues. These oil revenues declined with the worldwide recession, and Mexico was left with a debt of more than $80 billion with few exports to make repayments. Banks and the U.S. government provided Mexico with new loans to forestall the repayment problem, but it was then clear that the world community had a major financial crisis on its hands that would have to be addressed. As Mexico renegotiated its loans, the crisis came to a head in other countries as well. By 1986, more than 40 countries in Latin America, Africa, and elsewhere had encountered severe financial problems. Except for the Philippines, countries in Asia largely escaped severe debt problems.

Comparisons of the external debt situation between 1980 and 2003 for low-income, middle-income, and several individual countries are presented in Table 18-1. For developing countries as a whole, external debts as a percent of GNP were almost three times as great in the 1980s as they were in 1970. By 1989, developing countries owed more than $1.3 trillion. Debt service was running more than $100 billion per year. Twelve of the 17 countries identified by the World Bank as heavily indebted were in Latin America and the Caribbean. Africa's debt of more than $110 billion was three times the value of all its annual exports. Cote d'Ivoire provides an example of the severity of the debt problem: with a population of 11.7 million in 1989, it owed $15.4 billion or $1,300 per citizen in a country with an annual per capita income of $790. Forty-one percent of the country's export receipts were needed just to service the debt.

Since the 1980s there have been gradual debt reductions in several countries, but many other countries have continued to experience high

TABLE 18-1. INDICATORS OF EXTERNAL DEBT FOR DEVELOPING COUNTRIES

Country or country group	Total external debt as a percent of gross national income				
	1970-1975	1980	1990	2000	2003
Low income	10.2	16.4	41.0	56.3	35
Middle income	18.6	31.9	39.9	36.5	32
Argentina	20.1	48.4	61.7	56.0	115
Brazil	16.3	31.2	25.1	39.0	54
Morocco	18.6	53.3	97.1	49.0	47
Philippines	20.7	53.8	69.3	64.0	81

Source: World Bank, *World Development Report* (New York: Oxford University Press, various years).

debt levels (e.g., Argentina, the Philippines). Some attempts have been made to forgive debts of several of the most highly indebted, least developed counties, but debt problems have proven persistent. In addition, a number of Latin American and East Asian countries experienced other short-term financial crises in the 1990s, as discussed below.

Effects of the Debt Crisis. When a country attempts to reduce its external debt, domestic consumption must be cut to free up resources to produce goods that can be exported to earn foreign currency for debt service. Reductions in import demand are needed to save foreign exchange. Not all of the reduced spending affects traded goods. Some of it falls on non-traded goods and services when labor and capital shift to the production of traded goods for export.

Within the country, prices of traded goods must rise relative to wages and other prices to encourage the production of traded goods and to discourage domestic consumption. Exchange rate devaluation is the typical means of bringing about these adjustments in relative prices. Devaluation, however, takes time to have the desired effect. Thus, policymakers typically find ways to reduce their imports in the short term by means such as imposing tariffs or import quotas. Because some of the imports are raw materials or producer inputs, economic growth is often slowed as well.

Spending cuts and devaluations are painful since they inherently involve reductions in real income for the country. The cuts usually include reductions in basic social services that help the poor. The devaluations effectively cut real wages. As the currency is devalued, the country has to give up more in terms of domestic resources to earn each unit of foreign currency. The country is essentially selling its labor and other resources more cheaply on world markets.

Many developing countries had overvalued exchange rates prior to the debt crisis; thus, adjustments were needed irrespective of the crisis. The devaluations raised the prices of many agricultural exportables and importables, thus helping the farm sector. However, the resulting higher food prices hurt the poor in particular. The magnitude of this debt-induced hurt is difficult to judge, because several of these countries would have had to adjust their economies even without the debt crisis. But there is little doubt that the poor in developing countries have shouldered a large burden in adjusting to the crisis.

There is another way in which the debt crisis has affected the macroeconomic performance of debtor countries. Cuts in government spending have induced recessions that reduced government revenues. When countries can no longer borrow enough abroad to meet shortfalls, they often print money. Printing money usually increases inflation. Devaluation and import restrictions contribute to these inflationary tendencies.

The result of economic restructuring or structural adjustment in several indebted countries has been trade surpluses (achieved primarily through reduced imports), a net transfer of capital from developing to developed countries, and declines in domestic investment. Markets in developing countries for exports from developed countries shrank in the 1980s. U.S. agriculture, for example, was severely hurt by the debt problem at that time. Many countries that had been major markets for U.S. commodities became unable to import these food products.

At the time the debt crisis first hit, there was a major concern over the impending peril to the world financial system. The fear was that such countries as Mexico, Brazil, and Argentina would default on their loans, causing large commercial banks to go bankrupt. The threat to the banking community eventually receded, as threatened banks reduced their outstanding claims on developing countries, and increased the revenues they set aside to guard against disruptions in debt service.[7]

The threat to the poor in developing countries, however, has only receded slowly, and in many countries not at all. In parts of Latin America and the Caribbean, real wages have gradually increased, but not in all countries. In Sub-Saharan Africa, per capita incomes have continued to stagnate. Governments in many developing countries have cut their education and health budgets. Not all of these declines were due to debt problems, but many were. The rise in poverty and the reduction in social services have led to increased hunger and malnutrition in some countries. Environmental problems have increased as well, as countries exploit resources to meet current food and foreign exchange needs.

Solutions to Debt Problems. External debt problems of developing countries impose costs on both debtors and creditors. Risk of a commercial banking crisis in developed countries has receded, but many heavily indebted, developing-country economies continue to struggle. How can external debts be reduced, and who should bear the burden of adjustment? Several debt-reduction plans have been proposed and in some cases attempted, but the burden of adjustment to debt reduction has fallen primarily on the poor in debtor countries. Let us consider potential solutions to the problem.

---

[7] See Benjamin J. Cohen, "What Ever Happened to the LDC Debt Crisis?" *Challenge*, vol. 34 (May–June 1991), p. 49.

The first potential solution is for the countries to default on the loans. Total default would have the advantage of relieving pressures to cut government spending and to export more to service the debt. The disadvantages are, first, that the creditors could seize debtors' overseas assets.[8] Second, the creditors might seize payments to firms that attempt to export to the debtor and payments made by firms that attempt to import from it. Thus, the country would lose some potential gains from trade. Third, default might leave a country less able to borrow again for several years. This combination of disadvantages has meant that no countries have totally defaulted on their loans, although some countries have stopped payments or made only partial payments for a period of time (e.g., Peru, Brazil, Argentina).

When considering solutions to debt problems, it is important to separate the two different groups of countries whose governments have large debt problems. One group consists primarily of low-income, mostly African countries that owe money largely to governments or to multilateral lending agencies. The second group is composed of the heavily indebted countries, primarily in Latin America, that owe money mainly to commercial banks.

Both groups have high levels of debt, but otherwise their circumstances are different. The low-income African countries possess limited domestic resources such as oil or minerals, do not own much abroad, have had slow income growth for reasons primarily unrelated to debt, and have continued to receive new loans in excess of debt service. The countries that owe most of their debts to commercial banks, by contrast, own more resources (for example, Mexico, Venezuela, Nigeria, and Ecuador have oil reserves), have a great deal of wealth abroad in many cases, have had economic growth rates substantially reduced by their debt, and in most cases have paid more to creditors since 1982 than they have received back in new loans.

Because the lowest-income debtors owe mostly to governments, the creditor countries can mandate debt relief or restructuring without interfering in private international capital markets. Creditors can respond to the debt crisis in ways consistent with their humanitarian beliefs or, more likely, their overall foreign policy objectives. Low-income debtor countries can turn to the Paris Club for help in resolving debt issues (see Box 18-4). Because many loans to African countries are at below-market interest rates (subsidized), rescheduling these loans by extending the repayment period can significantly reduce the burden to

---

[8] See Jonathan Eaton, "Debt Relief and the International Enforcement of Loan Contracts," *Journal of Economic Perspectives*, vol. 4 (Winter 1990), pp. 43-56 for more details.

BOX 18-4
THE PARIS CLUB

The Paris Club is a forum for negotiations on countries' debts to government creditors. The Club, formed in 1956 in response to Argentine debt difficulties, has no set membership. The participants in any Paris Club negotiation are the debtor government and its creditors, who traditionally meet under the chairmanship of a senior French treasury official.

All creditors are treated equally in Paris Club rescheduling negotiations. Debtor countries approaching the Paris Club are usually required to conclude an agreement with the IMF for an IMF loan and an IMF-approved program for restructuring economic policies. An example of IMF conditions for a structural adjustment program would be reductions in government spending and fewer restrictions on exports.

Source: P. Krugman and M. Obstfeld, *International Economics* (Cambridge: Massachusetts Institute of Technology Press, 1988), p. 596.

---

the debtor. Recently, partial debt forgiveness for some of the poorest countries has occurred and more has been pledged. The Enhanced Heavily Indebted Poor Country (HIPC) debt relief intitiative, established by the World Bank and IMF in 1996, has reduced debt by more than $50 billion for 28 HIPCs, reducing the current value of the external debt of those countries by more than two-thirds. A Multilateral Debt Relief Initiative (MDRI) agreed to by G-8 countries in 2005 will provide up to 100 percent debt relief to all HIPCs if fully implemented. Most of the relief will be from IDA funds (see Chapter 19 for more details on the World Bank).

The solutions to debt problems for the heavily indebted countries that have primarily commercial debts are different from those for the HIPCs because whatever solution is arrived at must operate within the context of international capital markets that include commercial banks. Any solution will affect the distribution of the debt burden among debtors, private creditors, and the public in creditor countries.[9]

Several potential solutions to the commercial debt problem have been proposed, and some partially implemented. Most proposals involve a combination of debt rescheduling and restructuring of economic policies within the debtor nations. Other proposals include debt-for-equity swaps, cash buybacks of debt, and debt-for-conservation swaps. Debt rescheduling involves extending the repayment period for the loans, altering interest rates, forgiving part of the principal, or some

---

[9] Ibid.

combination of the three. Efforts to restructure economic policies (structural adjustment programs) involve reducing exchange rates to discourage imports and to encourage exports, cutting government spending and otherwise liberalizing the economy through reduced government intervention in markets and marketing.

In the 1980s, two successive U.S. Treasury secretaries developed plans that involved debt renegotiation to convert short-term loans into long-term loans, structural economic changes by debtor countries under IMF supervision, some new loans, and provisions for commercial banks to reduce principal and interest for selected borrowers in return for various forms of financial guarantees by the IMF, World Bank, and creditor governments. Agreements for modest debt reduction were reached with Mexico, Bolivia, Uruguay, Venezuela, Costa Rica, and the Philippines.

Most countries' debt sells at a discount on a secondary market in which the debt can be shifted from bank to bank or to other institutions. The debt sells at a discount because creditors believe they will not be repaid in full. For example, each dollar of Peru's debt sold for about 5 cents on the secondary market in 1991. Debtor countries can sometimes buy back part of their debt with cash or by swapping government-owned assets (such as stock in publicly owned companies). Buying back the debt seems to make sense because the value of the debt on the secondary market is only a fraction of the face value of the loan. There have been few buybacks and swaps, however, because countries lack the cash, are uneasy about foreign ownership of their assets, and the secondary value goes up when they attempt to buy back the debt. In a few cases (for example, in Costa Rica), outside groups bought up and eliminated part of the debt in exchange for government assurances of protecting rainforests or other natural resources. This type of activity is called a debt-for-conservation (nature) swap.

Rescheduling debts over a longer period of time at a fixed but below-market interest rate would eventually solve the debt problem because countries could grow out of their debt. Commercial banks would be better off if loans were rescheduled because they have little hope of fully collecting their loans under current loan schedules. The debtor countries would clearly be better off. Few loans have been rescheduled this way for three reasons. First, banks are not under pressure to reduce the debts as the banks are no longer in danger of insolvency as a result of loan defaults.

Second, no single bank has an incentive to act alone. Debt reduction, like domestic bankruptcy, needs an institutional setting to bring it about. Even when it is in the collective interests of the banks to reduce the debt, each bank has an incentive to insist on full payment of its own loans. If one bank does grant a concession to lower the interest rate or principal, it becomes more likely that other banks will collect their loans. Hence each bank waits around for other banks to voluntarily reduce the interest rate or principal owed so they can get what is known as a "free ride." This free-rider problem exists for debt-equity swaps, cash buybacks, and other proposed solutions as well.

Third, developed countries have been reluctant to play too large a role in debt relief for fear of large budget expenditures. While there are strong humanitarian grounds for debt relief through Paris Club negotiations for the poorest countries, the arguments carry less weight for debt relief in Latin America if that relief comes at the expense of foreign assistance to even poorer countries in Africa and Asia. Therefore the world continues to muddle along with very gradual debt reduction.

Regardless of the method used to reduce the debts, it would be enhanced by lowering trade barriers to developing country exports. These barriers make it difficult for the countries to acquire foreign exchange for debt service. For this reason, the WTO negotiations may have a role to play in solving the debt problem.

Effects of Structural Adjustment Programs on the Poor

Virtually all the debt-reduction plans, at least up until the mid-1990s, included provisions for structural adjustment programs in the debtor countries. Countries without serious debt problems have also been forced to undertake adjustment programs as a condition for new loans or grants or because of independent financial crises. One question is the extent to which these programs have hurt the nutritional and health status of the poor. A second question is whether the programs can be structured in a way that they could achieve their objective of stimulating economic growth without harming the poor.

Structural adjustment programs, which typically involve currency devaluation and reduced government involvement in markets, have direct effects on prices, employment, income, and government spending. Unfortunately, measurement of these effects on the poor is hampered by lack of information about what would have occurred without the programs. Mexico, for example, undertook a substantial structural adjustment program in the 1980s. Incomes dropped substantially, but eventually inflation did as well. Subsequent economic growth has allowed the government to target new social programs at the poor. In Costa Rica, real wages declined by 40 percent from 1977 to 1982. After the country undertook a structural adjustment program, real wages rose back to the previous level by 1985. Do these experiences indicate that structural adjustment programs do not hurt the poor? No, but the alternative may have been worse.

Any kind of adjustment to repay debt involves lower incomes and higher food prices, and almost inevitably hurt the poor. In most cases, the adjustment period is too long to ignore the pain even if the pain is only temporary. Those hurt the most tend to be the urban poor and the rural landless. Farmers may benefit even in the short run due to higher agricultural prices. However, targeted food and health programs are needed as a safety net even if social services in general are temporarily reduced. In countries where social spending targeted toward the poor remained high — such as Costa Rica, Bolivia, and Chile — indicators of health and nutritional status were affected very little by the adjustment.

Structural adjustment programs may help the poor in the long run. Many existing distortions in developing countries benefit the wealthy and hurt the poor. Unfortunately, poorly designed programs hurt the poor by removing certain social programs or distortions that had been put in place to offset other distortions that favor the rich. Recognizing this problem, beginning in the mid-1990s, the World Bank and the IMF, together with major lending countries, gradually moved away from structural adjustment requirements toward debt forgiveness in exchange for government commitments toward a well-defined poverty reduction strategy. The HIPC initiative, referenced above, limited debt forgiveness to those governments that follow specific procedures in drawing up a well-defined national Poverty Reduction Strategy Paper (PRSP). Unlike structural adjustment programs, most of the PRSPs call for increased public-sector activity, particularly in areas thought to benefit the poor such as public health and rural infrastructure. In mid-2004, the World Bank announced it was ending its structural-adjustment lending program.

Financial Crises in Latin America and Asia

In the 1990s, a series of shorter-run financial crises occurred in Latin America (1994-1995), East Asia (1997), Russia (1998), and Brazil (1998-1999). The impacts of the crises spread to other countries and regions. There were some similarities among the crises. In most cases increased private capital flowed into the countries shortly before the crises, including both bank lending and private investments. The IMF gradually relaxed its rules on capital flows and encouraged capital movements in the 1990s. Real exchange rates had generally appreciated as well, especially in Mexico and Thailand. When investors became nervous, they pulled their money out and the governments were forced to let their currencies depreciate. Problems worsened when neighboring countries were forced to depreciate their currencies because investors pulled out their money. As capital dried up in the affected countries, investment stalled and the countries went into deep recessions. In some cases the countries had problems with deficit spending or inflation before these crises, but in many cases had not.

The crises demonstrated that completely deregulated capital flows carry both benefits and costs. Advantages to the borrowers include resources to finance investments with high social returns and to compensate for balance-of-payments problems and recessions. The disadvantages are that foreign investors might pull their money out quickly, thereby destabilizing the economy. Also, the money may go toward projects that are too risky if the investors think that the government or the IMF will bail them out if there is a problem. In addition, capital flows can affect the exchange rate. If capital suddenly starts to flow out, the government must choose between higher interest rates or depreciation of the exchange rate.[10]

Governments can reduce the chances of financial crises by stronger regulation of domestic banking and financial institutions, and improving information flows with respect to economic and financial conditions. The IMF can assist by helping devise solutions in times of crisis while providing some financial assistance when private funds are not available. The IMF has to be sophisticated in its ability to distinguish between countries that are being fiscally irresponsible from those that are financially sound but are suffering sudden capital outflows due to temporary regional or global events.

Governments cannot simultaneously fix the value of the exchange rate and use macroeconomic policy tools to offset economic problems if capital is allowed to flow in or out of the country freely. Therefore some countries choose to have a flexible exchange rate with relatively free capital flows and attempt to manage their macroeconomic policies. Others choose to fix their exchange rates and institute some controls on capital flows to minimize the danger of financial crises. This combination allows them to manage their macroeconomic policies. A third group of countries decide to fix their exchange rates and allow free capital flows, but give up the ability to influence their macroeconomies. The latter countries are usually small ones with major trading partners to which they tie their currency. They also want to encourage strong foreign capital investment, and therefore do not want to institute capital controls.

[10] Joseph Joyce, "The IMF and Global Financial Crises," *Challenge* (July–August 2000), p. 98.

SUMMARY

Macroeconomic policies have a strong influence on prices, marketing margins, and hence on incentives for economic agents. A macroeconomy may be described in terms of aggregate demand, supply, or income. Policies in developing countries are frequently aimed at the supply side of the economy. Both fiscal and monetary policies influence inflation. Developing countries often go into debt because of many pressing needs and limited tax revenues.

Governments use foreign exchange rates, interest rates, and wage rates to influence trade, investment, and incomes. Many developing countries overvalue their exchange rates, a policy that discourages exports and encourages imports. They do so in part to keep prices down. They often subsidize interest rates and set minimum wages for the urban formal sector. Agricultural and land prices are influenced by macroeconomic policies.

Governments pursue particular macroeconomic policies to stimulate economic growth, distribute income, correct debt problems, reduce inflation, and so on. Policies are influenced to a large extent by urban lobbies. Forces external to the country also come into play. Well-integrated capital markets and bloc-floating exchange rates tie economic policies of developing to developed countries more so than was the case in the 1950s and 1960s.

While it is natural for governments in developing countries to borrow to finance investment, massive borrowing during the 1970s, followed by high interest rates and tight money in the early 1980s, led to a severe debt crisis. Many of the loans in Latin America were from commercial banks, and many of the loans in Sub-Saharan Africa were from official sources. Countries were forced to adjust their economies by exporting more, importing less, and reducing government spending in order to pay off debts. Attempted solutions to the debt crisis have been slow to reduce LDC debts. Repayment of many loans has been rescheduled over a longer period of time, and bank loan reserves have been increased, reducing the threat to world financial markets. Much of the burden of adjustment continues to fall on the developing countries themselves. Banks are hesitant to reduce principal or interest rates because there is no coordinating body to exact a comprehensive reduction that would eliminate the free rider problem. Cash buybacks, debt-for-equity swaps, and debt-for-conservation swaps have been used but are not widespread enough to solve the problem. Structural adjustment programs have often hurt the poor in the short run, suggesting a need for safety-net programs and increased debt forgiveness.

In recent years, several developing countries have experienced short-run financial crises in which private capital has flowed out rapidly, causing severe economic downturns. Capital controls are a possible remedy for capital outflows, but come at the cost of reduced foreign investment. Some countries with flexible exchange rates choose to allow free capital flows, but then attempt to manage their macro-policies to offset the dangers of the sudden capital flows.

IMPORTANT TERMS AND CONCEPTS

Balance of payments	Free rider
Bloc-floating exchange rate	International capital market
Capital flight	Fiscal policy
Cash buybacks	Macro-prices
Current account deficit	Minimum wage
Debt crisis	Monetary policy
Debt-for-conservation swaps	Money supply
Debt-for-equity swaps	Paris Club
Debt relief	Secondary market
Debt rescheduling	Structural adjustment program
External debt	Urban lobby
Financial crisis	

Looking Ahead

International relations between more developed and less developed countries are influenced in major ways by foreign assistance programs. In the following chapter we discuss the various types of foreign assistance, motivations for the aid, and effects on the less and more developed countries.

QUESTIONS FOR DISCUSSION

1. What are the three ways a macroeconomy can be described so as to arrive at gross domestic product (GDP)?
2. What do we mean by a country's "fiscal policy"?
3. What are the two primary monetary policies that may be used to finance a government deficit, and what are their effects?
4. What are the major macro-prices that governments often try to set?
5. Why do countries overvalue their currencies, and what is the effect of overvaluation?
6. What are the advantages of high versus low interest rates?
7. How are wage rates determined, and what are the advantages and disadvantages of minimum wage laws?
8. How are land prices affected by macroeconomic policies?

9. Why do governments pursue particular macroeconomic policies?
10. What is a structural adjustment program, and what are its effects?
11. How does a bloc-floating exchange rate system differ from a fixed exchange rate system?
12. How are interest rates, capital movements, exchange rates, and trade interconnected?
13. How might a macroeconomic policy suppress the comparative advantage of a country in producing a particular good?
14. Why is it natural for developing countries to borrow from developed countries?
15. Describe the major causes of the debt crisis.
16. Why have many heavily indebted countries devalued their currencies?
17. Why has voluntary rescheduling of debt servicing by commercial banks not resolved the debt crisis?
18. What are the advantages and disadvantages of cash buybacks of debt? Of debt-for-equity swaps? Of debt-for-conservation swaps?
19. Why are the urban poor often hurt more by structural adjustment programs than are semi-subsistence farmers?
20. What are the pros and cons of a developing country defaulting entirely on its debts?
21. What were the causes of financial crises in Asia and Latin America in the 1990s?
22. Who are the HIPCs?

RECOMMENDED READING

Cohen, Benjamin J., "What Ever Happened to the LDC Debt Crisis?" *Challenge*, vol. 34 (May–June 1991), pp. 47-51.

Joyce, Joseph, "The IMF and Global Financial Crises," *Challenge*, vol. 43 (July–August 2000), pp. 88-107.

Rogoff, Kenneth, "International Institutions for Reducing Global Financial Instability," *Journal of Economic Perspectives*, vol. 13 (Fall 1999).

Timmer, C. Peter, Walter P. Falcon, and Scott R. Pearson, *Food Policy Analysis* (Baltimore, Md.: Johns Hopkins University Press, 1983), ch. 5.

CHAPTER 19

Capital Flows and Foreign Assistance

Everywhere one turns in global poverty reduction efforts, high-minded rhetoric provides tattered veneer over deficient funding.
— Jeffrey Sachs[1]

This Chapter

1. Examines the nature of public and private capital flows to developing countries, including the rationale for and major types of foreign assistance to agriculture
2. Discusses the types, the objectives, and the positive and negative effects of food aid programs in less-developed countries
3. Identifies means for improving the effectiveness of foreign assistance.

DEVELOPMENT ASSISTANCE PROGRAMS RELATED TO AGRICULTURE

Flows of capital into developing countries can benefit them, given their shortage of capital relative to labor. Private capital flows, however, may not be sufficient to meet development needs for a variety of reasons. Restrictions on investments and other forms of capital flows in developing countries create risks for private investors, as do political uncertainty and long gestation periods for projects. Many key forms of infrastructure have attributes of public goods, and economics tells us that it is difficult to charge for use of public goods. All these factors reduce the willingness of the private sector to undertake investments and, hence, can slow the flow of capital into developing countries. The absence of sufficient incentives to invest can also result from incomplete development of international capital institutions. Foreign development assistance (aid) is one possible

---

[1] Jeffrey Sachs, "A New Global Consensus on Helping the Poorest of the Poor," *Annual World Bank Conference on Development Economics*, 1999.

solution to help the resulting capital imbalance, including assistance to the agricultural sector.

Foreign aid in support of agriculture in developing countries has taken many forms, and the nature and magnitude of its effects have generated considerable debate. Multiple objectives drive all foreign aid programs, with the result that the distribution of aid among different countries often bears little relation to need as manifested by hunger, poverty, or presence of market failure. Hence, we begin this chapter by examining the rationale for foreign assistance.

RATIONALE FOR FOREIGN CAPITAL FLOWS AND ASSISTANCE

From a donor's perspective, the rationale for foreign aid in general, as well as for aid to agriculture, rests on humanitarian (moral or ethical), political (strategic), and economic (commercial) self-interest grounds.[2] Several variants of the humanitarian argument have been made based on compensation for past injustices, uneven distribution of global natural resources, and a moral obligation to help the least-advantaged members of society.[3] The premise is that the emergence of international economic and political interdependencies has extended the moral basis for distributive justice from the national to the international sphere. Foreign assistance to agriculture can benefit one of the largest and poorest sectors in most developing countries.

The political self-interest rationale is based on the notion that aid will strengthen the political commitment of the recipient to the donor(s). A quick examination of the distribution of U.S. foreign assistance by country makes it clear that strategic political considerations have been an important motivation for aid (see Table 19-1). For example, in the 1980s and 1990s roughly 20 to 25 percent of all U.S. development assistance went to Israel and Egypt. Most recently, Iraq has become the largest aid recipient. And a small country like Peru received more U.S. development assistance than a large country like India.

The argument that aid serves a country's economic self-interest is based on the notion that aid increases exports from and employment in the donor country. For example, producers of food grains in the United States benefit from food aid to the extent that it increases total quantities demanded. Food aid can open up markets to a country's exports by

---

[2] See Anne O. Krueger, "Aid in the Development Process," *World Bank Research Observer*, vol. 1 (January 1986), pp. 57–8; and see Vernon W. Ruttan, *United States Development Assistance Policy* (Baltimore, Md.: Johns Hopkins University Press, 1996), ch. 2.

[3] Vernon W. Ruttan, "Solving the Foreign Aid Vision Thing," *Challenge*, vol. 34 (May-June 1991), p. 46.

TABLE 19-1: U.S. OFFICIAL DEVELOPMENT ASSISTANCE (ODA) TO TOP 10 RECIPIENTS (U.S. $millions, 2000)

1970-1971		1988-1989		2000-2001		2003-2004	
Country	ODA	Country	ODA	Country	ODA	Country	ODA
India	1,625	Israel	1,553	Russia	815	Iraq	2,157
Vietnam	1,227	Egypt	1,180	Egypt	789	Congo, D.R.	758
Indonesia	912	Pakistan	485	Israel	555	Egypt	724
Pakistan	584	El Salvador	410	Pakistan	428	Russia	69
Korea	526	India	236	Ukraine	240	Jordan	62
Brazil	421	Philippines	224	Colombia	223	Afghanistan	596
Turkey	421	Pacific Islands	199	Jordan	169	Pakistan	557
Colombia	351	Guatemala	186	Yugoslavia	155	Colombia	506
Israel	199	Bangladesh	174	Peru	154	Israel	495
Laos	187	Honduras	174	Indonesia	154	Ethiopia	472
Total above	6,452	Total above	4,821	Total above	3,682	Total above	7,588
Total ODA	11,689	Total ODA	12,426	Total ODA	11,163	Total ODA	18,217

Source: OECD database.

initiating commercial contacts. In general, foreign aid to agriculture can improve nutrition and stimulate economic growth, thereby, in low-income countries, stimulating demand for agricultural imports and, by extension, donor exports. Much foreign assistance is tied to the purchase of goods such as food or equipment from the donor. These purchases directly benefit producers in the donor countries.

This complex set of reasons for foreign assistance means that foreign aid does not always go to where need is greatest. The fact that aid is given in part for donor self-interest purposes would seem to impose on donors some obligation to ensure that the distribution and types of foreign assistance provided do not harm the recipients. We return to this issue below after examining the types and amounts of foreign development assistance.

Foreign Aid in the Context of Other Capital Flows

Foreign aid is not the largest type of capital flow to developing countries. Much larger flows occur privately through portfolio investment, foreign direct investment (FDI), and individual remittances. Portfolio investment is the purchase of shares or bonds, and can be an important source of capital for middle-income developing countries with growing financial markets. Foreign direct investment is the use of company-owned facilities and operations, which is particularly helpful if it involves the transfer of proprietary technologies and business methods. Individual remittances occur when migrants send money back to their families, as a gift or for investment. Such remittances are the fastest-growing kind of capital flow, and reached about $90 billion in 2003 (see Box 19-1).

Foreign aid is different from private investment or remittances in that, by definition, it uses government or charitable funds to serve a public purpose. Total foreign assistance encompasses economic development assistance plus military assistance and export credits. Often private funds from voluntary agencies are included. Foreign development assistance, as the term is used in this chapter, excludes the military-related component and export credits, while the term *official development assistance* (ODA) excludes private fund transfers as

---

BOX 19-1
REMITTANCES AS A DEVELOPMENT TOOL

Remittance transfers are generally small quantities of money sent through wire transfers or banks, or hand-carried to individuals, usually family members, in developing countries. Remittances have grown dramatically over the past decade to the point where, for some countries, they represent the largest source of capital inflow. Their importance differs by region, and they are relatively more important in Central America, the Caribbean, and some countries in Southeast Asia. In Latin America, remittances from the United States alone exceed $30 billion annually.

Because of the growing importance of remittance flows to developing countries, steps are needed to improve the efficiency by which remittances are transferred and their ultimate impacts on development. Efficiency can be improved by lowering the costs of and risks associated with remitting money, which will involve cooperation of international financial institutions. Steps are also needed to improve the enabling environment in recipient countries through reforms of banking systems, more transparent rules of access, encouraging acceptance of small-scale deposits, and so on. Incentives can be given for participation in the formal financial sector; these incentives will help mobilize savings of remittance recipients and other potential small-scale customers. Steps to encourage use of remittance funds for private productive investments will channel these funds into capital accumulation and away from short-term consumption. Well-defined legal and regulatory frameworks will help build the confidence of remitters to make productive investments and lower risks of losing their investments. Many of these steps would have the side benefit of mobilizing all forms of small-scale savings and investments; this micro-finance has been shown to facilitate broad-based growth.

Source: Samuel Munzele Maimbo and Dilip Ratha, *Remittances: Development Impact and Future Prospects* (Washington, D.C.: World Bank, 2005).

well. To qualify as any type of foreign assistance, the resources transferred must be sent from donor(s) to a recipient without a commensurate return flow of resources. There may be goodwill, political support, and so on, but direct payments are not made in return.

At one extreme, foreign development assistance can occur as loans at near-market interest rates. At the other, this assistance can be an outright grant. In the middle, the assistance can be a loan at a concessional (below-market) interest rate or with a maturity period longer than that commercially available.[4] Foreign development assistance may also come in the form of food aid or as technical assistance to provide needed expertise. To be classified as ODA by the Development Assistance Committee of the Organization for Economic Cooperation and Development (OECD), the assistance must have at least a 25 percent grant element.[5] The grant element is defined as the excess of the loan or grant's value over the (current) value of repayments calculated with a 10 percent interest rate.

The amount of U.S. and world ODA in recent years is presented in Table 19-2. The growth and decline of assistance flows is influenced strongly by international conflicts, such as the war on terror and the cold war. Declines in U.S. ODA during the 1990s illustrate the influence of the end of the cold war as well as the effects of budgetary constraints in donor countries. Increases after 2001 in part reflect the costs of the war on terrorism and of rebuilding Iraq.

Foreign assistance to agriculture is a portion of total ODA and includes such diverse components as aid used for agricultural research and extension, irrigation projects, rural roads, agricultural education and training, flood control projects, health improvement programs, integrated rural development projects, and agricultural policy assistance. It is difficult and perhaps not appropriate to separate out agricultural from nonagricultural aid. Foreign exchange and budget support directed at a country as a whole can indirectly benefit agriculture, as can food aid funneled through food-for-work programs to improve rural infrastructure. Support for infrastructure and education benefits all sectors of the economy.

---

[4] See Krueger, "Aid in the Development Process."

[5] OECD is an organization of Western industrialized nations designed to promote economic growth and stability in member countries and to contribute to the development of the world economy. Member countries are Australia, Austria, Belgium, Canada, Denmark, Finland, France, Germany, Greece, Iceland, Ireland, Italy, Japan, Luxembourg, the Netherlands, New Zealand, Norway, Portugal, Spain, Sweden, Switzerland, Turkey, the United Kingdom, and the United States.

TABLE 19-2: UNITED STATES AND WORLD OFFICIAL DEVELOPMENT ASSISTANCE (ODA), 1960-2004
(U.S. $millions, 2000)

Year	Total ODA	U.S. ODA	U.S. as % of Total
1960	22,256	13,137	59.0
1965	28,879	17,904	62.0
1970	24,835	11,665	47.0
1975	35,514	11,150	31.4
1980	49,248	13,420	27.2
1985	41,845	13,683	32.7
1990	65,578	14,108	21.5
1995	64,351	8,045	12.5
1996	59,578	10,044	16.8
1997	50,948	7,226	14.2
1998	54,035	9,115	16.9
1999	57,623	9,347	16.2
2000	53,735	9,955	18.5
2001	51,121	11,163	21.8
2002	54,825	12,410	22.6
2003	64,694	15,280	23.6
2004	69,011	17,923	26.0

Source: OECD database, Development Cooperation Report 2000; United Nations Statistics Division 2003.

Development Assistance Programs

Since World War II, the number of development assistance programs has risen greatly. Initially the United States was the major provider of economic aid. The relief and recovery efforts in Western Europe and East Asia shortly after the war were followed by a U.S. development assistance program for developing countries, called the Point Four Program. That program grew out of the fourth point in President Harry S. Truman's inaugural address of January 20, 1949. Truman called for a "bold new program for making the benefits of scientific advances and industrial progress available for the improvement and growth of underdeveloped areas."[6] The program provided technical assistance to Taiwan, South Korea, and other countries in Southeast Asia, the Middle East, and the less developed countries of Europe. Point Four effectively marked a shift in focus of U.S. foreign assistance from Europe to developing countries. This program was followed by the Development Loan Fund in 1957, which was combined with a technical assistance program

[6] Harry S. Truman, "Inaugural Address of the President," Department of State Bulletin 33, Washington, D.C. (January 1949), p. 125.

called the International Cooperation Administration in 1961 to form the U.S. Agency for International Development (USAID).[7] USAID is currently the major development assistance agency of the United States.

During the 1950s, assistance was increasingly extended by the United Kingdom, France, the Netherlands, and Belgium to their former colonies. The list of donors grew during the 1960s and now includes most members of OECD and many members of the Organization of Petroleum Exporting Countries (OPEC).[8] The United Slates provided roughly 60 percent of total ODA in the early 1960s, but this proportion dropped to about 13 percent by 1995 and then grew to about 26 percent by 2004. Even though the United States gives more ODA than any other country, in recent years it has ranked at or near the bottom among OECD countries in terms of the ratio of ODA to GNP, a rough measure of the ability to "afford" aid. In 2003 it gave 0.15 percent while the weighted average across all OECD countries was 0.25 percent and simple average was 0.41 percent. In 2002, the OECD countries pledged to contribute 0.7 percent of their GNP to ODA as a means of achieving the United Nations Millennium Goals. In order to achieve this goal, ODA will need to grow dramatically.

Each donor country has its own internal agency for administering its bilateral (country-to-country) assistance programs. Some of the money and technical assistance is funneled through specific projects and other assistance through broader programs. Much of direct assistance to agriculture comes through specific projects. While the agricultural component of total aid is substantial, support for agricultural assistance has declined recently for several countries, particularly in Latin America.

Not all development assistance is publicly funded or administered. Non-governmental organizations (NGOs) are becoming increasingly significant sources of development assistance. Examples of NGOs are CARE, the Red Cross, Lutheran Relief Service, Catholic Relief, Oxfam, Save the Children, the Gates Foundation, and a host of others. There is a wide range of types and sizes of NGO institutions. Many, but not all, are focused on community or microeconomic development. NGO projects tend to be small and designed to benefit the poorest of the poor. Some of the efforts are long term and development oriented while others provide short-term relief in crisis situations.

---

[7] See Elizabeth Morrison and Randall B. Purcell, *Players and Issues in U.S. Foreign Aid* (West Hartford, Conn.: Kumarian Press, 1988) for additional historical details.

[8] OPEC is a group of countries devoted to seeking agreement among themselves regarding selling prices and other issues related to oil exports. OPEC members include Algeria, Ecuador, Gabon, Indonesia, Iran, Iraq, Kuwait, Libya, Nigeria, Qatar, Saudi Arabia, the United Arab Emirates, and Venezuela.

TABLE 19-3: OFFICIAL DEVELOPMENT ASSISTANCE
BY COUNTRY OF ORIGIN
(U.S. $millions, 2000)

Country	1970	1980	1990	2002	2004
Australia	749	1,254	1,183	925	1382
Austria	71	335	488	457	652
Belgium	443	1,119	1,101	1,021	1370
Canada	1,281	2,021	3,058	1,937	2393
Denmark	219	904	1,450	1,570	1910
Finland	–	207	1,048	448	618
France	3,370	7,825	8,869	4,985	7995
Germany	2,216	6,706	7,826	5,155	7073
Greece	–	–	–	284	609
Ireland	–	56	71	382	553
Italy	545	1,284	4,204	2,225	2343
Japan	1,694	6,304	11,230	8,870	8358
Luxembourg	–	9	31	138	227
Netherlands	727	3,064	3,143	3,249	3995
New Zealand	–	135	118	119	198
Norway	136	914	1,492	1,680	2075
Portugal	152	–	176	271	970
Spain	–	305	1,195	1,547	2403
Sweden	433	1,809	2,485	1,687	2551
Switzerland	112	476	929	898	1301
United Kingdom	1,653	3,486	3,266	4,569	7392
United States	11,284	13,420	14,108	12,410	17924
Total DAC countries	25,306	51,318	67,469	54,825	74138

Source: OECD database; Development Co-operation Report 2000; United Nations
  Statistics Division 2003.

All NGOs are non-profit, privately run institutions. While many, such as CARE, receive substantial public support, most depend heavily on private voluntary financial contributions. The largest NGOs are based primarily in developed countries, but carry out their activities in developing countries. Several NGOs have religious affiliations.

Multilateral Assistance Programs

In addition to publicly supported bilateral aid and assistance from NGOs, a variety of multilateral (multi-country) organizations provide significant financial and/or technical support to developing countries (see Table 19-4). The World Bank was created in 1944, followed shortly afterward by several other United Nations-supported agencies. A brief

TABLE 19-4. MAJOR DEVELOPMENT ASSISTANCE PROGRAMS

Multilateral Financial Network

1 World Bank
2 InterAmerican Development Bank
3 African Development Bank
4 Asian Development Bank
5 European Development Fund
6 United Nations Development Programme (UNDP)
7 World Food Programme (WFP)
8 United Nations Children's Fund (UNICEF)
9 International Fund for Agriculture and Development (IFAD)
10 Organization of Petroleum Exporting Countries (OPEC) Fund
11 Caribbean Development Bank

Bilateral Financial Assistance

1 U.S. Agency for International Development (USAID)
2 Japanese Overseas Economic Cooperation Fund (OECF)
3 Agencies in Germany, France, United Kingdom,
   and many other countries
4 Non-governmental Organizations (NGOs): Foundations
   (e.g., Gates, Rockefeller), Cooperative for American Relief
   Everywhere (CARE), Catholic Relief, many others

Technical Assistance Network

1 United Nations Development Programme
2 Food and Agriculture Organization of the United Nations (FAO)
3 Bilateral assistance programs listed above
4 International Agricultural Research Centers (IARCs)
5 Many others, including private assistance

description of the major U.N. financial and technical assistance agencies is presented in Box 19-2.

The World Bank is not a commercial bank, but an international organization that combines donor funds with investors' deposits to make relatively low-interest loans to developing countries. It has more than 180 country members and consists of three major arms that together represent the largest source of long-term multilateral economic development assistance. The first arm, the International Bank for Reconstruction and Development (IBRD), established in 1945, makes long-term loans at interest rates related to its own cost of borrowing, mostly for large-scale projects. The second arm, the International Development Association (IDA) established in 1960, makes loans only to the poorest countries. More than 90 percent of IDA lending goes to countries with an annual per capita income under $500. Loans from IDA have long repayment periods and concessional interest rates. The third arm, the

---

BOX 19-2
MAJOR UNITED NATIONS AGENCIES
FOR FINANCIAL AND TECHNICAL ASSISTANCE
TO DEVELOPING COUNTRIES

The United Nations Development Programme (UNDP) is the central funding and coordinating mechanism within the United Nations for technical assistance to developing countries. The United Nations Fund for Population Activities (UNFPA) helps countries gather demographic information, undertake family planning projects, and formulate population policies and programs. The United Nations Children's Fund (UNICEF) provides technical and financial assistance to developing countries for programs that benefit children and for emergency relief for mothers and children. The purpose of the Food and Agriculture Organization (FAO) is to raise nutrition levels and standards of living by improving the production and distribution of food and other commodities derived from farms, fisheries, and forests. It also helps countries with food emergencies. The World Food Programme's (WFP) purpose is to stimulate economic and social development through the use of food aid and to provide emergency food relief. The World Health Organization (WHO) conducts immunization campaigns, promotes and administers research, and provides technical assistance to improve health systems in developing countries. The United Nations Education, Scientific, and Cultural Organization (UNESCO) promotes international intellectual cooperation in education, science, culture, and communications. The UNDP, UNFPA, WFP, and UNICEF are funded through voluntary contributions, public and private, while FAO, WHO, and UNESCO are funded primarily through assessments on member nations with some additional voluntary contributions and other sources of funds.

Source: Elizabeth Morrison and Randall B. Purcell, *Players and Issues in U.S. Foreign Aid* (West Hartford, Conn.: Kumarian Press, 1988), pp. 41–5.

International Finance Corporation (IFC), is a profit-making enterprise and is funded by capital from its 130 member countries. It makes loans to the private sector, to mixed (public/private) enterprises, and to government-owned agencies that channel financial assistance to the private sector. Two other arms of the World Bank are the Multilateral Investment Guarantee Association (MIGA) and the International Centre for the Settlement of Investment Disputes. Since the late 1950s, regional development banks have been formed for Latin America, the Caribbean, Asia, and Africa. These banks make both concessional and non-concessional loans.

Trends in Development Assistance

The net financial receipts of developing countries have been volatile over time. Total Official Development Finance increased gradually from the 1960s until the mid-1990s, then dropped (after taking into account inflation and exchange rate differences), and recently has risen again. Private lending increased dramatically during the 1970s, but virtually dried up in the 1980s, causing total financial flows to decrease during the 1980s. In the 1990s, private flows into developing countries again increased significantly, but then dropped off again after 1999. The ODA portion of total financial flows to developing countries was highest (above three-quarters) for the least developed countries, next highest (more than half) for all low-income countries, about one-quarter for lower-middle-income countries, but only about 10 percent for upper-middle-income countries. ODA accounted for only about 1 to 5 percent of GNP in the low-income countries. A brief snapshot of total public and private capital flows to developing countries since the 1980s is presented in Table 19-5. The volatility is evident.

Since the 1950s, the composition of development assistance has changed dramatically.[9] As noted, the relative contribution of the United States has declined substantially. Multilateral ODA has grown in

TABLE 19-5. TOTAL NET RESOURCE RECEIPTS OF DEVELOPING COUNTRIES
FROM ALL SOURCES ($U.S. billions, current $)

	1987-88 average	1999	2000	2001	2002	2003
I. Official Development Assistance (ODA)	43.8	53.2	53.7	52.4	58.3	69.0
II. Other official flows	3.0	15.6	−4.3	−1.6	−.04	−1.1
III. Private flows at market terms	21.5	116.0	78.1	49.7	6.3	30.5
IV. Net grants by NGOs	4.1	6.7	6.9	7.3	8.8	10.2
Total net flows in current $	72.5	191.5	134.5	107.9	73.3	108.5
Total net flows at 2002 prices and exchange rates	87.2	182.6	134.0	112.0	73.3	96.0

Source: OECD Database

[9] See John Mellor, "Foreign Aid and Agriculture-led Development," in *International Agricultural Development*, 3rd edn, ed. Carl K. Eicher and John M. Staatz (Baltimore, Md.: Johns-Hopkins University Press, 1998), pp. 55–66.

importance — from 6 percent of the total in the 1960s to more than 25 percent currently. In the early post-World War II years through the 1960s, development assistance focused primarily on agriculture. Assistance was given to agricultural extension and community development (in the 1950s and early 1960s) and to agricultural research and infrastructure such as irrigation (in the 1960s). This assistance helped contribute to the Green Revolution and agricultural productivity growth in many parts of the world. Donors were buoyed by these successes and began in the 1970s to turn their attention to "second-generation" problems such as provision of services to the poor, integrated rural development programs, and natural resource management. By the 1980s it began to become clear that in many countries progress toward economic development was limited by inappropriate macro-economic policies, and the focus of donors switched to support for policy reforms such as market liberalization, more open trade and foreign exchange regimes, and tax reforms. As discussed in Chapter 18, these "structural adjustment policies" were promoted by the large multilateral donors such as the World Bank and the International Monetary Fund (IMF) and were supported by bilateral aid agencies. They were designed to improve growth prospects, but often had adverse effects on the poor.

In the 1990s, the focus of many donors switched to poverty reduction and environmental management programs. Research showed that many of the "top-down" programs of the 1970s and 1980s failed because many of the intended beneficiaries were not consulted during program design and, hence, felt no sense of "ownership." Aid program design became more participatory during the 1990s; specific programs included community-driven development, community-based natural resource management, and participatory research projects. Most recent evidence has shown that aid is most effective when it is accompanied by good policy and management in recipient countries and when it is used to provide a comprehensive package of assets. Sachs discusses different forms of capital and assets — business, human, infrastructure, natural assets, knowledge and public institutions — that, when packaged together, lead to growth and poverty reduction. When subsets of these packages are provided, the results are less likely to be positive.[10]

During all these changes in composition and form of development assistance, one thing has been relatively constant: the decline in public support for agriculture. Partly due to the success of the Green Revolution, partly due to changing donor interests, support for agricultural infrastructure, research, and extension has fallen since 1960. These

[10] Jeffrey Sachs, *The End of Poverty* (New York: Penguin, 2005), esp. ch. 13.

declines have been especially harmful in Africa, where Green Revolution technologies were never really suitable and agricultural output has stagnated since 1980.

Effects of Foreign Assistance

The economic effects of development assistance on recipients are the regular subject of debate in the popular press and among policy-makers. Such debates need to be informed by good information, so studies examining aid's effectiveness should be closely scrutinized. The effects of aid can be assessed at the project, the sector, or the national levels. At the project level, rates of return have been calculated for individual projects such as in irrigation, flood control, roads, and agricultural research. For example, the World Bank has estimated that the rate of return on its IBRD loans from 1960 to 1980 was 17 percent and on its IDA loans, 18 percent.

Given the high rates of return to many aid projects, questions are raised about why private lending has not been forthcoming, or why concessional loans are needed. The answer is that returns to aid are spread among the recipient population in ways that a private company could not capture to repay its investors. For example, aid that helps a small child avoid malnutrition or attend school can generate benefits far in excess of its cost, but those benefits cannot be seized by a lender. Indeed, aid is most effective when it focuses on precisely these kinds of public services, which will not be provided by private firms. The developing country's own governments cannot provide enough of these public goods because they lack the tax base or administrative capacity to finance them, either from current revenues or to repay its own loans. As a result, public-sector development assistance from richer to poorer countries has an important role to play in making the world economy more efficient as well as more equitable.

At the sector or national levels, development assistance can augment domestic savings, help provide foreign exchange, and minimize adverse impacts of needed policy reforms. These effects can stimulate growth. Many studies have attempted to assess the impact of development assistance at the national level as it affects savings, investment, or growth. Generally, the results have been positive but, in many cases, inconclusive.[11] The fundamental problem is that aid typically flows to countries in trouble and so is usually associated with bad economic

---

[11] For more details, see Constantine Michalopoulos and Vasant Sukhatme, "The Impact of Development Assistance: Review of Quantitative Evidence," in *Aid and Development*, ed. Anne O. Krueger, Constantine Michalopoulos, and Vernon W. Ruttan (Baltimore, Md.: Johns Hopkins University Press, 1989), ch. 7.

outcomes, even though without aid the outcome might have been worse. Poor countries face many different kinds of problems, from natural disasters and disease to wars and corruption. A widely cited view suggests that aid has positive effects on growth only in countries with good fiscal, monetary, and trade policies, but recent studies suggest that aid has helped even in countries with weak governments.[12]

The small size of aid relative to other capital sources, as described above, has undoubtedly contributed to many of the inconclusive findings about the effects of aid on growth. It is also difficult to measure the effects of aid without breaking it down into its different types and the different reasons for which it is given. For example, aid for infrastructure may have different impacts on growth than does aid for policy reform, or for emergency food relief. Emergency food relief programs may be very effective at reducing hunger and human suffering associated with short-term problems, but their effects on aggregate growth may be hard to measure.

Despite a relatively weak link between aid and aggregate economic growth, for reasons outlined above, aid has helped reduce poverty and improved the quality of life of the poor. Targeted aid has helped eradicate smallpox, put polio on the brink of eradication, reduced death from diarrheal diseases, and reduced the incidence and severity of many illnesses.[13] These results will, over time, contribute to economic growth and development, but their effects are indirect and hard to measure.

Few studies have attempted to evaluate the effects of development assistance to the agricultural sector. One study for 98 countries did find a positive effect of foreign assistance to agriculture from 1975 to 1985, particularly in Asia. However, the effects of aid to agriculture in Latin America and the Middle East were non-significant.[14] Investment in research and dissemination of new technologies may be among the most effective kinds of aid for reducing poverty, especially in Africa where the level of such investment has been lowest and where rural poverty is particularly severe.[15]

---

[12] See Craig Burnside and David Dollar, "Aid, Policies and Growth," *The American Economic Review*, vol. 90 (September 2000), pp. 775–86; and Carl-Johan Dalgaard, Henrik Hansen, and Finn Tarp, "On The Empirics of Foreign Aid and Growth," *The Economic Journal* vol. 114, no. 496 (June 2004), pp. F191–F216.

[13] See Jeffrey Sachs, *The End of Poverty* (New York: Penguin Press, 2005); Joseph Stiglitz, "Overseas Aid is Money Well Spent," *Financial Times*, April 14, 2000, p. 20; and Shalendra Sharma, "The Truth About Foreign Aid," *Challenge*, vol. 48 (July–August 2005), pp. 11–25.

[14] See George W. Norton, Jaime Ortiz, and Philip G. Pardey, "The Impact of Foreign Assistance on Agricultural Growth," *Economic Development and Cultural Change*, vol. 40 (July 1992), pp. 775–86.

Effects on Donors

Many donor countries believe that development assistance is effective in stimulating growth in developing countries, but ask whether the effects on their own countries are negative. In other words, they question whether their economic self-interest is served by aid. For example, farm groups and the farm press in the United States frequently express concern that foreign aid may be generating foreign competition. Several studies have assessed whether foreign aid to agriculture does indeed hurt U.S. farmers. These studies have found that for particular commodities in particular countries at particular stages of development, foreign competition is increased as a result of aid. However, they also found that agriculture as a whole in donor countries is helped by foreign aid to agriculture in developing countries.[16]

The reason why farmers in donor countries often benefit from aid to agriculture in developing countries is that agricultural growth in LDCs increases incomes, and these incomes, in turn, stimulate food demand. Middle-income countries, particularly, still have relatively high population growth rates and high income elasticities of demand for food. Demands shift toward higher-quality grains and livestock products as incomes rise. Consequently, when agricultural production rises in these countries, if domestic economic policies permit that production growth to stimulate other sectors of the economy, the result is an expansion in food demand that must be met partially through food imports (see Box 19-3). Of course, when countries eventually reach higher income status, their growth in food demand slows. If and when most of the currently developing countries reach that status, trade will be governed by comparative advantage and trade-distorting policies. Development assistance to agriculture will no longer be an issue.

In summary, while empirical evidence is not entirely conclusive about the effects of development assistance in general or to agriculture in particular, it appears that positive but modest gains are likely for both recipients and donors. It is unlikely that large gains will be realized except in a few small countries, because aid usually represents a small portion of ODA.

---

[15] See Colin Thirtle, Lin Lin, and Jenifer Piesse, "The Impact of Research-led Agricultural Productivity Growth on Poverty Reduction in Africa, Asia and Latin America," *World Development*, 31, vol. 12 (December 2003), pp. 1959–75.

[16] See, for example, Alain de Janvry, Elisabeth Sadoulet, and T. Kelley White, "Foreign Aid's Effect on U.S. Farm Exports," U.S. Department of Agriculture, Economic Research Service, Foreign Agricultural Economic Report Number 238, Washington, D.C., November 1989; and James P. Houck, "Link between Agricultural Assistance and Agricultural Trade," *Agricultural Economics*, vol. 2 (October 1988), pp. 158–66.

BOX 19-3
COUNTRIES WITH SUCCESS IN AGRICULTURE
OFTEN INCREASE FOOD IMPORTS

The World Bank's *World Development Report 2000/2001* lists nine countries whose agricultural sectors grew at an annual rate of 4.9 percent or higher from 1990 to 1999. The report also lists cereal imports for those countries in 1990 and in 1999. In every case, cereal imports increased despite the relative successes in agriculture.

Country	Average annual growth rate (percentage 1990–1999)[1]		Cereal imports (thousands of metric tons)[2]		Percent increase in cereal imports
	Agriculture	Overall GDP	1990	1999	1985–1999
Albania	6.2	2.3	168	364	117
Cameroon	5.3	1.3	384	402	4.7
Chad	4.9	2.3	30	50	67
Mauritania	5.2	4.1	198	501	153
Myanmar	4.9	6.3	7	110	1,471
Nicaragua	5.4	3.2	170	202	19
Peru	5.8	5.4	1,643	2,556	56
Vietnam	4.9	8.1	227	711	213
Yemen	5.0	3.0	1,314	2,628	100

Sources: [1] World Bank, *World Development Report 2000/2001* (New York: Oxford University Press, 2000).
[2] The FAOSTAT Database.

FOOD AID

Food aid to developing countries has been an important dimension of foreign assistance since the mid-1950s. The importance of food aid has, however, fallen over time. In the 1960s, it represented more than 20 percent of total ODA, but by 2000 it had fallen to around 2 percent. Food aid still constitutes a relatively large share of U.S. ODA, but for the U.S. this share is also falling. About 45 percent of food aid is used to provide emergency relief in times of severe food shortages, 30 percent supports specific development projects, and the remaining 25 percent is program or non-project food aid. The latter is given as hard currency assistance to buy imports of food, usually cereals, most often from the donor country.

The role and effects of food aid have been controversial because of its many purposes. While food aid fulfills a humanitarian and development mission, it also provides a means for donor countries to dispose

Food aid helped alleviate hunger in Ethiopia during the 1980s
(photo: Mesfin Bezuneh).

of surplus commodities and to develop new markets. As with any foreign aid, food aid serves the foreign policy objectives of donors. While this multiplicity of objectives has added instability over time to food aid allocations, it has also strengthened the political support for maintaining food aid programs within the donor countries.

Critics of food aid have argued, among other things, that unrestricted cash donations would be preferable to food. While it is clear that recipients would prefer cash, many donors treat food aid as an addition to, rather than a component of, their economic assistance. It is highly unlikely that donor budgets would be expanded by the value of food aid if the latter were eliminated.

Food aid is provided both bilaterally and multilaterally. The United States has been the largest source of food aid since the enactment of the Agricultural Trade Development and Assistance Act of 1954, commonly referred to as Public Law (P.L.) 480. The U.S. share of total food assistance has been about 60 percent in recent years; the European Community has contributed about 15 percent, Japan 6 percent, and other countries the rest.[17] About a quarter of food aid is funneled through multilateral organizations, primarily the World Food Programme of the United Nations.

---

[17] Food aid once represented about one-half of U.S. grain exports, but in recent years it has declined to less than 10 percent.

History of Food Aid

The history of food aid to developing countries is marked by shifting emphases on its multiple objectives. During the period from 1959 to 1965, the United States and Canada were particularly concerned about disposal of farm surpluses, developing markets for farm products, and providing emergency food relief. Most of the aid provided during this period was in grain, but several other products were given, including tobacco. In 1961, an amendment was added to P.L. 480 to permit food to be used for economic development instead of being restricted to emergency relief. Improved export markets, led by demand growth in developing countries, reinforced the objective that food aid helps developing markets.

The era from 1966 to 1972 was a period of heavy use of food aid for emergency relief, particularly in drought-stricken areas of South Asia. Self-help of recipients was also promoted during this period. The European Community and Canada increased their shipments of food aid for emergency relief in this period. The 1966 to 1972 period might be called the idealistic era of food aid.

Unfortunately, any idealism with respect to food aid programs was pretty much destroyed by the cutbacks in food aid that followed food price increases in 1972 to 1975. The United States had depleted its grain surplus by exporting commercially to the Soviet Union and other countries. From 1972 to 1973, U.S. commercial grain exports doubled, and the volume of food aid fell in 1974 to its lowest level since the enactment of P.L. 480. Furthermore, during this period half of all U.S. food aid went to South Vietnam and Cambodia as a result of U.S. involvement in those countries.

In 1975, the U.S. Congress instituted more humanitarian and development criteria for receiving food aid by passing the International Development and Food Assistance Act (see Box 19-4). This legislation called for increased food aid to the poorest countries. The remainder of the 1970s also saw increasing food aid quantities from EC countries. However, the use of food aid for political purposes also increased after 1975. For example, U.S. food aid to Bangladesh declined from 1.15 million tons in 1975 to 0.34 million tons in 1985, while food aid to Egypt increased from 0.58 million tons to 2.00 million. This increase to Egypt was directly linked to the Camp David Peace Agreement signed with Israel in 1979. Food aid quantities increased in the mid-1980s in response to severe drought problems in Ethiopia, the Sudan, and other Sub-Saharan African countries. In the 1960s, most food aid went to Asia and Latin America. By the mid-1980s Sub-Saharan Africa was absorbing as much food aid as the much more populous Asia.

BOX 19-4
THE UNITED STATES P.L. 480 FOOD AID PROGRAM

Since 1954, most U.S. food aid activities have been coordinated under P.L. 480. Numerous amendments and extensions have been added to the original act, but currently the major provisions fall under the three following titles:

Title I — was formerly the most important component of P.L. 480, but by the early 2000s, it had shrunk to about $100 million per year compared to more than $50 billion in agricultural exports from the United States. Recipient governments buy grain on credit with interest rates of 3 percent or less over 20 to 40 years, repayable in local currency. These governments can sell the grain internally and use the profits for development. The lower interest rates and long repayment period mean that almost 70 percent of the food aid loan is a grant.

Title II — involves gifts of food for emergency relief and for economic development, and in 1991 it surpassed Title I as P.L. 480's largest component, now accounting for more than 85 percent of the program. The food is given to and distributed by private agencies such as CARE, who use the food for infant-feeding programs and for mother- and child-health programs in addition to emergency distribution. In a recent year about 70 percent of CARE's budget was P.L. 480. Shipping and labor are paid for by the U.S. government. Food given under Title II is also used in food-for-work programs.

Title III — involved using food aid in government-to-government programs to support economic development, but has not received funding since 2001.

Other U.S. food aid programs exist, but food aid is now dwarfed by agricultural export credit programs that support commercial grain exports.

Types of Food Aid Programs Today

Emergency food aid grabs most of the headlines as it relieves crises associated with droughts in Ethiopia, the Sudan, and North Korea, and flooding in Bangladesh and other parts of Asia. Emergency food aid has also played a significant role in feeding refugees from Afghanistan, Iraq, and other countries in recent years. This short-term food aid is essential for reducing acute hunger problems. The possibilities of using food aid to foster long-term development, however, are more closely linked to program or project food aid.

Program food aid is, in many respects, similar to more general financial assistance, as it provides currency to buy imports, in this case food that can be sold or otherwise distributed in the domestic market.

This aid fosters the development of marketing linkages with the donors, it helps the recipients save foreign exchange, and the funds generated by the sales can be used for development. Some donors participate in determining how the funds generated by commodity sales are used. Donors may insist that funds be used for investments in the agricultural sector or to support specific policy changes affecting agriculture. Some of the recent food aid shipments to Sub-Saharan African countries were intended to soften the adjustments to structural changes in their economies.

Project food aid is aimed at meeting specific development objectives. Projects tend to be multi-year, to be targeted at nutritionally vulnerable individuals or groups, and may involve food in exchange for work on the project. Donor and recipient countries agree on who will be targeted by the project, the amount of food each individual receives, the delivery system for the food, and the design, implementation, and monitoring of the project activities.[18] Most of the projects involve the rural sector and can vary in size from a few hundred thousand dollars to $100 million or more. Food aid projects often involve forestry development, soil conservation and watershed management, resettlement projects, training, development of irrigation works, and construction and maintenance of rural roads.

Effects of Food Aid

The positive and negative effects of food aid on recipient countries have been studied and debated for many years. On the positive side, food provides real resources that can be used to expand investment and employment. Food aid can have a disproportionate but positive effect on disadvantaged groups, notably by supporting specific nutrition or food-for-work projects or by providing food to the poor for free or at subsidized prices. Food can be used to help recipient governments support storage and stabilization schemes to provide a small buffer against poor production years.

Food aid can also have adverse effects on the recipients. These potential adverse effects can occur in a number of ways: (1) disincentive effects on local agricultural production through reduced prices because of greater supply, (2) dependency effects because the government can substitute food aid for agricultural development programs, and (3) the uncertainty of food aid quantities from year to year.

---

[18] See Robert Chase, "Commodity Aid for Agricultural Development," in *Trade, Aid, and Policy Reform,* ed. Colleen Roberts (Washington, D.C.: World Bank, 1988), pp. 199–204.

The disincentive issue has been examined empirically in several studies.[19] In theory, additional supplies could depress food prices and discourage production. Some empirical studies have found this to be the case, but other studies have not. The disincentive effect is minimized if food aid is given or sold to those who otherwise could not afford the food. Transferring food is like transferring income. The quantity of the aid compared to the country's overall food production is important. For example, it appears that there has been a disincentive effect in Egypt due to the large quantities of aid shipped, but it is extremely difficult to sort out the impact of food aid from the many policy-induced distortions. Even when food aid reduces prices, it is likely to have a beneficial effect on the poor, who generally purchase more food grains than they sell.[20]

The idea that food aid creates dependency has not been examined as frequently. Food aid is no different from other aid in that, by providing resources, it may lead to less effort to raise revenues domestically or to promote agricultural development. Conditions are usually placed by donors on program aid that minimize this possibility. A second part of the dependency argument is that, over the long run, food aid leads to more food imports and changes in preferences away from domestically produced foods. Some evidence shows that this preference effect may be occurring, although it is difficult to separate changes induced by food aid from those that occur because of income growth and other trade.

Food aid can be used in a positive way by recipients to further both agricultural and overall economic development. Emergency food aid will always be variable and it can play a major life-saving role during short-term emergencies. It appears that the potential positive development role of food aid has not been fully exploited, although some efforts are under way to improve its development contribution.

Most donor countries find public opinion is generally supportive of food aid, especially when it is used in visible programs to prevent starvation. The future, however, for food aid is uncertain due to budget tightening in donor countries, reduced price supports for agriculture as a part of the opening of global markets, and questions about its effectiveness as a development tool. Stronger multi-year commitments are certainly needed if food aid is to be more effective in promoting development.

---

[19] See S. T. Maxwell and H. W. Singer, "Food Aid to Developing Countries: A Survey," *World Development*, vol. 7 (1979), pp. 223–47, for a summary of results of 21 studies.

[20] See James Levinsohn and Margaret MacMillian, "Does Food Aid Harm the Poor? Household Evidence from Ethiopia," National Bureau of Economic Research, Inc., NBER Working Papers: 11048 (2005).

ADMINISTRATION OF FOREIGN ASSISTANCE

Foreign assistance is administered through a wide array of bilateral and multilateral public agencies and through private organizations (Table 19-4). Once the aid reaches the recipient country, the government usually plays an important role in implementing the program or project for which the aid is intended. If basic infrastructure is missing, incentive structures are severely distorted, or the government bureaucracy is highly inefficient, effective administration of economic assistance programs can be extremely difficult.

How Aid is Administered

Most bilateral assistance agencies such as USAID maintain offices in the recipient countries staffed with people who provide the link between the home office (and the policies of the donor government) and the local officials. Operating with a budget established for that country by the donor, programs and projects are designed, implemented, and evaluated by personnel in that office, by people sent from the donor country, and by local counterparts in the recipient country.

Bilateral foreign assistance often comes with strings attached. These strings can involve policy changes by the recipient country or simply a requirement that goods and services procured with the aid money be obtained from the donor country. In other cases, the donor might make the provision of aid contingent on policy reforms or political changes.

Multilateral aid is administered by the appropriate international organization, which usually maintains an office in the recipient country. Assistance is often earmarked for specific purposes, and projects are designed and evaluated by teams of experts. Multilateral donors such as the World Bank often have more leverage in policy discussions than bilateral donors both due to the greater resources typically involved and because they are "owned" in part by the recipient countries that are also Bank members. In addition, multilateral donors may be regarded as having less political bias than bilateral donors.

Foreign assistance, especially food aid, may be channeled through non-governmental organizations. These NGOs generally have to spend an enormous amount of time and resources raising money. NGOs based in the United States receive about a quarter of their resources from the government, but the remainder comes as donations from private donors.

Food aid is used in Kenya to help pay labor for road construction
(photo: Mesfin Bezuneh).

Improving Aid Effectiveness

The effectiveness of foreign assistance in promoting development could be improved by changes on the part of both donors and recipients. Bilateral assistance would be more effective with reduced tying of aid to procurement of donor goods and services, by longer-term commitments, by reduced bureaucracy in many cases, by increased coordination with other donors and across programs, and by increased attention to the development and poverty alleviation goals and less to narrow political goals. Multilateral organizations need increased flexibility in many of their programs that support technical assistance. Many of the U.N. agencies need better evaluation and budgeting procedures. Many NGOs need to improve their relations with local governments. They need to refine their ability to monitor and evaluate projects. Interactions with official development agencies also can be improved in many cases.

Coordination of the activities of various bilateral, multilateral, and NGO aid sources could help increase aid effectiveness. This coordination would probably have to be provided by a multilateral agency working with the local government.

Many recipient countries could make more effective use of foreign assistance with more enlightened macroeconomic and sectoral economic policies. The differences in the relative effectiveness of aid to Cote d'Ivoire and Ghana in the 1960s and 1970s helps make this point. Ghana was significantly ahead of Cote d'Ivoire by most development

405

measures in the late 1950s. Cote d'Ivoire followed policies that guided rather than controlled the private sector. Ghana pursued policies of import substitution and intervened strongly in almost every sector of the economy. The two countries received similar levels of aid but, by the early 1980s, Cote d'Ivoire had surpassed Ghana in most indicators of development.[21] As a result of such experiences, some donor countries restrict a portion of their assistance to countries that demonstrate a viable institutional environment to make good use of the resources.

Streamlining the bureaucracies in many developing countries could significantly improve the effectiveness of foreign aid. In many cases bureaucracies are a colonial legacy designed to improve accountability. However, they often merely reduce government efficiency and provide opportunities for corruption.

Foreign assistance can play a significant role in agricultural and overall economic development. With stronger multi-year commitments on the part of donors, and policy and bureaucratic reform on the part of recipients, greater returns can be realized by both groups from foreign assistance programs. There are several examples of aid being used successfully to soften the adverse impacts of macroeconomic and policy reforms. Using aid to create safety nets for needed reforms can satisfy humanitarian, economic, and political goals.

SUMMARY

Foreign development assistance in support of agriculture in developing countries has been substantial, has taken many forms, and has generated considerable debate. The rationale for foreign aid rests on humanitarian, political, and economic self-interest grounds. As a result, aid does not always go to where need is greatest. Economic aid may come in the form of financial assistance, food aid, or technical assistance. Some foreign aid comes as loans at below-market interest rates. Foreign aid to agriculture includes aid for agricultural research and extension, irrigation projects, rural roads, agricultural policy assistance, and many other items.

The United States is the largest bilateral donor, but the share of total ODA coming from the United States has declined over time. The United States currently gives the second smallest percentage of its GDP of any OECD donor. Aid is also funneled through multilateral agencies and NGOs.

Food aid to developing countries has been an important dimension of foreign assistance since the mid-1950s. It provides emergency

---

[21] Vernon W. Ruttan, "Solving the Foreign Aid Vision Thing," *Challenge*, vol. 34 (May–June 1991), p. 43.

relief in times of severe food shortages, and supports specific development projects and programs. Food aid provides a means for donor countries to dispose of surpluses, develop new markets, and pursue foreign policy objectives.

Aid effectiveness could be improved if there were increased coordination among donors and more attention was paid to needs than to politics. Developing countries could improve aid effectiveness by improving the policy environment, streamlining bureaucracies, and demanding more focused and long-term donor commitments.

IMPORTANT TERMS AND CONCEPTS

Bilateral aid

Concessional interest rates

Dependency effect of aid

Disincentive effect of aid

Economic self-interest

Food aid

Foreign development assistance

International Bank for
    Reconstruction and
    Development

International Development Association

International Finance Corporation

Multilateral aid

Nongovernmental organization

Official development assistance

Point Four Program

Public Law 480

U.S. Agency for International
    Development

World Bank

World Food Programme

Looking Ahead

This chapter concludes the section of the book concerned with macroeconomic and international issues affecting development. The book concludes in the next chapter with a discussion of how the various components required for agricultural development can be combined in an overall strategy. An assessment of future development prospects is provided, and suggestions are made for how you as individuals can contribute to solving the world food-poverty-population problem.

QUESTIONS FOR DISCUSSION

1. What is the rationale for foreign development assistance?
2. What are the major types of foreign development assistance?
3. What are some of the major effects of foreign development assistance on recipients and donors?
4. Distinguish between bilateral and multilateral aid.
5. Give several examples of foreign aid to agriculture.
6. How do NGOs differ from official sources of foreign development assistance?

7. What are the three major arms of the World Bank and how do they differ?
8. Which country is currently the largest bilateral donor of foreign aid?
9. Why might foreign development assistance help U.S. farmers?
10. What are the objectives of food aid?
11. What is the case for and that against food aid?
12. How have food aid programs changed over time?
13. What is the difference between program and project food aid?
14. How do the three titles of P.L. 480 differ?
15. How might the effectiveness of foreign development assistance be improved?

RECOMMENDED READING

Houck, James P., "Link Between Agricultural Assistance and Agricultural Trade," *Agricultural Economics*, vol. 2 (October 1988), pp. 158–66.

Krueger, Anne O., "Aid in the Development Process," *World Bank Research Observer*, vol. 1 (January 1986), pp. 57–78.

Mellor, John, "Foreign Aid and Agriculture-Led Development," in *International Agricultural Development*, 3rd edn, ed. Carl K. Eicher and John M Staatz (Baltimore, Md.: Johns Hopkins University Press, 1998), pp. 55–66.

Ruttan, Vernon W., "Why Foreign Assistance?" *Economic Development and Cultural Change*, vol. 37 (January 1989), pp. 411–24.

Sachs, Jeffrey, *The End of Poverty* (New York: Penguin, 2005), esp. ch. 13.

Sharma, Shalendra, "The Truth About Foreign Aid," *Challenge*, vol. 48 (July–August 2005), pp. 11–25.

Lessons and Perspectives

Progress in reducing hunger has been uneven across regions and countries...reducing hunger and malnutrition will require strengthening governance of the food and agriculture system at the global, country, and local levels...scaling up public investment for agricultural and rural growth, taking targeted steps to improve nutrition and health, and creating an effective global system for preventing and mitigating disasters....We must push ourselves even further to develop and implement solutions and policies to achieve food and nutrition security for the poorest of the poor and those most afflicted by hunger.
— Joachim von Braun[1]

This Chapter

1. Summarizes how the various components required for agricultural development can be combined to increase agricultural productivity and stimulate economic growth and development
2. Discusses how principles highlighted in this book can be used to assess future prospects for agricultural development in developing countries
3. Suggests ways that individuals can contribute to reducing the food-poverty-population problem.

AN INTEGRATED APPROACH TO AGRICULTURAL DEVELOPMENT

It is easy to be pessimistic about prospects for solving poverty and hunger problems in developing countries. Most countries in Sub-Saharan Africa have stagnated for decades, and disease problems such as HIV/AIDS have made a bad situation worse. Latin American countries suffered setbacks on the path to development during the 1980s, and have improved only gradually since that time. Several Asian countries have

---

[1] Joachim von Braun, Director General, International Food Policy Research Institute, from remarks prepared for the CGIAR Annual General Meeting, Marrakesh, Morocco, December 6, 2005.

grown rapidly over the past 25 years, but population growth remains rapid in many already densely populated Asian countries, while water becomes scarcer. Concerns for the global environment have focused attention on the growing problem of resource degradation in all developing regions. Over the past several years, numerous policy prescriptions have been suggested, yet none has been universally successful. Import substitution policies, domestic and trade policy liberalization, land reform, foreign aid, education, privatization, investment in large-scale industries, integrated rural development projects, farming-systems research, and many other solutions have been offered. Some of these suggestions have contributed to the development process; others have not. Blame for slow progress is often laid at the doorstep of the developed countries, sometimes with justification.

While economic development has been painfully slow and uneven, there is certainly room for guarded optimism. Globally, the percentage of people living in poverty has fallen significantly since the 1960s, as has the absolute number of malnourished people despite the growth in population. Several lessons have been learned about what it takes to stimulate agricultural and overall economic development. One of these lessons is that there are no panaceas. Development requires a mix of technical and institutional changes that work best in combination. The exact mix varies between countries, and policies appropriate for one environment may not necessarily be so for another. A second lesson is that developing countries are primarily responsible for their own development, but interdependence in trade and capital flows means that developed-country policies can assist or retard that development. Developed countries have forgiven some debt obligations of the most highly indebted and lowest-income countries, and recently pledged to do more. A few years ago, the United Nations and its member states set a series of millennium development goals for 2015 that would significantly reduce poverty, hunger, and disease while promoting education, gender equality, and environmental sustainability.[2] Some progress has been made toward achieving those goals, even if their total attainment seems unlikely.

In *Economics of Agricultural Development*, you have examined the dimensions of world food-income-population problems (see top of Figure 20-1). You have considered the interconnections among these problems and their linkages to health, nutrition, literacy, and the environment. There is enough total food in the world at the moment, but hunger is caused by distributional problems that are, in many cases, related

---

[2] See Jeffrey Sachs, *Investing in Development: A Practical Plan to Achieve the Millennium Development Goals*, the Millennium Project (London: Earthscan Publishing, 2005).

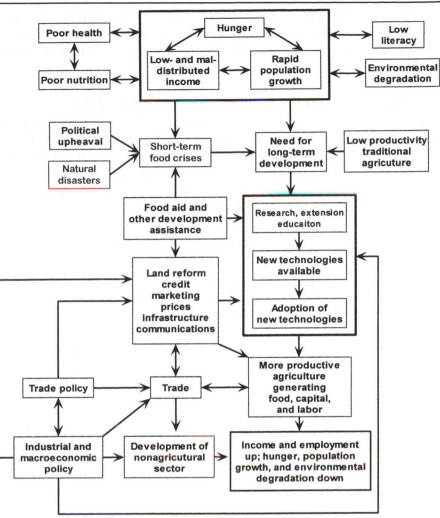

Figure 20-1. The hunger problem and the solution process.

to poverty. There are short-term food crises and long-term or chronic malnutrition. You have considered economic development theories, the role of agriculture in those theories, and the nature of existing agricultural systems. You have learned that developing-country farmers tend to be relatively efficient at what they do, but have low productivity because of their limited access to resources, their existing technological and institutional environments, and pervasive risks. Having learned something about the dimensions of the problem, the role of agriculture in economic development theories, and the nature of agriculture in developing countries, you then examined several components of the development process. Let us review below the interrelationships among those components, and assess where the need is greatest for additional insights with respect to the development process.

Technical and Institutional Change in Agriculture

In the 1950s, many development experts felt that the keys to agricultural development were capital investment and the transfer of technologies from more developed countries. By the 1960s and 1970s, it was clear that technology transfers and capital investment had a role to play, but that many other factors were equally as or more important. Differences in resource bases across countries meant that indigenous research and extension were vitally important. Education was required if countries were to produce, adapt, transfer, and receive new technologies. By the 1990s, countries that were successful in agricultural development had put in place a research- and technology-transfer system that included: (1) indigenous agricultural research stations and educational institutions, (2) a mechanism for linking those stations to on-farm research and testing, and (3) ties between the national research system and the international agricultural research and training network.

As important as indigenous capacity for agricultural research and extension is, the last several years have also demonstrated the need for a whole series of institutional reforms related to the agricultural sector. These reforms have been proven important not only because of their influence on production incentives and the distribution of economic gains, but also because of their influence on the types of technologies produced and adopted. Land reform, improved credit policies, marketing system development, nondiscriminatory pricing policies, rules on intellectual property protection, and incentive systems to reduce environmental externalities are examples of the institutional changes that may be required.

New technologies are not gifts of nature, and institutional changes do not magically appear. New technologies require research investments, and the levels and types of technologies produced and subsequently adopted are influenced by changes in relative prices of inputs and outputs. Institutional changes are also induced by changes in relative prices and by technical change.

The logic of the induced technical and institutional change theories and their apparent empirical verification in several countries, particularly in East Asia, give cause for optimism. However, the failure of many countries to follow a path of sustained development has forced economists to broaden the induced innovation theory. This broadening has come about by incorporating transactions costs and collective action into the theory.

Transactions costs refer to the costs of information, of adjusting a fixed-asset base, and of negotiating, monitoring, and enforcing contracts. The fact that information is not perfect and is costly to acquire, and that people are willing to exploit their situation at the expense of others, has received particular attention. If one group has greater access to information than another, that group can act collectively to press for policies or new technologies that benefit it at the expense of others.

If a small but wealthy elite with large landholdings finds it cheaper to acquire information and act collectively, it may press for technical and institutional changes for personal benefit at the expense of the masses. The elite may press for changes that not only distribute income in its favor, but reduce the agricultural growth rate because the resulting technologies and policies may not be appropriate for the resource base in the country.

Factors that can help reduce transactions costs are those factors that reduce information costs. Improved roads and communications infrastructure are examples; new communications technologies are having dramatic impacts even in remote areas. Education is critical. Land reform can help in many countries, as can institutional change to enhance contracting, to improve the legal infrastructure, and to provide certain types of regulations.[3] Freer markets to provide efficient price signals to individual farmers should also help.

Markets are generally held to be the best means of providing signals to actors. In developing countries, however, market imperfections due to transactions costs, unequal asset distributions, and other factors are the norm. Some government involvement is legitimate and necessary to reduce these imperfections and allow the markets to work. Government involvement can be justified to provide "public goods," to create equal access to opportunities, and to achieve equity outcomes consistent with society's wishes.

Macroeconomic and International Institutional Changes

In the 1950s and 1960s (and to some extent before and after), several economists in developing countries recommended policies that discouraged agricultural exports and encouraged production of goods that would substitute for imports. The argument for these policies was based on potential or perceived exploitation of developing countries by more developed countries. However, countries that integrated more closely

---

[3] For example, regulations may be needed to reduce externalities. These regulations may be particularly important for avoiding environmental externalities.

Young boys in Guatemala (photo: Nancy Alexander).

into international markets tended to develop more rapidly than did those that closed or isolated their economies. The more rapid development was due in part to the lower level of rent-seeking behavior and corruption as well as the increased efficiency gains from trade and specialization.

Some of the countries that discriminated against agricultural trade encouraged capital-intensive imports, causing capital-intensive industries to develop in labor-abundant countries. These industries placed a drag on economic development because human resources, freed up by increases in agricultural productivity, were underemployed. This tendency to develop capital-intensive industries may have been induced in part by transactions costs and collective action, but also by a perceived need to imitate more developed countries.

For the past 25 years, several developing countries have suffered from heavy external indebtedness. This problem has forced some countries to reform their policies in ways that are intended to spur longer-run economic growth. The debt overhang is so large for some Latin American and African countries, however, that without debt reduction on the part of those holding the loans, development will remain impeded for years to come. Official debt can be rescheduled through the Paris Club and by multilateral organizations. Some debt reduction for the lowest-income heavily indebted countries has recently occurred, and more has been promised. However, commercial debt reduction will only be achieved very slowly without improved international oversight to reduce the free-rider problem that currently exists.

Continued efforts by developing countries to reduce overvalued exchange rates and to phase out policies such as export taxes will be required to stimulate agricultural growth. Phasing out export taxes, however, will necessitate new mechanisms for generating government revenues, such as land or income taxes. Such taxes become somewhat more feasible as information flows improve in a country. New institutions will be needed, along with increased government responsibilities. Reduced trade restrictions by developed countries as a result of negotiations under the auspices of the WTO would also help developing countries. Increased regional economic integration among developing countries may also play a positive role in some cases.

Capital Flows and Foreign Assistance

Capital flows have provided a two-edged sword for many countries, particularly in Asia and Latin America. Capital inflows have helped stimulate investments and growth, particularly in East Asia, but have led to financial crises when outflows occur rapidly over a short period of time. The crisis that spread through East Asia beginning in 1997 had devastating impacts on human welfare and set back progress toward development in affected countries. Developing countries must each decide upon the appropriate mix of regulations for capital flows, fixity of exchange rates, and freedom to adjust macroeconomic policies. Flawed decisions can lead to economic instability and stagnation.

Economic development assistance can help relieve short-term food crises and can contribute to longer-term development. Emergency food aid is essential for averting famine following natural disasters and major political upheavals. Longer-term financial aid could help to reduce the debt problem in several countries and provide real resources for development.

Aid effectiveness could be improved by longer-term commitments and increased donor coordination. Less tying of aid to factors such as procurement from donor sources, but increased tying of aid to institutional changes that eliminate distortions or reduce transactions costs would help.

Coordinated international action has been successful in dealing with specific development problems. International support for agricultural research led to productivity increases that enhanced food security, reduced famines in highly populated areas, and helped alleviate rural poverty. Worldwide immunization efforts, coordinated by the World Health Organization, have significantly reduced deaths due to common childhood diseases. Concerted efforts to provide food to famine victims have reduced famine mortalities. Similar international coordination could be effective in reducing debts, providing assistance for policy reforms, and for other specific actions. Because development needs and the impacts of different interventions vary from country to country, international actions to promote growth have tended to be less successful than those used to address specific short-run problems.

In summary, it is clear that many pieces are needed for a country to solve its development puzzle. Enhanced information flows are vitally important for agricultural development. More labor-intensive industrial growth is needed in several countries if the employment problem is to be solved.

ASSESSING FUTURE PROSPECTS

Several countries in Asia have grown at relatively rapid rates for almost two decades (with some short-term financial instability), but masses of impoverished people still live in Asia. Latin American countries that grew in the 1960s and 1970s stagnated in the 1980s and most of the 1990s. Most Sub-Saharan African countries have grown very slowly, stagnated, or declined for the past 40 years. Hunger problems persist despite increased food production, per capita, in the world over the past 50 years. Poverty rates have, however, fallen. Environmental problems have grown worse in several countries as well. What does the future hold for reducing hunger, poverty, population growth, and environmental problems? Let us consider some of the underlying forces at work.

Supply and Demand for Food

The real price of food in the world has trended slightly downward for several years as supply growth has outstripped demand growth. The major long-run supply shifters are new technologies, while the major demand shifters are population and income growth. As we look to the future, population will continue to grow, but the *rate* of population growth will continue to fall. Incomes have increased rapidly in several Asian countries, including China with its massive population base. Asia has two-thirds of the world's population and, as a region, the best chance of continued supply increases due to research-induced technical change. Food production per capita will likely continue to increase in Asia as supply shifts outpace demand shifts, although the magnitude of this increase is a bit uncertain (Box 20-1). There will be increased diversification away from rice, however, as diets change with higher incomes. In particular, demands for animal proteins are likely to increase, with important implications for livestock production and marketing systems.

416

BOX 20-1
THE PROSPECTS FOR CEREAL TECHNOLOGIES

The spectacular burst in yield potential from new varieties of rice and wheat that began the Green Revolution has not been repeated. Rice yields on experimental farms have not grown dramatically since the introduction of IR-8 in 1966. However, the difference between yields on the best farms and the yields on experiment stations has shrunk dramatically since 1970, particularly in Asia. This reduced difference is due to widespread irrigation, high application of fertilizer, and good management. Future gains in rice production must come increasingly from rain-fed upland and deep-water areas, unless new biotechnologies provide yield breakthroughs.

The prospects for wheat and maize are more optimistic, even in the shorter run. The Centro de Mejoramiento de Maiz y Trigo (CIMMYT) reports a continuing increase in the yield potential of wheat of about 1 percent annually. Substantial progress has been made toward breeding in disease resistance, especially against wheat leaf-rust. Yield growth for wheat in less favorable conditions has been less spectacular. High-yielding varieties for low-rainfall marginal areas are limited, and there are virtually no new varieties for the lowland humid tropics. Major breakthroughs in these areas may pave the way for a technology-driven boom in wheat yields. Maize shows the most promise. There is a large gap between experiment station and farmers' yields, and weed control seems to be the critical problem. Human-based solutions to weed problems provide opportunities for increased employment while increasing maize yields.

In terms of genetic engineering and other biotechnologies in general, the outlook is promising but uncertain. Many improvements, such as increased insect/disease and drought resistance, are on the horizon, but public fears about biotechnologies in developed countries have slowed down the development and spread of these technologies in developing countries as well.

In Latin America, increased food production per capita is likely, but not at a rapid rate, as the debt problems continue, limiting public investments in agriculture. Population growth rates have already declined from their peaks of earlier decades, facilitating this per capita increase. More urbanization and income growth will have implications for food demands and a changing face of marketing and trade in food.

Unfortunately, many Sub-Saharan African countries will continue to experience disease and stagnant per capita growth in food production. Some increased investments in education and in agricultural research systems have been realized, but population growth rates are still high. If small income increases can be realized, population growth may decelerate, but environmental problems appear to have already degraded the resource base in parts of the Sahel to the point of reducing productivity. Aids and malaria remain serious health problems. Recently the Gates Foundation has joined other public and private organizations in a concerted attempt to solve these serious health problems, but solutions will be difficult and will take time.

Institutional Changes

Improved information technologies and infrastructure development have improved information flows in some developing countries. These improvements may create pressures for political and institutional changes, changes that offer favorable opportunities for development. Reduced transactions costs that result would induce the development of technologies that are better suited to the relative resource scarcities of the countries. More market-oriented policies may continue to create efficiency gains, as they have in Asia. There is evidence that some governments in Latin America and Africa have laid the groundwork for these types of gains as well.

The willingness of more developed countries to provide foreign assistance and international institutional changes to help poor countries is constantly in flux. Indifference was growing during the 1990s among policy-makers in the United States and many other developed countries. The fall of communism in Eastern Europe and the breakup of the Soviet Union a few years ago reduced political pressures on Western governments to help developing countries for the purpose of keeping those countries out of Soviet influence. The increased aid burden to the Soviet Republics and Eastern Europe diverted assistance away from other regions. The heightened focus on the terrorism threat since September 11, 2001, has caused many nations to focus their resources more on countries posing security threats than on attacking more broadly the root causes of poverty and hunger.

The relatively wealthy countries of the world must resist isolationist temptations. Terrorist attacks and threats may help stimulate countries eventually to seek longer-term solutions to problems abroad. A long-term goal of promoting democracy and freedom can only be attained through steps to reduce poverty and build economic opportunity. Many security, income, hunger, and environmental problems require a supra-national decision-making process. In order to strengthen the United Nations agencies that could make these decisions, developed countries will need to increase their contributions to official development assistance.

People in developing countries can benefit from grass-roots help.

HOW YOU CAN HELP

You as individuals can do a great deal to help solve hunger, poverty, ill-health, environmental degradation, and other development problems. Some of you can get involved directly through working for grass-roots organizations in developing countries. The Peace Corps is an example in the United States, but there are many others. For those from developed countries, spending time living and working in a developing country can greatly improve your understanding of development problems. We are each captive of the pictures in our mind, and living in a developing country provides a more accurate picture of the world.

Getting directly involved in influencing the fortunes of others can bring you a feeling of significance or satisfaction. The frustrations of working with desperately poor people are many. If you are not an optimist, you may not want to try. However, if you are adventurous, flexible, and somewhat persistent, you may want to consider working at a grass-roots level in a developing country.

Some of you can obtain a graduate education to become animal scientists, plant breeders, plant pathologists, entomologists, agricultural economists, soil scientists, microbiologists, or some other type of agricultural scientist needed to help solve world food, income, and environmental problems. Employment opportunities exist for rewarding careers at universities, in international agricultural research centers, national research centers, and private firms. Until the world's population stabilizes, the battle to keep world food production increasing at roughly 2 to 3 percent per year will continue.

Most of you will take very different career paths, but the opportunity always exists to contribute to solving poverty problems through financial contributions to private voluntary organizations. All of you can strive to keep informed about what is happening in the world outside your state and country. You can try to keep politicians informed and let them know that you support foreign assistance contributions to countries where needs are greatest.

SUMMARY

In this chapter, but also in the whole book, we have stressed the interrelatedness of hunger, population, and poverty problems. There are no panaceas, but a set of interconnected pieces to a development puzzle. We have learned over the years what many of these pieces are. In this chapter we stressed particularly the importance of enhanced information flows if broad-based development is to occur. Open economies, employment-based industrial policies, and development policies that do not discriminate against agriculture are essential. For developed countries, now is the time for renewed commitment to finding solutions to development problems.

IMPORTANT TERMS AND CONCEPTS

Agricultural scientist
Enhanced information flows
Feeling of significance
Grass-roots organization process

Interdependence
No panacea
Supply and demand shifters
Supra-national decision-making

QUESTIONS FOR DISCUSSION

1. Why might there be room for guarded optimism with respect to future agricultural and economic development?
2. Describe the interconnectedness among the pieces that can contribute to solving the development puzzle.
3. How has the theory of induced technical and institutional innovation been broadened in recent years and why?
4. What factors can help reduce transactions costs?
5. Why have relatively open economies grown more rapidly than relatively closed economics?
6. What factors will determine the long-run future price of food in the world?
7. Why do enhanced information flows offer favorable prospects for development?

8. What might you as an individual do to help solve hunger, poverty, and other development problems?

RECOMMENDED READING

Hunger Task Force, *Ending Hunger: It Can Be Done*, the Millennium Project (London: Earthscan, 2005).

Pinstrup-Andersen, Per, and Rajul Pandya-Lorch, *The Unfinished Agenda: Perspectives on Overcoming Hunger, Poverty, and Environmental Degradation* (Washington, D.C.: International Food Policy Research Institute, 2001).

Runge, C. Ford, Benjamin Senauer, Philip G. Pardey, and Mark W. Rosegrant, *Ending Hunger in Our Lifetime: Food Security and Globalization* (Baltimore, Md.: Johns Hopkins University Press, 2003).

Glossary of Selected Terms

absolute advantage: When one country's cost of producing a good is lower than the cost in other countries.

agricultural extension: The process of transferring information about improved technologies, practices, or policies to producers, consumers, or policymakers.

agricultural productivity: Level of agricultural output per unit of input.

balance of payments: Difference between receipts from all other countries and payments to them, including all public and private transactions.

biased technical change: The process of adoption of new means of production that use one factor more intensively than other factors, holding all other things constant.

bilateral: Two-party or two-country, such as aid from one country to another.

biotechnology: A set of tools, including traditional breeding techniques, that alter living organisms, or parts of organisms, to make or modify products; improve plants or animals; or develop micro-organisms for specific uses. Modern biotechnology includes use of recombinant DNA, monoclonal antibodies, and novel bio-processing techniques, among others.

birth rates and death rates: The number of births or deaths per 1,000 population in a year.

buffer stocks: Supplies of a product that are stored and used to moderate price fluctuations. These stocks are sold during periods of rising prices, and purchased when prices fall.

capital accumulation: Investment.

common property: Property for which the rights of use are shared and ownership is not private but shared by all.

comparative advantage: Ability of a country to produce a good or service at a lower opportunity cost than can another country. The theory of comparative advantage implies that a country should devote its resources not to all lines of production, but to those it produces most efficiently.

concessional: Subsidized, usually used with respect of interest on loans.

consumer/producer surplus: Consumer surplus is the area below the ordinary demand curve and above the price paid; it is a measure of well-being. Producer surplus is the area above the supply curve and below the price paid; it is a measure of returns to fixed factors of production.

debt rescheduling: Extending the repayment period for loans, altering interest rates, forgiving part of the principal, or some combination of the three.

demographic transition: The historical shift of birth and death rates from high to low levels in a population. Death rates usually decline before birth rates, resulting in rapid population growth during the transition period.

discount rate: The value used to determine the present value of future cash flows arising from a project or an investment.

economic development: Improvement in the standard of living of an entire population. Development requires rising per capita incomes, eradication of absolute poverty, reduction in inequality over the long term, and increased opportunity of individual choice.

economic or structural transformation of an economy: The increase in the size of the nonagricultural sector relative to agriculture that occurs in all economies as economic growth occurs.

elasticity: A measure of the percentage response of one variable (for example, quantity demanded) to a 1 percent change in another variable (for example, price).

experiment station: A center or station at which scientists conduct research.

external debt: Debts owed by the government in one country to creditors in another country.

externality: An economic impact of an activity by an individual or business on other people for which no compensation is paid. Externalities may be positive or negative and are often unintentional.

foreign assistance or foreign aid: Includes financial, technical, food, and military assistance given by one or several countries to another country. This assistance may be given as a grant or subsidized loan.

foreign exchange rate: The number of units of one currency that it takes to buy a unit of another currency.

free rider: An individual or business that receives the benefits of the actions of another individual or business without having to pay for those benefits.

free trade area: A block of countries that agree to lower or eliminate tariffs and other trade barriers among themselves, but each country maintains its own independent trade policy toward nonmember nations.

fungibility of credit: The degree to which money loaned for one purpose can be used for another.

General Agreement on Tariffs and Trade (GATT): Multilateral agreement, originally negotiated in 1947, for the reduction of tariffs and other trade barriers. The agreement provides a forum for intergovernmental tariff negotiations.

globalization: The increasing integration of economies around the world, particularly through trade and financial flows. Also refers to the movement of people and knowledge across international borders.

Green Revolution: The dramatic increases in wheat and rice harvests that were achieved in the late 1960s, primarily in Asia and Latin America, following the release of fertilizer- and water-responsive, high-yielding, semi-dwarf varieties of those crops.

high-yielding variety: Varieties of plants that have been improved through agricultural research so that they yield more per amount of input than the traditional varieties.

human capital: The level of education, skills, knowledge, health, and nutrition of an individual or a population.

import substitution: Actions by a government to restrict imports of a commodity to protect (from international competition) and encourage domestic production of the good.

induced innovation theory: A theory that hypothesizes that technical change is induced by changes in relative resource endowments and by growth in product demand; institutional change is induced by changes in relative resource endowments and by technical change.

institutions: Organizations or rules of society. Government policies, regulations, and legal systems are examples.

integrated pest management: The coordinated use of biological, cultural, and chemical pest control practices to reduce insects, diseases, and weeds. The purpose is to control pests in both an economically and an ecologically sound manner.

intellectual property rights: Laws regulating the copying of inventions, identifying symbols, and creative expressions. These laws encompass four separate and distinct types of intangible property — patents, trademarks, copyrights, and trade secrets.

international agricultural research centers (IARCs): The set of agricultural research centers supported by a group of public and private funding sources. These centers provide improved technologies and institutional arrangements to help developing countries increase their food production. Funding is coordinated by the Consultative Group on International Agricultural Research (CGIAR).

international capital market: The transfers of capital (money) among countries in response to short- and long-term investment opportunities.

international commodity agreement: A formal agreement among the major producing and consuming countries of a commodity that specifies a mechanism for stabilizing price. An agreement may specify import and export quotas for each country.

International Monetary Fund (IMF): An international financial institution designed to: (1) promote international monetary coordination; (2) foster international trade; (3) facilitate stabilization of exchange rates; (4) develop mechanisms for multilateral transactions between members; and (5) provide resources for enhanced international financial stability.

land reform: An attempt to change the land tenure system through public policies.

land tenure: The rights and patterns of control over land.

less developed country (developing country) (LDC): Generally refers to countries in which per capita incomes are below $6,000, although a few countries with higher incomes consider themselves to be less developed or developing.

market failure: When markets fail to efficiently organize production or allocate goods in a way that maximizes social welfare.

micro-finance: Small-scale provision of credit, savings, and insurance services, usually to the very poor.

moneylender: An informal lender whose business it is to lend money to borrowers, usually at high interest, with little or no collateral or paperwork.

money supply: Currency plus money that can be easily withdrawn from checking or savings accounts.

monopoly power: When a single seller or united group of sellers has the power to alter the market price as opposed to having to just accept the market price.

monopsony: A market with a single buyer.

multilateral: Refers to many countries as opposed to two countries (bilateral). Examples are multilateral aid, multilateral trade, and multilateral agreements.

multiple exchange rates: When a country sets different rates between its currency and foreign currencies depending on the class of imports. May be used to control foreign exchange by limiting certain types of imports.

Official Development Assistance (ODA): Foreign assistance that excludes military related assistance, export credits, and private fund transfers while having at least a 25 percent grant element. The grant element is defined as the excess of the loan or grant's value over the (present) value of repayments calculated with a 10 percent interest rate.

opportunity cost of capital: The rate of return on the best alternative use for the funds. It is the cost of alternative investments forgone when a particular investment is made.

overvalued exchange rate: When the official value of a currency is too high given the exchange rate that would otherwise prevail in international money-markets given the supply and demand for the country's currency.

parastatal: An institution, such as a marketing board, that is used by a government to control the production, distribution, international trade, and domestic price of a product. This product might be an agricultural good or an input such as fertilizer.

production function: Describes, for a given technology, the different output levels that can be obtained from various combinations of inputs or factors of production.

production possibilities frontier: The trade-off between the maximum amount of two goods that can be produced in a country given existing production technologies and the available productive resources.

protectionism: A reaction by an industry or a country to foreign competition. That reaction is usually manifest through tariffs, quotas, or other means of reducing imports to shield domestic producers.

public goods: Goods or services that are non-rival (consumption or use by one person does not preclude consumption by another) and non-exclusive (a person cannot be excluded from consumption or use, except at prohibitively high costs).

scale-neutral technology: A technology that can be employed equally well by any size firm.

social cost: The total value of resources used in production of a good, including the value of externalities, which are not borne by the producer of the good or reflected in the market price.

structural adjustment program: Government program aimed at adjusting the economy to reduce imbalances between aggregate supply and demand. Structural adjustment programs typically involve: devaluation of the foreign exchange rate to increase exports and reduce imports, reduced government spending, and removal of many government policies that distort prices, including barriers to trade.

426

subsidized (concessional) credit: Loans made with interest rates below the rates prevailing in the market.

sustainable development: Development that meets the needs of the present without compromising the ability of future generations to meet their own needs.

tariff: A tax or duty placed on goods imported into a country.

technology: The method for producing something. New technologies are often embedded in inputs, for example, seeds or machines. Hence higher-yielding seeds or more efficient machines are often referred to as improved technologies. Technological progress occurs when more output is obtained from the same quantity of inputs. Technology transfer occurs when methods (perhaps embedded in materials) from one location are applied in a second location.

terms of trade: The relationship between the prices of two goods that are exchanged; for example, the price of an export good relative to the price of an import good. When the price of an export good increases relative to the price of an import good, the terms of trade have increased for the export and are said to be favorable.

trade preferences: Refers to favorable tariff treatment accorded by one country or group of countries to exports of certain other countries.

transactions costs: The costs of adjustment, of information, and of negotiating, monitoring, and enforcing contracts.

World Bank: The major multilateral-funded organization that makes loans to developing countries. It contains the International Finance Corporation, the International Bank for Reconstruction and Development, and the International Development Association.

World Trade Organization (WTO): The international institution created in 1994 to replace the GATT, and strengthen enforcement of international trade rules and the settling of trade disputes.

Author Index

Page numbers followed by b indicate material in boxes.
Page numbers followed by f indicate figures.
Page numbers followed by n indicate footnotes.
Page numbers followed by t indicate material in tables.

Subject Index

Page numbers followed by b indicate material in boxes.
Page numbers followed by f indicate figures: an ft means a combined figure/table.
Page numbers followed by n indicate footnotes.
Page numbers followed by t indicate material in tables.
The word *agricultural* may be abbreviated to ag.